APPLIED MATHEMATICS Made Simple

The Made Simple series
has been created
primarily for self-education
but can equally well
be used as
an aid to group study.
However complex the subject,
the reader is taken
step by step,
clearly and methodically,
through the course. Each volume
has been prepared by
experts,
using throughout the
Made Simple technique of teaching.
Consequently the gaining
of knowledge now becomes
an experience to be enjoyed.

Accounting	English
Acting and Stagecraft	French
Additional Mathematics	Geology
Advertising	German
Anthropology	Human Anatomy
Applied Economics	Italian
Applied Mathematics	Journalism
Applied Mechanics	Latin
Art Appreciation	Law
Art of Speaking	Management
Art of Writing	Marketing
Biology	Mathematics
Book-keeping	New Mathematics
British Constitution	Office Practice
Chemistry	Organic Chemistry
Childcare	Philosophy
Commerce	Photography
Commercial Law	Physics
Company Administration	Pottery
Computer Programming	Psychology
Cookery	Rapid Reading
Cost and Management	Russian
Accounting	Salesmanship
Dressmaking	Soft Furnishing
Economics	Spanish
Electricity	Statistics
Electronic Computers	Typing
Electronics	

APPLIED MATHEMATICS Made Simple

Patrick Murphy, M.Sc., F.I.M.A.

Made Simple Books
W. H. ALLEN   London
A division of Howard & Wyndham Ltd

Made and printed in Great Britain by
Richard Clay (The Chaucer Press), Ltd., Bungay, Suffolk
for the publishers W. H. Allen & Company Ltd.,
44 Hill Street, London W1X 8LB

First edition, September 1971
Revised and reprinted June 1975

ISBN 0 491 00579 2 Paperbound

Foreword

This book provides an elementary study of the classical subject matter of Dynamics, Statics and Hydrostatics using the new scheme of SI units. The standard and content of the work covers C.S.E. and 'O' level G.C.E. examinations in Applied Mathematics and Mechanics as well as the relevant parts of the syllabuses for Physics and General Science courses related to Engineering, Building, and Agriculture.

Vectors are employed in Chapters 6, 7, and 8 on Relative Motion and Forces in Equilibrium, and 'A' level G.C.E. students in Mathematics could profit from a reading of these chapters in particular.

The book has also been written for the home study reader who is interested in widening his mathematical appreciation or simply reviving forgotten ideas and the author hopes that the style of presentation will be found sufficiently attractive to recapture those who may at one time have lost interest. No knowledge of the calculus is required and all the necessary preparation for the wide use of trigonometry is to be found in the early part of Chapter 6.

The method of introducing Applied Mathematics has long been a matter for argument. Some teachers prefer to come from the familiar ideas of applied forces in Statics while others prefer the more exciting introduction of Kinematics and the laws of motion. Thereafter some would complete Statics before Dynamics or vice versa.

In this book if the student reads the chapters in numerical order he will commence with the preliminary ideas in statics and work his way through Dynamics and Statics until he reaches the subject of Hydrostatics—treated separately in the last three chapters of the book. By this means he should come to appreciate the unity of the ideas in Dynamics and Statics. The author recommends this method of approach, while leaving the student free to select his own sequence of reading, which may emphasize Dynamics or Statics according to choice.

Following the introduction of the SI scheme of units the opportunity has been taken to emphasize the concepts of weight and mass by the use of absolute units throughout.

Finally, my special thanks are due to B. Knight, Ph.D., who offered so much helpful comment on the manuscript and to my students Jennifer Bull, Lynne Clayton, and Peter Shaw who so patiently read the proofs and approved the exercises.

PATRICK MURPHY

Table of Contents

FOREWORD   v

1 MECHANICS   1
  (1) Rigid Bodies   1
  (2) Characteristics of a Force   2
  (3) Measuring Force with a Spring   4
  (4) Mass and Weight   6
  (5) The Unit of Force   8
  (6) Newton's Third Law   9
  (7) Simple Force Diagrams   10
  (8) Bodies in Equilibrium   11

2 PARALLEL FORCES AND RIGID BODIES   13
  (1) Turning Effect of a Force   13
  (2) The Beam Balance   18
  (3) Calculation of Moments   20
  (4) Resultant of Two Parallel Forces   22
  (5) Centre of Gravity   28
  (6) Composite Figures   31
  (7) Solid Figures   33

3 KINEMATICS   35
  (1) Speed and Velocity   36
  (2) Speed–Time Formulae   38
  (3) Distance–Time Graphs   39
  (4) Non-uniform Speeds   44
  (5) Speed–Time Graphs   46

4 MOTION WITH UNIFORM ACCELERATION IN A STRAIGHT LINE   50
  (1) Acceleration   50
  (2) Equations of Motion   55
  (3) Acceleration due to Gravity   59
  (4) Falling Bodies and Equations of Motion   62

5 NEWTON'S LAWS OF MOTION   66
  (1) Momentum   66
  (2) The First Law   67
  (3) The Second Law   68
  (4) Impulse of a Force   72
  (5) The Third Law   72

6 AN INTRODUCTION TO VECTORS   83
  (1) Revision of Trigonometry   83
  (2) Sine rule and Cosine rule   91
  (3) Vector Quantities   96
  (4) Vector Addition   101
  (5) Resultant of Three Vectors   104

7 VECTOR ALGEBRA AND COPLANAR FORCES   107
  (1) Addition and Subtraction   108
  (2) Distributive Law and Associative Law   109
  (3) Vectors in Elementary Geometry   113
  (4) The Polygon of Forces   115
  (5) Non-concurrent Vectors   118
  (6) Resolution of Vectors   120
  (7) Resolving Non-concurrent Vectors   126
  (8) Locating the Line of Action   131

8 RELATIVE MOTION   135
  (1) Latitude and Longitude   141
  (2) Finding Latitude   143
  (3) Finding Longitude   143
  (4) Course Calculations   147
  (5) Moving Reference Point   153
  (6) Positions of Closest Approach   156

9 PROJECTILES   162
  (1) Vector Components   163
  (2) Acceleration   165
  (3) Equations of Motion   166
  (4) General Equations for Projectile Motion   169
  (5) Motion Down an Inclined Plane   175
  (6) Motion Down a Chord of a Circle   177
  (7) Lines of Quickest Descent   178

10 FRICTION   179
  (1) Force of Friction   180
  (2) Experimental Investigation of Static Friction   183
  (3) Motive Force at an Angle to the Slope   190
  (4) Toppling before Sliding   193
  (5) Stable and Unstable Equilibrium   195
  (6) Freely Hanging Load   197

11 RIGID BODIES IN EQUILIBRIUM UNDER THE ACTION OF COPLANAR FORCES   201
  (1) Point of Concurrency   201
  (2) Force Normal to a Surface   206
  (3) Frictional Force and Normal Reaction   209
  (4) Equilibrium of a Body under the Action of More than Three Forces   212

12 WORK, ENERGY, AND POWER   217
  (1) Work   217
  (2) Kinetic Energy   218
  (3) Motion at an Angle to the Force Producing it   220
  (4) Potential Energy   225
  (5) Conservation of Energy   227
  (6) Momentum   232
  (7) Power   233

viii Table of Contents

13 MACHINES 236
    (1) Mechanical Advantage 236
    (2) Levers 237
    (3) Wheel and Axle 243
    (4) Velocity Ratio 244
    (5) Inclined Plane 249
    (6) Pulleys 252
    (7) The Law of the Machine 254

14 HYDROSTATICS I: DENSITY, RELATIVE DENSITY, AND BUOYANCY 257
    (1) Density 258
    (2) Mixtures 262
    (3) The Relative-density Bottle 265
    (4) Archimedes' Principle 267
    (5) The Hydrometer 271
    (6) Buoyancy and Equilibrium 273

15 HYDROSTATICS II: LIQUID PRESSURE 276
    (1) Fundamental Principles 277
    (2) Pascal's Law 278
    (3) Pressure due to Weight of Liquid 279
    (4) Hydraulic Machines 281
    (5) Horizontal Thrust in a Liquid 284

16 HYDROSTATICS III: BAROMETERS AND BOYLES' LAW 287
    (1) Atmospheric Pressure 287
    (2) Pressure of a Gas 292
    (3) Hare's Apparatus 292
    (4) Boyle's Law 294
    (5) Pumps 297
    (6) The Siphon 298

TABLES 301
    Logarithms 302
    Antilogarithms 304
    Natural Sines 306
    Natural Cosines 308
    Natural Tangents 310
    Squares 312
    Square Roots 314
    Reciprocals 318

SI UNITS AND ABBREVIATIONS 320

ANSWERS 321

INDEX 343

CHAPTER ONE

MECHANICS

This book is concerned with three main branches of classical applied mathematics. They are **Statics**, the study of forces on bodies at rest; **Hydrostatics**, the study of forces on liquids and gases at rest, and **Dynamics**, the study of the motion of a body together with the forces which cause the motion. Very often the single title of **Mechanics** is used to refer to the study of forces on bodies whether at rest or in motion. Since an introduction to statics contains many familiar ideas and experimental possibilities, we shall start with a study of the relations between forces which keep a body at rest. First let us consider what we mean by the terms 'body', 'force', 'rest', and 'motion'.

(1) Rigid Bodies

A **body** in applied mathematics is anything to which we can apply a force. It may be this book, a pen, a screwdriver, a screw, a tyre being fitted to a wheel, a saucepan, a door, a letterbox flap, a zip, a paper-clip, a rubber balloon, even a jelly. Such a variety of shapes and sizes will introduce into our problems some difficulties which we would like to avoid, especially in this elementary stage. The main difficulty lies in the distortion which takes place as a result of the applied forces. A body may bend, sag, expand, or compress in a manner which is too difficult to predict and may obscure the basic principles involved in applied mathematics. For example, the use of steel girders instead of wooden bookshelves is an expensive way of minimizing the possible sag in the shelves. The point of this example is that it illustrates the possibility of choosing a body for which the distortion is so small that it may safely be ignored. We are therefore allowed to speak of such things as a 'horizontal shelf of length 1·5 m': we ignore completely any sag that occurs; for our purposes the shelf rests in a horizontal plane and its length remains at 1·5 m.

However, on some occasions the forces will be so great that even a steel girder will suffer a distortion which can no longer be ignored: examples occur in bridge building and 'skyscraper' construction. In order to simplify our work we therefore introduce the idea of a **rigid body**, i.e. a body which does not undergo any distortion resulting from the forces which act upon it.

DEFINITION: *A rigid body is such that the distance between any two of its points remains unchangeable.*

This is of course an idealized definition, but it does approximate very closely to the real situation when the forces on the body are such that any distortion which takes place is small enough to be ignored. Experience alone will tell us what we mean by 'small enough'.

The idea of being 'at rest' appears to be straightforward enough until we remember that our planet not only travels around the sun but also spins on its own axis. With everything on earth moving in this way, what can we mean by saying that a body is at rest? For our purposes, to be at **rest** means to be in an unchanging position relative to the surface of the earth.

We are not here concerned with the length of time for which the body remains at rest, provided that its position does not change at all while we are studying it. For example, a house is at rest and so is the car in the garage and the swing in the garden—although we know that one remains in the state of rest much longer than the others. Clearly, motion is a state of not being at rest, so we describe **motion** as changing position relative to the surface of the earth.

A change in position is called a **displacement**. Thus a body originally at O, the point of intersection of the South to North and West to East lines, which then moves to a point P, as illustrated in Fig. 1, has its displacement

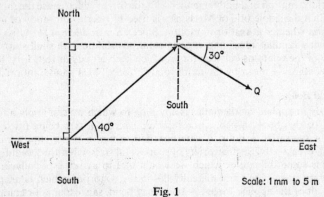

Fig. 1

completely represented by the length and direction of the line OP. The use of the arrowhead indicates the sense of the direction of the displacement. If the body moves on to Q, the length and direction of PQ in Fig. 1, together with the arrowhead, completely represents the further displacement of the body from P to Q.

(2) Characteristics of a Force

A **force** is defined as *that which changes or tends to change a body's state of rest or motion.* It must be noted that this statement only defines force: it does not say how force should be measured.

Thus we may use a force either to bring a body to rest or to set it in motion. If we see a body changing its direction of motion, we know that a force must be acting on it. Similarly, if we see a body at rest, we know that the total force on the body is zero. Our first ideas about forces arise from personal actions such as pushing, pulling, pressing, and lifting in moving furniture about a room. We can usually remember the method by which a particularly heavy or awkward piece of furniture can be most conveniently moved and we realize intuitively that the force concerned has certain characteristics, i.e. it is easier if we lift at one particular corner or pull in a certain direction.

The **characteristics of a force** are as follows:

(*a*) the *point of application,* i.e. whereabouts on the body the force is applied;

(*b*) the *direction* and sense of the force, i.e. at what angle to the floor we direct the force;

(*c*) the size or *magnitude* of the force.

When we speak of the direction of a force, we are plainly considering that the force acts along a straight line passing through the point of application; this straight line is referred to as the **line of action** of the force.

The examples illustrated in Fig. 2 show how forces may be represented in diagrammatic form. The artistic details are clearly superfluous, and a diagram such as Fig. 2 (*d*) is an adequate picture of the details of the force exerted by the 'stick' man on the chest of drawers in Fig. 2 (*c*).

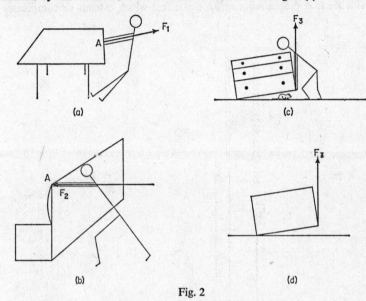

Fig. 2

By choosing a suitable scale we could also make the length of the line represent the magnitude of the force. To obtain the magnitude of a force we must have some means of measuring or comparing forces with a basic force, and it is for this purpose that we now turn to a discussion of weights, the most common of all forces, and their measurement by the spring balance.

Exercise 1

1. Draw a diagram to illustrate from Fig. 1 the displacements

    (*a*) from O to Q
    (*b*) from P to O
    (*c*) from Q to P
    (*d*) Using the scale of Fig. 1, measure the distances from O to P, P to Q, and O to Q.

2. $ABCD$ is a square of side 10 m. Describe the displacement of a body from A to C and from C to B.

3. Fig. 2 (*d*) was an abstract diagram to represent the action in Fig. 2 (*c*). Draw similar diagrams to represent the action in Figs. 2 (*a*) and 2 (*b*).

4. Draw diagrams similar to Fig. 2 (*d*) to represent the action of jacking up a car (i) from the rear, and (ii) from the side.

(3) Measuring Force with a Spring

Coil springs are made to resist compression or extension and are commonly found in car suspensions, engine valves, bed mattresses, retractable ball-point pens, etc. The quality of a spring is judged by its ability to continue returning to its original shape and size after the load has been removed, and the fact that a car coil suspension or a bed mattress will last ten years or more is evidence that such quality exists. Since that which extends or compresses

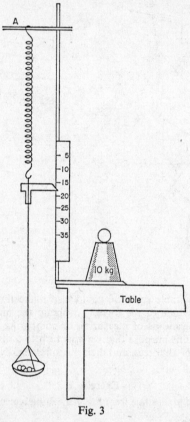

Fig. 3

a spring will be called a **force,** we shall measure the force by the **extension** it produces. The interesting feature of the spring is that any extension or compression is directly proportional to the force required to produce that extension or compression. In other words, the extension produced by a force T is doubled by the force $2T$, trebled by the force $3T$, and so on. Alternatively, the extension produced by a force T is halved by the force $\frac{1}{2}T$, quartered by $\frac{1}{4}T$, and so on.

It is not difficult to determine the relation between the force on a coil spring and the extension it produces. The apparatus consists of a coil spring approximately 0·10 m in length suspended from a fixed point A, as illustrated in Fig. 3. To the end of the spring is attached a pointer which moves up or down over a fixed scale in order to register the length of the extension. For convenience we shall extend the spring by suspending a known number of pennies from its free end. By counting the number of pennies we shall be able to test whether the force they exert on the spring is directly proportional to the extension.

On the free end of the spring is hung a pan or similar container which is sufficient to keep the spring taut. Each reading in the experiment is made with the pan and its contents at rest. The zero mark is noted corresponding to the starting position with the pan empty. Subsequently the load is increased by adding enough extra pennies to give a measurable extension of approximately 5 mm at a time. After taking five or six readings the spring is gradually unloaded and the readings taken again as indicated in the table below. A graph is plotted of the load against the extension it produces, as suggested in Fig. 4.

No. of Coins in Load	Extension Reading (mm)		
	Loading	Unloading	Average
10	9	10	9·5
15	14	15	14·5
20	19	20	19·5
25	24	25	24·5
30	30	30	30

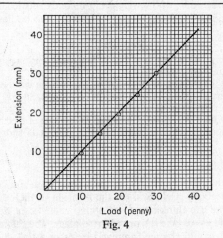

Fig. 4

On plotting the readings it will be found that they lie approximately on a straight line passing through the origin. This indicates that equal increases in the load produce equal increases in the extension, a relation which is called **Hooke's Law** for the spring. If we continue to increase the load, the spring will eventually become over-extended and the pointer will not return to the zero position when the pan is unloaded. As soon as this happens we say that the spring has passed its **elastic limit.** After this limit has been passed, smaller and smaller increases in the load will produce the same increase in the extension—until the spring breaks. Clearly Hooke's Law will only hold if the spring is within its elastic limit.

In the experiment we saw that a force was exerted on the spring by placing a body in the pan. Had we used a different spring, the same body would still have exerted the same force but would probably have produced a different extension. This force associated with the body is called the **weight** of the body. For the present all we can say is that the magnitude of the weight of a body is the magnitude of the force which that body exerts on a spring on which it hangs at rest. We have yet to find the direction of this force called weight.

Examining the graph in Fig. 4 we see that a body which produces an extension of 19·5 mm is exerting the same force on the spring as 20 coins. We can say therefore that the magnitude of the weight of this body is the same as the weight of 20 pennies. In other words, we are relating the weight of the body to the weight of the pennies. However, even though we could use the graph to calibrate the spring in pennies, this is not a satisfactory unit to use for the measurement of weight: among other defects, it suffers from the fact that different countries have different coinages.

It is found that the magnitude of the weight of a body varies from place to place, being less at the Equator than at the poles, less at the top of a mountain than at its foot. The difference is very small and is ignored for most practical purposes. Some ideas of this small difference can be gained from the fact that 214 pennies at the poles would exert approximately the same force on a spring as 215 pennies at the Equator. The important fact to remember is that if the magnitude of the weight of one body is twice that of another at the same place, it will remain so everywhere the two bodies are taken together.

Exercise 2

1. Using the extension/load graph of Fig. 4, find the extension produced by loads of (*a*) 17 pennies, and (*b*) 23 pennies.

2. Assuming that the spring does not reach its elastic limit, use the graph of Fig. 4 to find (*a*) the extension produced by a load of 35 pennies, and (*b*) the load which would produce an extension of 41 mm.

3. An experiment on a coil spring shows that an increase of 8 pennies in the load produces an extension of 5 mm. Find the load which extends the spring by (*a*) 30 mm, (*b*) 7·5 mm.

(4) Mass and Weight

We have seen that the magnitude of the weight of a body depends upon two things: (*a*) the body itself, and (*b*) the position of the body in space or on earth. In (*a*) we are thinking of some characteristic of the body which remains unchanged as we take it from place to place. The shape could change (putty), the colour could change (unmixed paint in a tin), the temperature could change (a packet of frozen peas), or the feel of the body could change (the surface could melt). It is impossible to think of anything which would not change in some way, except the quantity of matter in the body. This latter characteristic of the body will remain constant everywhere; we agree to call it the **mass** of the body. Thus it is common knowledge that an astronaut may become weightless when in space but *his mass remains the same*. It should be noted that mass is not a force.

We remarked earlier that if, at any one place, body A exerts twice the force on a supporting spring as body B, then we can say that the weight of A is twice the weight of B wherever the two bodies are taken together. Since we have now concluded that the weight of a body is dependent on its mass to the extent that if we double the mass we must double the weight (i.e. weight is proportional to mass), it follows that we may compare the masses of two bodies simply by comparing their weights, provided we do the comparison at the same place. It would be no use finding the weight of one body at the North Pole and the other body at the Equator for the purpose of comparing their masses. A more concise statement of the relation between the weights and masses of the two bodies is given by

$$\frac{W_1}{W_2} = \frac{m_1}{m_2}$$

where the letters and subscripts have their obvious meaning.

All methods of measurement consist of comparing the object to be measured with an internationally agreed standard unit. The SI unit of mass is the kilogramme (abbreviated to kg). This is equal to the mass of the international prototype of the kilogramme which is kept in the Department of International Weights and Measures at Sèvres near Paris.

All we need to do is to obtain a copy of the prototype and we shall then be able to measure the mass of any body. Let us see how this can be done.

Using the apparatus shown in Fig. 3, though not necessarily using the same spring, we first suspend the body and note the extension it produces; call this e_1. We then replace the body with the known 1-kg mass and note the extension produced; call this e_2. As we know that the extensions are proportional to the weights which produce them, we can write

$$\frac{W_1}{W_2} = \frac{e_1}{e_2}$$

where W_1 is the weight of the body and W_2 is the weight of the 1-kg mass.

But, since

$$\frac{W_1}{W_2} = \frac{m_1}{m_2}$$

we now have

$$\frac{m_1}{m_2} = \frac{e_1}{e_2}.$$

However, the second mass we placed in the container was 1 kg, i.e. $m_2 = 1$ kg.

Therefore

$$\frac{m_1}{1} = \frac{e_1}{e_2}$$

or

$$m_1 = \frac{e_1}{e_2} \text{ kg}$$

If we could obtain a spring such that $e_2 = 1$, the scale at the side of the spring could be calibrated to read off, in kilogrammes, the mass of any body placed in the container. Unfortunately the weight of a body or the force it exerts on the spring varies from place to place, so it would be impossible to claim that the load of 1 kg would always produce an extension of $e_2 = 1$ as just suggested. We conclude therefore that the spring may be used to *compare* masses but certainly not measure them everywhere on the surface of the earth. This highlights the difference between mass and weight: the mass of a body is constant, but the weight of the body varies. The weight of a body is a force whose direction is still to be discussed. Mass is not a force.

(5) The Unit of Force

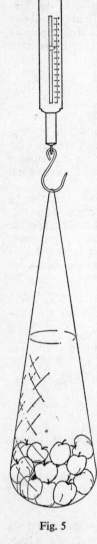

Fig. 5

If a weight *W* causes an extension *e*, then any force of magnitude *F* which produces an equal extension *e* will be such that $F = W$. It follows therefore that the extension of a spring can be used to measure any force no matter how it is applied, whether by a hanging load, a strong arm, or any other means. All we need is a unit for the calibration of the spring. We have already mentioned the possibility of calibrating such a spring in terms of penny units and in terms of kilogramme units. Neither of these methods is satisfactory because when used as a load on the spring the same masses exert different forces at different places on the earth. What we need is a unit of force which is independent of our position, or, to put it more poetically, is always absolutely true. Such a unit will be called an **absolute unit of force.** The absolute unit of force which we shall use in this book is the **newton** (abbreviated to N). Thus, if we calibrate a spring in newtons we shall ensure that we are not measuring force with a unit which is itself changing according to its position in space.

DEFINITION: *A newton is the force required to give a mass of 1 kg an acceleration of 1 metre per second per second in the direction of the force.*

We shall need to study some dynamics before we can appreciate this definition fully. However, this does not stop us measuring our forces in newtons and employing spring balances calibrated to give readings in newtons, as illustrated in Fig. 5.

The following examples give some idea of the magnitude of the newton: this book has a weight of approximately 4 N; the average teacup weighs approximately 1 N, and the average new-born baby weighs approximately 40 N. Appropriately enough, the standard eating apple weighs approximately 1 N.

Exercise 3

1. State whether the following quantities measure a force *F* or a mass *m*:

(i) 14 kg, (ii) 10 N, (iii) 9 kg, (iv) 1·7 kg, (v) 2·3 N, (vi) 0·1 N.

(In the next two questions you are only expected to work intuitively just to get some 'feeling' about the unit of force.)

2. If a force of 1 N gives a mass of 1 kg an acceleration of 1 metre per second per

second in the direction of the force, what acceleration will the following forces give a mass of 1 kg in the direction of the forces?

(i) 4 N, (ii) 40 N, (iii) 3 N.

3. What acceleration would a force of 20 N give to a mass of

(i) 4 kg, (ii) 0·5 kg?

4. The weight of a new penny is twice the weight of a new half-penny. If the mass of a new half-penny is 1·78 grammes, what is the mass of a new penny?

5. A new ten-penny coin has a mass of 11·30 grammes. The mass content of the coin is 75% copper and 25% nickel. Find the mass of copper and the mass of nickel in the coin.

(6) Newton's Third Law

Take two different springs and carry out experiments like the one on p. 5 in order to find the relation between the loads and corresponding extensions for each spring. By measuring the extensions we are able to find the magnitude of the force being exerted on the springs in terms of the number of pennies required to produce the extension.

Let us suppose that the same increase in load extends spring L by 7 mm and spring R by 5 mm. We now place them on a horizontal board, note their natural (i.e. unloaded) length, couple them together and stretch them apart to fix them on two nails A and B as shown in Fig. 6. It is found that the ratio of the extensions is still 7:5, for example, if the extension of L is 14 mm then the extension of R is 10 mm. Thus indicating that each spring exerts a force of equal magnitude on the other.

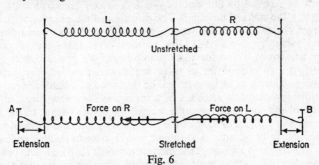

Fig. 6

This is an example of Newton's Third Law, which states that *action and reaction are equal and opposite*. For example, if a body L exerts a force (call this the **action**) on body R, then body R will always exert a force (call this the **reaction**) of equal magnitude but opposite direction on body L.

If we pull the springs in Fig. 6 farther apart until the extension of L is 21 mm, for example, it will be found that the extension of R is 15 mm, indicating again that each spring is exerting a force of equal magnitude on the other.

We could, however, make our experiment more direct by taking two spring balances which have already been calibrated in units of force (they are then known as **dynamometers**). On coupling them together and stretching them apart as shown in Fig. 6, the force in each spring can be read off immediately. If the force in L is 6 N, then the force in R will also be 6 N.

Exercise 4

1. A load of 6 pennies will extend a spring L by 7 mm, and a load of 8 pennies will extend another spring R by 5 mm. If the springs L and R are coupled together as in Fig. 6 and the extension of L is 21 mm, find the extension of R and the penny load equivalent of force on both springs.

2. A force of 3 N produces an extension of 4 mm in a spring L and 9 mm in a spring R. If the two springs L and R are coupled together as in Fig. 6 and the extension of spring L is 20 mm, find the extension of R and the force on both springs.

3. If each spring in Question 2 had registered a force of 18 N, what would have been the extension of each spring?

4. Is it necessary to confine both springs in Fig. 6 to the horizontal? Would the same force be registered in both springs if the horizontal board in Fig. 6 was tilted? Try to suggest a reason to justify your answer.

5. The weight of a body is 6 N and the weight of each of two spring dynamometers is 2 N. If the body is suspended from one spring which in turn is suspended from the other spring, what is the reading on each of the dynamometers, assuming that they are each calibrated in newtons?

6. What would be the readings in Question 5 if the dynamometers were interchanged?

(7) Simple Force Diagrams

A plumb-line consists of a small heavy body suspended at rest by a cotton thread. The direction of the thread is said to be **vertical** at the place where the line is set up. In the area covered by, say, a house or a factory, all plumb-lines will be parallel to one another—a fact which bricklayers use when constructing the walls of buildings. The direction of the plumb-line pointing away from the earth is called vertically upwards and the reverse direction is called vertically downwards. For reasons of strength and durability, builders use strong twine for their plumb-lines. We prefer cotton thread because it is light enough for us to ignore its mass: indeed, any thread or string whose mass may be ignored will be described as a **light string.**

Consider a plumb-line attached to a spring, as shown in Fig. 7 (*a*). The extension of the spring registers the magnitude of the weight of the body. By comparison with the weight of the body, the weight of the thread will be small enough to be ignored. The thread is said to be **in tension,** and we are seeking to indicate in a diagram the manner in which the forces act. The force of the tension in the string must act along the direction of the string, so this force acts vertically upwards on the body. Since the body is at rest, and since the only other force acting on the body is its weight, it follows that the total force on the body is zero and the weight must act vertically downwards in the opposite direction to the tension in the string.

DEFINITION: *The weight of a body acts vertically downwards with the magnitude of the force it exerts on any support which keeps the body at rest.*

Fig. 7 (*b*) illustrates the method of representing the forces on the body. Thus W is the magnitude of the weight, and the adjacent arrowhead indicates its direction. T is the magnitude of the tension in the string, and the adjacent arrowhead indicates the direction of T. Notice that we show T and W acting in the same straight line; since the body is at rest, this also tells us that $T = W$. Observe also that the force diagram indicates directions with the arrowheads, and the magnitudes by the letters T and W. No scales have been used to indicate magnitude by the length of arrow.

Fig. 7 (c) shows the same body at rest on a horizontal table. The weight W of the body acts on the table and therefore produces a reaction of the table on the body in the same straight line. The reaction is represented by R together with the adjacent arrowhead. Since the body is at rest, $R = W$ from Newton's Third Law.

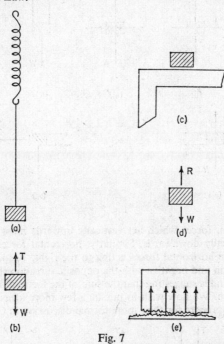

Fig. 7

The difference in the two situations of Fig. 7 should be noted. In Fig. 7 (b) there is clearly only one possible direction for the upward force, namely the direction of the thread. In Fig. 7 (e) we can see that, by the very nature of surface contact, the body touches the table at more than one point, and at each contact point there must be an upward force. What we agree to do is consider all these forces added together and replaced by the one force given by R in Fig. 7 (d). We shall discuss this procedure further in the section on parallel forces.

Exercise 5

1. Fig. 8 shows a small heavy body of weight W at rest under the action of forces which are either horizontal or vertical. The magnitude and/or direction of some of the forces have been omitted. Working intuitively, insert these missing forces.

2. In Question 1, if $W = 10$ N, $T = 15$ N, $P = 4$ N and $X = 12$ N express all the missing forces in newtons.

(8) Bodies in Equilibrium

When a body is at rest under the action of a set of forces we say that the body is in **equilibrium**. Since the forces concerned must have a special relation with each other in order to keep the body at rest, the set of forces is also described as being in equilibrium. That is, we find it convenient to apply the description 'in equilibrium' to both the body and the set of forces acting thereon. In our discussion on simple force diagrams we applied common sense to deduce the results in Fig. 8. Clearly our first thoughts are that, if the body

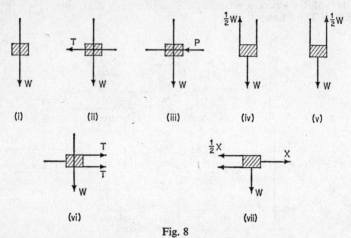

Fig. 8

is in equilibrium, forces which act vertically upwards must balance forces which act vertically downwards. Similarly, horizontal forces acting to the left must balance horizontal forces acting to the right. Associating positive with one direction and negative with the opposite direction shows that what we are really doing is suggesting that the sum of the forces along any straight line must be zero. We now wish to pursue a few more ideas at an intuitive level in order to prepare for their more formal discussion in Chapter Two.

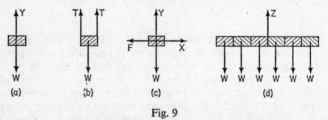

Fig. 9

Consider the force diagrams in Fig. 9, which show a body of weight W resting in equilibrium in each case except (d). In Fig. 9 (a) the sum of the forces is $Y - W = 0$, since the body is in equilibrium. We therefore deduce that $Y = W$. In Fig. 9 (b) the sum of the forces is $2T - W = 0$, and we deduce that $T = \frac{1}{2}W$; but note that we have assumed from a common-sense point of view that the two parallel forces T and T have the sum of $2T$, a result to be discussed at greater length in Chapter Two.

Comparison of Figs. 9 (a) and 9 (b) suggests that Y is equivalent to the combined effect of the two forces T and T since it produces the same result of keeping the body in equilibrium, i.e. we may replace T and T by Y to produce the same result. Since this is true, we say that Y is the **resultant** of the two parallel forces T and T. To summarize: the resultant of two parallel forces of equal magnitude T is a parallel force of magnitude $2T$.

In Fig. 9 (c) we have added two extra horizontal forces to those of Fig. 9 (a). Since Y and W kept the body at rest it follows that F and X cannot disturb the equilibrium, so their sum is zero. Therefore $X - F = 0$ and $X = F$.

In Fig. 9 (d) a new body has been made by using adhesive of negligible weight to stick together six smaller bodies each of weight W. Treating this as one body of weight $6W$ at rest, we know from the discussion above that $Z - 6W = 0$ and $Z = 6W$. This clearly implies that a force of $6W$ in the same line as Z but in the opposite direction is the resultant of the six weights of W each. More important, it suggests that if we consider a large body as being made up of smaller bodies, then all the separate weights of these smaller bodies may be summed and replaced by the one single force equal to the weight of the whole body and acting through a particular point in the body, a point called its **centre of gravity**.

We can avoid any doubts arising about the position of the centre of gravity by considering the body to be so small that it exists as a single particle of matter concentrated at a point. In this case we speak of a **particle** of mass m, regarding the mass as concentrated at a single point in space.

Exercise 6

1. A man approaches a weighing machine, takes off his coat, puts it over his arm, and then weighs himself. Comment.

2. In a cart pulled by a very tired donkey, a farmer is seen to be holding two buckets of pig food. When asked why he did not put them on the floor of the cart he replied: 'Because the donkey is tired.' Comment.

3. The baggage allowance set by an airline is 15 kg. A holidaymaker packs a case by placing each garment or object on the bathroom scales, making a note of its mass and keeping a running total until it reaches 15 kg including the case. Suggest a better way.

CHAPTER TWO

PARALLEL FORCES AND RIGID BODIES

(1) Turning Effect of a Force

So far we have only applied forces to those bodies which have been so small that the turning effect of the forces may be ignored. Now that we are dealing with larger rigid bodies this ability of a force to turn or rotate a body will need to be considered. The turning effect of a force is experienced in such actions as opening doors or gates, lowering tip-up theatre seats, playing on a see-saw, using a carpenter's brace-and-bit drill, and so on. What we need to discover is whether it is possible to measure this turning effect in some way.

Watching children on a see-saw, we soon notice how a heavy body near the pivot can balance a lighter body farther away from the pivot. A closer study of this balancing act reveals that there might be a relation between weight and position roughly expressed by two children sitting 1 m from the pivot being balanced by one child 2 m from the pivot. More precisely, we feel that a weight of $2W$ which is 1 m from the pivot will balance a weight of W placed 2 m from the pivot. We can draw diagrams such as Fig. 10 to represent these ideas.

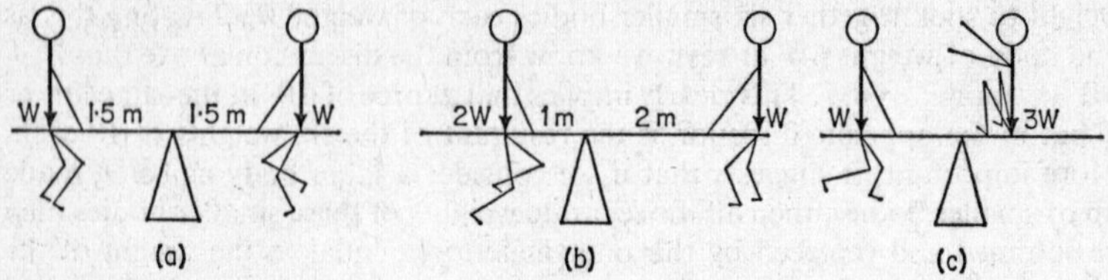

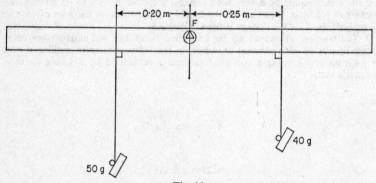

Fig. 10

Perhaps the reader could hazard a guess about the respective positions in Fig. 10 (*c*). Certainly we must now realize that the turning effect of a force about a pivot is dependent on the magnitude of the force and its distance from the pivot. We can begin to confirm this by the following demonstration.

A metre ruler or any similar graduated stick has a 4-mm-diameter hole drilled on its centre line and nearer to one edge than the other, as illustrated in Fig. 11. The pivot, or **fulcrum** as it is called, may be any fixed horizontal

Fig. 11

support such as a nail in the wall or a knitting needle passed through the hole just drilled. With luck, the ruler may balance in the horizontal position; if it will not balance, gummed paper should be stuck on one end until it does come to rest horizontally. We now need a number of standard 50-g and 20-g masses which will be suspended from the ruler by a cotton thread whose mass is small enough to be ignored in obtaining the following results.

Consider a mass of 50 g suspended from one arm of the ruler 0·2 m from the fulcrum to be balanced by a mass of 40 g suspended from a point of the other arm until the ruler is once again in equilibrium in a horizontal position. By trial and error we eventually find that the mass of 40 g has to be placed

0·25 m from the fulcrum F. The relation between the two masses and their distances from F is given by

$$50 \times 0·2 = 40 \times 0·25.$$

Further experiment with other masses and their positions confirms the relation as shown in the following table, in which three results are left for

Left Arm		Right Arm	
Mass, g	Distance, m	Mass, g	Distance, m
50	0·4	100	0·2
40	0·3	50	0·24
20	0·3	50	0·12
20	0·4	40	?
50	0·3	?	0·15
60	?	100	0·24
m_1	l	m_2	r

completion by the reader. The relation is generalized from the last entry in the table by

$$m_1 l = m_2 r$$

or,
$$\frac{m_1}{m_2} = \frac{r}{l} \qquad \text{(i)}$$

There is something curious about the results we have just obtained. We started by speaking about the turning effect of a force; yet here in equation (i) we have a relation involving mass, which is *not* a force. How has this come about? The answer is that the force acting on the ruler and producing the turning effect is the **weight** of the mass suspended, and acts vertically downwards along the cotton thread. Consequently, with the ruler in equilibrium in the horizontal position, the distance measured in each case was the perpendicular distance from the fulcrum to the line of action of the weight. Furthermore, since we have already seen (page 7) that

$$\frac{W_1}{W_2} = \frac{m_1}{m_2} \qquad \text{(ii)}$$

we may combine equations (i) and (ii) to yield

$$W_1 l = W_2 r \qquad \text{(iii)}$$

Equation (iii) now leads us to the following definition.

DEFINITION: *The turning effect of a force about a fulcrum is called the moment of the force about the fulcrum. The moment of a force about a fulcrum is the product of the magnitude of the force and its perpendicular distance from the fulcrum.*

Clearly, if the line of action of the force passes through the fulcrum, the moment of the force about the fulcrum will be zero. Since the basic unit of force is the newton (N), and the basic unit of length is the metre (m), it follows that the basic unit for the measurement of the moment of force is the newton metre (N m). Thus a force of 1 N whose line of action is 1 m from P has a moment about P of 1 N m. Similarly a force of 60 N whose line of action is 0·3 m from P has a moment about P of 18 N m.

We need more information about the direction of the force before we can say whether the moments are clockwise or anticlockwise. In the above example the moment of the force W_1 about the fulcrum F is $W_1 l$. As we are looking at it we describe this moment as anticlockwise. The moment of the force W_2 about the fulcrum F is $W_2 r$, and is described as clockwise. The statement $W_1 l = W_2 r$ is thus saying that the anticlockwise moment of W_1 about F balances the clockwise moment of W_2 about F. This procedure of calculating the moments about F is called 'taking moments about F' for the forces acting on the body.

In each of the problems so far we have only dealt with systems of parallel forces so that the moments have been easy to calculate and the balancing of clockwise and anticlockwise moments has been equally easy to verify.

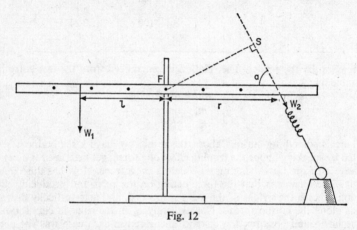

Fig. 12

It is a simple matter to adapt our existing apparatus to examine the case of two non-parallel forces acting on a beam. Holes are drilled symmetrically in the beam (Fig. 12) so that the force in the spring may be applied directly to it. The angle a at which the force W_2 is applied to the beam is arranged by the position of the heavy object to which the spring is fastened; finally the beam is balanced horizontally by positioning force W_1. The moment of W_2 about F is given by $W_2 \times FS$, where FS is perpendicular to the direction of W_2.

The results obtained show that

$$W_1 l = W_2 \times FS$$

But, by trigonometry,

$$FS = r \sin a$$

Hence

$$W_1 l = W_2 r \sin a$$

Example: Using the apparatus of Fig. 12 find the moment of the force $W_2 = 2$ N applied to the beam with $r = 0·3$ m and $a = 30°$. Find also where $W_1 = 3$ N must be applied in order to maintain the beam horizontally at rest.

SOLUTION: From the given information the perpendicular distance of F from the direction of W_2 is given by

$$FS = r \sin a = 0.3 \sin 30° = 0.15 \text{ m}.$$

Therefore the moment of W_2 about F is $2 \text{ N} \times 0.15 \text{ m} = 0.3 \text{ N m}$ clockwise. To balance this force we need

$$W_1 \, l = 0.3$$
$$3l = 0.3$$
$$l = 0.1 \text{ m } (Answer).$$

Exercise 7

1. A metre ruler is pivoted at F as shown in Fig. 11 and rests horizontally in equilibrium when unloaded. If a mass of 60 g is suspended from one arm at a point 0.4 m from F, calculate the position of each of the following masses which balance the ruler when suspended from the other arm:

(a) 100 g, (b) 200 g, (c) 120 g, (d) 240 g.

2. In Fig. 11 the hole which was drilled in the metre ruler had to be nearer to one edge than the other. In Fig. 13 the ruler is 'upside down' from its position in Fig. 11.

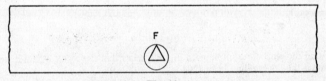

Fig. 13

Can the ruler still rest horizontally in equilibrium? What is the difficulty in trying to balance the ruler in this position?

3. A metre ruler rests horizontally in equilibrium when unloaded. A mass of 100 g is suspended from a point 0.48 m from the fulcrum to balance a mass of 120 g. Find the point from which the 120 g is suspended.

4. If in Question 3 the point of suspension of the 100-g mass is moved 60 mm nearer to F, what change is needed in the position of the 120-g mass in order to maintain the ruler in equilibrium in the horizontal position?

5. If in Question 3 the 100-g mass is decreased by 10 g, by how many grammes must the 120 g be decreased if both masses remain where they are with the ruler in equilibrium?

6. Calculate the moments about P of each of the following forces:

Force	Distance of Line of Action from P
10 N	0.5 m
20 N	300 mm
40 N	0.75 m
100 N	30 mm

7. $ABCD$ is a square of side 0.25 m. A force of 10 N acts along AB in the direction A to B. Calculate the moment of this force about the points:

(i) A, (ii) C, (iii) B, (iv) D.

8. Fig. 14 shows a uniform beam at rest in the horizontal position with F at its midpoint. Vertical forces are applied by means of two equal springs fixed to the beam at the points P and Q such that $PF = 2FQ$. Forces L and R are exerted on the springs,

and the corresponding extensions of the springs are x and y, the beam being maintained in the horizontal position. If a force of 1 N extends either spring through 0·01 m, complete the following table:

L	x	R	y
2 N	?	?	?
?	0·03 m	?	?
?	?	5 N	?
?	?	?	0·08 m

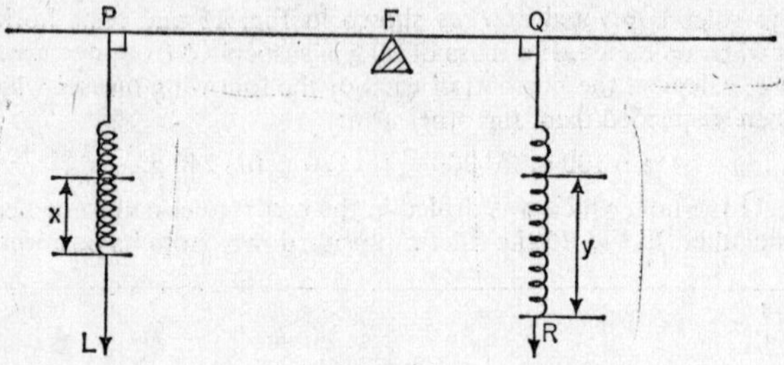

Fig. 14

9. Using the worked example with Fig. 12, show that the result would have been the same if the angle a had been 150°.

10. Calculate the position of W_1 in the worked example of Fig. 12 if angle a had been 45°.

(2) The Beam Balance

Looking back at the equations, p. 15,

$$\frac{W_1}{W_2} = \frac{m_1}{m_2} \quad \text{and} \quad \frac{m_1}{m_2} = \frac{r}{l},$$

we see that the apparatus of Fig. 11 enables us to compare the masses and weights of any bodies at any place. From these equations it is clear that if $l = r$, then $m_1 = m_2$, and $W_1 = W_2$. This result was suggested by Archimedes about 240 B.C. in the form: 'equal weights at equal distances are in equilibrium'. As the apparatus suggested in Fig. 11 is too crude for accurate work, manufacturers make a more sensitive instrument for comparing masses; this is called a **beam balance**.

A typical beam balance is shown in Fig. 15. The beam AB is supported on a steel knife-edge at F. At each end of the beam a scale pan (P and Q) is suspended from a knife-edge (A and B). In conjunction with the small plumb-line attached to the support column, the screws on the base-board enable the instrument to be levelled before readings are taken. When not in use, the beam AB rests on two stops at S. To take readings the beam is lifted clear of S by use of the control knob C.

Attached to the beam just under the fulcrum F is a long pointer whose central position on the scale indicates that the beam is horizontal, and whose direction of movement indicates which pan is carrying the lighter load. The construction is such that the two arms AF and FB are of equal length. Adjustments can be made by moving the counterweight nuts (n and m) so that, when both pans carry no load, the beam is in equilibrium in the horizontal position. The method of use is to place the body of unknown mass in pan *P* and then load pan *Q* with known masses until the beam balances horizontally; the sum of the masses in pan *Q* is then equal to the unknown mass in pan *P*. There is an unfortunate confusion of words in describing this operation which, although obtaining the mass of the body, is actually called 'weighing the body'; furthermore, the standard masses marked 20 g, 50 g, 100 g, etc., are sometimes referred to as 'weights'.

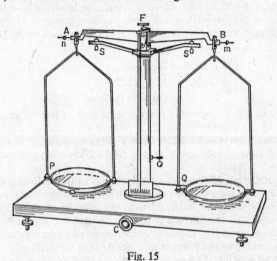

Fig. 15

Finally, it must be remembered that the beam balance determines the mass of a body in grammes or kilogrammes, a result which is independent of where the balance is used since the mass of a body is the same everywhere. The spring balance, however, determines the weight of the body, a result which *will* vary according to where the body is situated.

Wear and tear on the balance reduces its accuracy owing to displacement of the knife-edges, bending of the beam, or wear on the pans. Consequently the arms of the balance become unequal in length and the scale pans become unequal in weight or the beam becomes improperly balanced, and we can no longer rely on equal masses being able to balance one another. We can overcome these deficiencies by the simple method outlined in the following exercise.

Exercise 8

1. Assuming that our balance is faulty, we place the body in pan *P* and balance the beam by pouring sand or similar dry material into the pan *Q* until the pointer reaches some chosen mark on the scale. Complete the explanation of how to find the mass of the given body.

2. A body is placed in pan *P* and counterbalanced by 140 g placed in pan *Q*. Without using sand or any other masses, how would you discover whether 140 g was the true mass of the body?

3. Using three different masses of 4 g, 2 g, and 1 g, it is possible to balance any body whose mass is a whole number of grammes up to 7 g on an accurate beam

balance. Using the same balance find the minimum number of masses and their size which enable you to balance any body whose mass is a whole number of grammes up to (*a*) 15 g, (*b*) 31 g, (*c*) 63 g.

4. Using only the three masses 9 g, 3 g, 1 g, how is it possible to balance a body of mass 7 g on an accurate beam balance?

5. Using the three masses 9 g, 3 g, and 1 g only, show how to balance any body whose mass is a whole number of grammes up to 13 g.

(3) Calculation of Moments

For each case in which we have loaded the ruler or beam we have found that a calculation of the form 100 g × 0·24 m has been involved. If we examine such a calculation we see that it may be replaced for example by two others such that

$$100 \times 0·24 = 24$$
$$= 12 + 12$$
$$= (30 \times 0·4) + (60 \times 0·2)$$

In other words, 100 g suspended from a point 0·24 m from *F* produces the same result as 30 g suspended from a point 0·4 m from *F*, together with 60 g suspended from a point 0·2 m from *F*. This replacement is not unique, for we could have chosen $(50 \times 0·3) + (90 \times 0·1) = 24$, and many other replacements.

We shall make use of this type of calculation in the following examples, in which each beam is now drawn as a single horizontal line and is considered to be in equilibrium when balanced with the fulcrum at its midpoint. We know, of course, that the fulcrum or pivot is an axis, but when all the forces act in the same plane (i.e. coplanar forces) it is usual to simplify the discussion by referring to the fulcrum as a point; thus we shall refer to 'taking moments about a point'. The meaning is always obvious from the context of the problems and the diagrams which represent them.

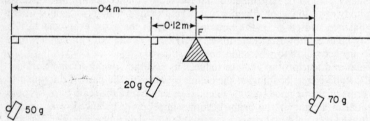

Fig. 16

Example: A beam which rests horizontally in equilibrium when unloaded with the fulcrum at *F*, carries loads as suggested in Fig. 16. Calculate the distance *r* from the fulcrum *F* to the point of suspension of the mass of 70 g.

SOLUTION: Since the beam is in equilibrium, the sum of the anticlockwise moments is equal to the sum of the clockwise moments about the fulcrum *F*. Therefore

$$(50 \times 0·4) + (20 \times 0·12) = 70r$$
$$20 + 2·4 = 70r$$
$$\frac{22·4}{70} = r$$
$$r = 0·32 \text{ m} \quad (Answer).$$

Example: A beam rests horizontally in equilibrium when unloaded with the fulcrum at its midpoint. A vertical downward force of 10 N is applied at one end 0·48 m from F. Find the two positions of the vertical force which will keep the beam horizontal when applied to the beam 0·16 m from F.

SOLUTION: Let the unknown force be T. Taking moments about F for the forces acting on the beam when in equilibrium we have,

$$T \times 16 = 10 \times 48$$
$$\therefore T = \frac{10 \times 48}{16} = 30 \text{ N.}$$

So far all we have shown is that we need a force of 30 N to supply the required anti-clockwise moment to balance the 10 N already present.

That there are two possible solutions can be seen from Fig. 17. The first solution (*a*) will tend to lift the beam off the fulcrum so that the axis must be threaded through the beam.

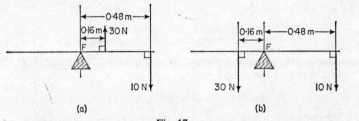

Fig. 17

In each example we have considered so far the beam has balanced horizontally in equilibrium when unloaded. However, experience shows that we may balance the beam horizontally about any point as fulcrum using only one force supplied by the suspension of a mass or body from the beam. Fig. 18 (*a*) shows how this can be done using a body of weight W.

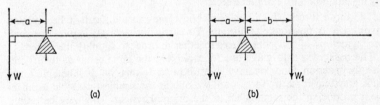

Fig. 18

We have already seen that, if a body rests in equilibrium on a fulcrum F, the sum of the anticlockwise moments about F balances the sum of the clockwise moments about F. When we look at Fig. 18 (*a*) we notice that we have omitted the force which will provide the clockwise moment about F, and this raises the question of where it is to come from. On page 13 it was suggested that we could consider the weight of a body as acting through one point called its centre of gravity. Continuing with this assumption, we consider the weight W_1 of the beam in Fig. 18 (*b*), to act through a point at a distance b from F so that, on taking moments about F for the forces on the beam, we get

$$Wa = W_1 b$$

It follows, therefore, that if a body has its centre of gravity outside the vertical line through the fulcrum, we shall have to consider the moment of the weight of the body about the fulcrum. There is, of course, a reaction force on the beam at the fulcrum F, and its direction in Fig. 18 is clearly vertically upwards. We would like to obtain the magnitude of this force. Perhaps the reader can guess the magnitude of this force in Fig. 18.

Exercise 9

1. Where on a horizontal beam must a 100-g mass be attached in order to produce the same turning effect as the following arrangement of masses:

(*a*) 20 g at 0·3 m + 30 g at 0·4 m on the same arm;
(*b*) 30 g at 0·2 m + 40 g at 0·3 m on the same arm;
(*c*) 100 g at 0·1 m + 20 g at 0·4 on different arms;
(*d*) 200 g at 0·3 m + 100 g at 0·4 m on different arms?

2. A steel beam 1 m in length rests horizontally in equilibrium when pivoted about its midpoint F. If a vertical downward force of 10 N is applied at a point 0·4 m from F, and a vertical downward force of 80 N is applied at a point 0·1 m from F on the same arm, find the points of application of the following forces in order to maintain the beam in equilibrium in the horizontal position:

$$(a)\ 30\ \text{N};\ (b)\ 40\ \text{N};\ (c)\ 36\ \text{N};\ (d)\ 48\ \text{N}.$$

3. The moment of a force about a pivot F is 24 N m. How far from the pivot F is each of the following forces in order to produce the same turning effect: (*a*) 8 N; (*b*) 48 N; (*c*) 60 N?

4. A beam of length 1 m would be in equilibrium if pivoted horizontally about its midpoint F. A weight $10W$ is applied to one end of the beam in order to balance it about a point 0·10 m from the midpoint. Find the weight of the beam.

(4) Resultant of Two Parallel Forces

The more forces which act on a body, the more difficult it becomes to investigate its possible equilibrium. To carry out such an inquiry we shall need a method of collecting forces together in order to simplify the system.

The basic idea is to make one force achieve the same result as two forces already in action. Once we have learnt how to do this, we shall be able to reduce a system force by force until we obtain the simplest possible arrangement producing the same effect as the original system. This may be only a single force if we are fortunate. If no forces are applied to a body, it will remain at rest; so we reason that, if a body remains at rest under the action of a set of forces, the simplest system which will produce the same effect is no force at all.

A system of two forces keeping a body in equilibrium is illustrated by the suspension of a body from a spring balance. Fig. 19 (*a*) shows a piece of wood of uniform thickness 3 mm and having three holes H_1, H_2, and H_3. It is suspended from H_1. There are only two forces acting on the body: the weight W of the wood and the tension T in the spring. The two forces are equal in magnitude; both act in the same vertical line, but they are in opposite directions. Since the piece of wood is in equilibrium, we must have $T - W = 0$ or $T = W$. The important fact here is that the weight of the body acts in a vertical line through H_1 when suspended as in Fig. 19 (*a*). We draw this vertical line on the surface of the wood and then repeat the procedure by suspending the body from H_3. Since we have two lines on the wood surface, we ask ourselves what significance can be given to their point G of intersection? Each line was the line of action of the weight, and it is clear that if the weight is to be considered as acting through a point, this point must be

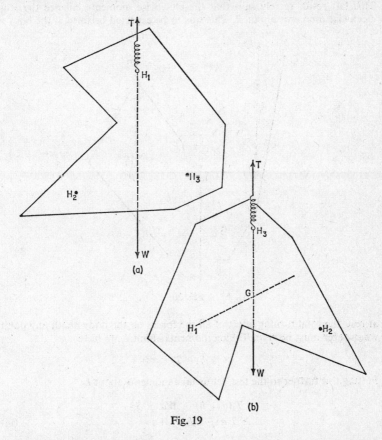

Fig. 19

the point of intersection G. We can put this to the test by suspending the body from H_2. It will be found that the vertical line passes through G again. Therefore the weight may reasonably be considered to act always through the point G. The point through which the weight of a body may be considered to act at all times is called its **centre of gravity**.

Let us now suspend our piece of wood by two spring balances as shown in Fig. 20. Both spring connections are kept vertical so that all three forces acting on the body are parallel to each other. In keeping with our idea that the sum of the forces acting on a body at rest must be zero, we find that

$$T_1 + T_2 - W = 0,$$
$$\text{or,} \qquad T_1 + T_2 = W \qquad\qquad\qquad \text{(i)}$$

But we can obtain more information than this by tracing the vertical lines of action of the forces to intersect a horizontal line at L, K, and R, as shown in Fig. 20. By measurement we confirm that

$$T_1 b = T_2 a \qquad \text{(ii)}$$

This last result merely says that the clockwise moments balance the anti-clockwise moments about K. This was to be expected because, if the body is

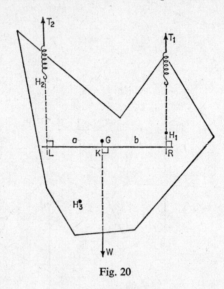

Fig. 20

at rest, the total turning effect of all the forces on the body about any point whatsoever must be zero. Taking moments about K we have

$$T_1 b - T_2 a = 0.$$

Putting this further to the test, let us take moments about L:

$$T_1(a + b) - Wa = S$$
$$\therefore \ T_1 a + T_1 b - Wa = S \qquad \text{(iii)}$$

But we have already seen that $T_1 + T_2 = W$, so that $T_1 a + T_2 a = Wa$. Substitution for Wa in equation (iii) yields

$$T_1 a + T_1 b - (T_1 a + T_2 a) = S$$

which reduces to $\qquad T_1 b - T_2 a = S$

On using the result (ii) we find that $S = 0$, which means that the sum of the moments taken about L is zero.

Returning to Fig. 20, we can say that the three forces T_1, T_2, and W are in equilibrium since they keep the body in equilibrium. In other words, the two forces T_1 and T_2 balance the weight W. We know from Fig. 19 that the simplest way to balance the weight W is to apply an equal and opposite force in the same vertical line through G, so it follows that this could also replace the two forces T_1 and T_2. We describe this force as the **resultant** of the two forces T_1 and T_2.

DEFINITION: *The resultant of two parallel forces acting on a rigid body is that single force which will produce the same effect on the body.*

The forces T_1 and T_2 in Fig. 20 are called **like** parallel forces. Two parallel forces in opposite directions are called unlike parallel forces, e.g. T_1 and W are unlike parallel forces. Here we have only dealt with vertical forces, but the relations obtained above hold for parallel forces in any direction. This can

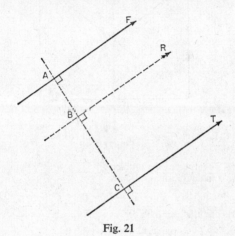

Fig. 21

be seen with the aid of Fig. 21. Draw any line to intersect at right-angles the line of action of the two like parallel forces F and T. Call the points of intersection A and C.

The magnitude of the resultant R of the two forces F and T is given by

$$R = F + T$$

and is parallel to F and T. The line of action of R is suggested to pass through B, and the position of B is given by

$$F \times BA = T \times BC.$$

We may appreciate this result by the following reasoning. If R is the resultant of F and T, then it must have the same moment as F and T about any point, including B. But the moment of R about B is zero, so the moment about B of F and T together must also be zero, which means that

$$-F \times BA + T \times BC = 0$$

or
$$F \times BA = T \times BC.$$

Observe that if $F = T$, then $R = 2F$ and $BA = BC$. Hence the resultant of two like parallel forces of equal magnitude is the same distance from each force. Note also that we have taken the anticlockwise moment as a positive quantity.

Example: The centre of gravity of a uniform rectangular board lies at the intersection of its diagonals. The dimensions of the board are 0·4 m × 0·3 m, and its weight is W. It is suspended by two vertical strings attached to one of the longer sides, 50 mm and 100 mm from opposite ends. Find the tension in each string when the longer sides are horizontal.

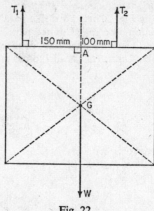

Fig. 22

SOLUTION: Let Fig. 22 represent the problem.
Since the board is in equilibrium, we have

$$T_1 + T_2 - W = 0 \qquad \text{(i)}$$

Taking moments about A, we have

$$100\,T_2 - 150\,T_1 = 0$$

or

$$2T_2 = 3T_1$$
$$T_2 = 1\cdot5T_1 \qquad \text{(ii)}$$

Substituting this result in equation (i)

$$T_1 + 1\cdot5T_1 = W$$
$$2\cdot5T_1 = W$$
$$T_1 = 0\cdot4W$$

Using equation (i) again, we get

$$T_2 = 0\cdot6W$$

So the tensions in the strings are $T_1 = 0\cdot4W$ and $T_2 = 0\cdot6W$ (*Answer*).

Example: A decorator uses a plank placed on two supports 3 m apart. The plank is uniform and placed so that its weight of 100 N acts midway between the supports. If the decorator's weight is 600 N, find the reactions at the supports when he stands (*a*) equidistant from each support, (*b*) 1 m from one support.

SOLUTION: Fig. 23 represents case (*b*), with R_A and R_B the vertical support forces on the plank.
In case (*a*) the total weight of 700 N is acting at the midpoint of the plank, so

$$R_A = R_B \text{ and } R_A + R_B = 700$$

Therefore

$$R_A = R_B = 350 \text{ N}   (Answer).$$

In case (*b*) we again have

$$R_A + R_B = 700$$

Taking moments about A for all the forces on the plank we get

$$R_B \times 3 - (100 \times 1.5) - (600 \times 2) = 0$$
$$3R_B = 150 + 1200$$

Therefore $$R_B = 450$$

Since $R_A + R_B = 700$, we can see that $R_A = 250$.
So the reactions are $R_A = 250$ N and $R_B = 450$ (*Answer*).

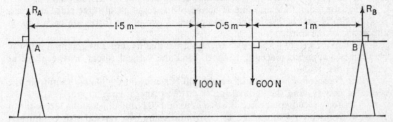

Fig. 23

Example: Find the magnitude, direction and line of action of the resultant of two like parallel forces of magnitude 100 N and 300 N, whose lines of action are 0.12 m apart.

SOLUTION: Fig. 24 represents the problem, with line AB drawn perpendicular to the lines of action of the given forces. We do not know where the resultant force is but we do know that it will be parallel to the other two: we therefore call it R and suggest that it intersects AB at C. We now suggest that $AC = x$, in which case

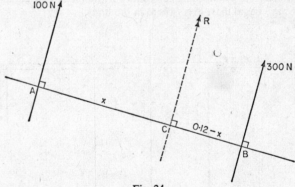

Fig. 24

CB must be $(0.12 - x)$. We know that $R = 100 + 300 = 400$ N. Since the resultant passes through C, the sum of the moments of the two forces 100 N and 300 N about C must be zero. Alternatively the moment of the force of 100 N about C must balance the moment of the force of 300 N about C. Therefore

$$100x = 300 (0.12 - x)$$
$$100x = 36 - 300x$$
$$400x = 36$$
$$x = 0.09 \text{ m}$$

Hence the resultant is a parallel force of 400 N acting at a distance of 90 mm from the force of 100 N.

Exercise 10

1. A uniform piece of plywood has two small holes drilled through it so as to be freely suspended from a horizontal pivot. Draw a sketch in the manner of Fig. 19 to indicate the vertical direction for each point of suspension when the piece is in the shape of (*a*) a square, (*b*) a rectangle, (*c*) a circular disc.

2. Two like parallel forces of magnitude 100 N and 200 N act along lines which are 0·36 m apart. Find the magnitude and line of action of their resultant.

3. Find the magnitude and line of action of the resultant of three parallel forces, each of 100 N and acting in the same plane, if two of the forces are each 1 m on either side of the third.

4. A plank of weight 80 N rests horizontally so that its centre of gravity is midway between two supports placed 2 m apart. Find the vertical forces on the plank at each support.

5. If in Question 4 a woman of weight 500 N stands on the plank 0·5 m from the centre of gravity, find the vertical forces on the plank at each support.

6. A uniform rectangular piece of wood of weight 40 N and dimensions 1 m × 0·8 m is suspended with its shorter edges horizontal using two vertical strings attached to the upper edge 0·1 m and 0·2 m from each end. Find the tension in each string.

7. What is the force at the fulcrum on the beam in (*a*) Fig. 17 (*b*), and (*b*) Fig. 18 (*b*)?

(5) Centre of Gravity

We recall that the centre of gravity of a body is that point at which the whole weight of the body may be considered to act. We use the abbreviation C.G. for centre of gravity. To simplify our work we frequently restrict our inquiries to **uniform** bodies only. By 'uniform' we mean that equal volumes of matter have equal mass everywhere in the body.

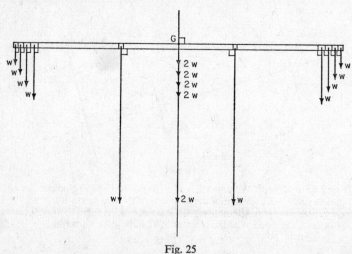

Fig. 25

Centre of gravity of a thin uniform rod. We consider the rod to be divided into small sections or elements of equal weight *w*. We now take two elements which are at equal distances from the centre of the rod, so that their weights give two like parallel forces of equal magnitude *w* as illustrated in Fig. 25.

We know that these weights have a resultant of $2w$ acting vertically downwards through G, the midpoint of the rod. We may therefore replace the two separate weights of w each by one of $2w$ through G. We continue in this manner until we have considered the weights of all the elements of the rod. This leaves us with forces of $2w$ all acting in the same vertical line through G and the sum of all these forces is the weight of the complete rod. We have shown therefore that the weight of a thin uniform rod may be considered to act at the midpoint of the rod.

Centre of gravity of a thin uniform rectangular board. Let the rectangular board be represented by $ABCD$ in Fig. 26. We divide the figure into elementary thin strips, such as PQ, parallel to AB.

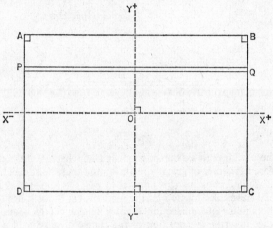

Fig. 26

Since the weight of each strip acts through its midpoint, all the weights of the separate strips act on the line $Y^- Y^+$ which is equidistant from AD and BC. Therefore the weight of the complete board must act through a point somewhere along the line $Y^- Y^+$. Similarly, if we divide the rectangle into strips which are parallel to AD, we deduce that the weight of the board must act through a point on $X^- X^+$. Since there is only one centre of gravity it must be at the intersection of $Y^- Y^+$ and $X^- X^+$, a result which we have accepted as intuitively obvious or obtained experimentally so far.

In applied mathematics it is customary to describe a thin body as a lamina. Thus when we speak of a 'rectangular lamina' instead of a 'rectangular board' we are saying that the thickness of the figure may conveniently be ignored in order to keep our calculations in two dimensions. We often take this one step farther and speak simply of finding the centre of gravity of a rectangle or a triangle or any other geometrical figure. The centre of gravity is often referred to as the **centroid** or the **centre of mass**: for our purposes these are all one and the same point. However, it is as well to mention one source of difference. We assume that the weights of the elements of any body are all parallel to each other. But we know that, if the body is very large, the verticals through two extreme points will not be exactly parallel since vertical lines meet at the centre of the earth. So the C.G., the point through which the total weight acts, may not be the same as the geometric centre of mass or centroid. Furthermore, the body may be so large that even the weights of equal masses may not be the same everywhere in the body.

Centre of gravity of a uniform triangular lamina. Again we divide the figure into strips, such as PQ in Fig. 27, parallel to one side. Since the strips are narrow, we may consider their weight to act at their midpoints. The weight of the complete triangle must be the resultant of all the weights of the strips acting at their midpoints.

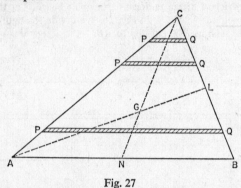

Fig. 27

But all these midpoints lie on the median CN, where N is the midpoint of AB. Therefore the centre of gravity of the lamina must lie somewhere on CN. Similarly, if we had chosen strips parallel to BC, the centre of gravity of the lamina would have been located on the median AL. Since it lies on both AL and CN, their point of intersection G must be the centre of gravity of the lamina. So we describe the position of the centre of gravity of a triangle as being at the point of intersection of its medians.

By geometry we find that for any triangle,

$$NG = \tfrac{1}{3}NC,$$
$$LG = \tfrac{1}{3}LA$$

and one often hears the phrase, 'the centre of gravity of a triangle is one-third of the way up the median'.

Exercise 11

1. By reasoning similar to that used for the rectangular figure on page 29, find the C.G. of a uniform lamina in the shape of a parallelogram.

2. A uniform lamina is in the shape of an equilateral triangle with a height of 0·6 m. How far is the C.G. from each vertex?

3. $ABCD$ is a rectangle with $AC = 60$ mm. Find the position of the C.G. of each of the triangles ABC and ADC.

4. A uniform lamina is in the shape of a trapezium $ABCD$ having AB parallel to CD. Explain why the C.G. must lie on the line joining the midpoint of AB and CD.

5. Explain how you would obtain the C.G. of a piece of uniform thin card of any shape.

6. A piece of cardboard in the shape of a circular disc has a smaller concentric disc removed. Where is the C.G. of the remaining ring or annulus? Given an example of another shape whose C.G. is not a point of the body.

(6) Composite Figures

We will now find the C.G. of a lamina whose shape consists of figures whose C.G. is already known. For example a shape made up of a triangle attached to a square or a rectangle with a triangular piece removed. Such figures are known as composite figures.

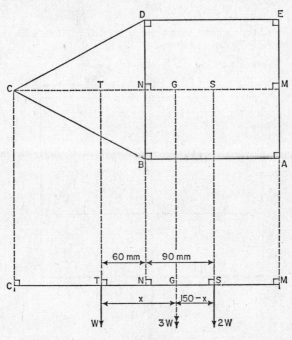

Fig. 28

Example: *ABCDE* in Fig. 28 is a uniform plane lamina, with *ABDE* a square of side 180 mm and *BCD* an isosceles triangle of height $CN = BD = 180$ mm. Find the C.G. of *ABCDE*.

SOLUTION: N and M are the midpoints of *BD* and *AE*, respectively; so, as the figure is symmetrical about the line *CNM*, the C.G. of the square and the C.G. of the triangle both lie on the line *CNM*.

(i) We now need to find the weights of each part of the figure. Since the lamina is uniform, the weight of each part of the figure is proportional to its area, i.e. doubling the area means doubling the weight. Since triangle *BCD* has the same base and height as the square, its area is half that of the square. If we let the weight of the triangle *BCD* be W, the weight of the square must be $2W$.

(ii) We have now reduced the problem to finding the position of the resultant of two parallel forces: W acting at the C.G. of the triangle, and $2W$ acting at the C.G. of the square. Both these C.G.s lie on the straight line *CNM*.

(iii) A side view of the forces is shown at the bottom of Fig. 28.

The weight W acts at T, where $NT = \frac{1}{3}NC = 60$ mm.

The weight $2W$ acts at S, where $NS = \frac{1}{2}NM = 90$ mm.

The resultant is $3W$ acting at G, such that $W \times GT = 2W \times GS$ (see page 25). We now suggest $GT = x$ and $GS = 150 - x$. Therefore

$$Wx = 2W(150 - x)$$
$$x = 300 - 2x$$
$$3x = 300$$
$$x = 100.$$

The C.G. of the complete lamina lies at point G on CNM such that $TG = 100$ mm or $NG = 40$ mm   (*Answer*).

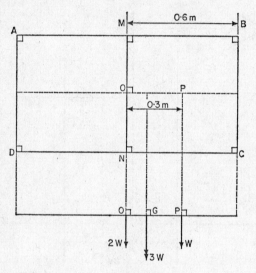

Fig. 29

Example: $ABCD$ in Fig. 29 is a rectangular piece of uniform cardboard with $AB = 1\cdot2$ m. A rectangular piece of the same cardboard is now fixed to $ABCD$ and covers the part $BMNC$, where M and N are the midpoints of AB and DC. Find the position of the C.G. of the new body.

SOLUTION: We may represent the weights of $ABCD$ and $BMNC$ as $2W$ and W respectively.

The weight $2W$ acts at O (the centre of $ABCD$) and the weight W acts at P (the centre of $BMNC$). OP is on the line midway between AB and DC.

The total weight of $3W$ will clearly act at the centre of gravity G on OP. The distance $OP = 0\cdot3$ m and we let $GO = x$. Therefore

$$2Wx = W(0\cdot3 - x) \qquad \text{(see page 25)}$$
$$2x = 0\cdot3 - x$$
$$3x = 0\cdot3$$
$$x = 0\cdot1.$$

So the C.G. of the complete figure is at G on OP such that $OG = 0\cdot1$ m.

Exercise 12

1. Repeat the worked example on page 31 for $BD = CN = 240$ mm.
2. Repeat the worked example on page 32 for $AB = 3$ m.
3. $ABCD$ is a uniform square piece of cardboard and E is the midpoint of DC. Another layer is now added on to triangle ABE. Find the C.G. of the new body if $AB = 90$ mm.
4. A rectangular piece of paper $ABCD$ is folded flat about the diagonal AC. Find the C.G. of the folded figure.
5. ABC is a straight line with $AB = 1$ m, $BC = 2$ m. Find the C.G. of three weights: $2W$ at A, $5W$ at B, and $3W$ at C.
6. $ABCD$ is a square with $AB = 1$ m. Find the C.G. of four weights: W at A and D, and $2W$ at B and C.
7. Find the C.G. of three equal weights W placed one at each vertex of the triangle ABC.

(7) Solid Figures

We now wish to extend the idea of the moment of a force into three dimensions by considering the effect of forces applied to solid figures. Consider a horizontal trap-door of weight 30 N which is in the form of a rectangle $ABCD$ hinged along one edge, AB, as shown in Fig. 30. Assuming the door

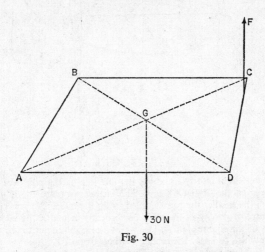

Fig. 30

is uniform, we deduce that the C.G. is at the intersection G of the diagonals AC and BD. If we attach a spring balance to any point on DC we soon discover:

(*a*) that the least force necessary to support the door is 15 N vertically upwards,
(*b*) that this least force is independent of where we attach the spring balance on CD.

With the least force F just supporting the door, we observe once again that the sum of the moments of the forces about the axis AB is zero. If the perpendicular distance of F from AB is $2l$, then the perpendicular distance of the weight from AB is l.

Taking moments about AB,

$$F \times 2l = 30 \times l$$

Therefore $F = 15$ N.

Example: A circular table-top of radius 1 m is supported on a horizontal floor by four vertical legs symmetrically placed at the vertices of a square $ABCD$ of side 1·2 m, as shown in Fig. 31. Find the least vertical force which will tip the table over when applied to the edge of the circular top if the total weight of the table is 200 N.

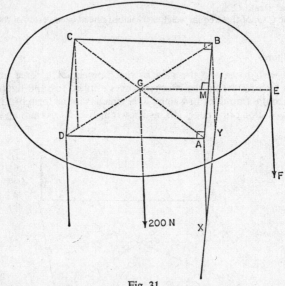

Fig. 31

SOLUTION: The table will clearly tip about the line joining the points of contact of two of the legs on the floor (XY say). The farther we place the required force from this line, the smaller the force needed to tip the table.

Let the centre of gravity of the table be vertically below G, the intersection of the diagonals of the square $ABCD$. Draw GME at right-angles to AB to give the point E from which to apply the force F farthest from XY.

Now $GM = 0·6$ m and $GE = 1$ m. Therefore $ME = 0·4$ m.

Taking moments about XY with the table just about to tip about XY we have

$$F \times ME = 200 \times MG$$
$$F \times 0·4 = 200 \times 0·6 = 120$$

Therefore $\qquad\qquad\qquad F = 300$

So the least force required to tip the table is 300 N (*Answer*).

Exercise 13

1. A uniform table with a circular top of radius 0·8 m is supported by four equal vertical legs symmetrically placed at the vertices of a square $ABCD$ of side 1 m as illustrated in Fig. 31. If the total weight of the table is 180 N, calculate the least

vertical force which may be applied to the edge of the table to tip it over when the. table rests on a horizontal floor.

2. A uniform square-top table of side 1 m is supported on a horizontal floor by four equal vertical legs placed at the midpoints of the sides. If the total weight of the table is 140 N, calculate the least vertical force applied to the top necessary to tip the table over, assuming that the C.G. is vertically below the centre of the top.

3. A circular table is supported on a horizontal floor by three equal vertical legs symmetrically placed on the rim of the circular top. If the weight W of the table acts through the centre of the top, find the least weight which will tip the table over when placed on the rim of the top.

4. What is the advantage of a three-legged stool over one with four legs?

5. What advantage is there in splaying the legs of a stool outwards?

CHAPTER THREE

KINEMATICS

Kinematics is the study of motion without reference to the forces which may cause that motion.

To appreciate some of the basic ideas of this section, let us suppose that we are standing at a point A in a field and that we receive instructions to walk a distance of 50 m in a straight line. Fig. 32 shows some of the alternative

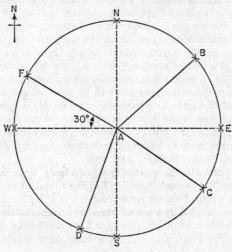

Fig. 32

routes available which satisfy the instructions; it indicates that we may arrive at any point on a circle centre A and radius 50 m.

In order to arrive at a definite point on the circle we must know not only how far to walk but also in what direction. For example, to get to E from A

we must walk 50 m in the East direction; to get to *F* from *A* we must walk 50 m in a direction W 30° N. We see from this example that to specify completely a change in position we must give: (*a*) a starting-point, (*b*) a direction, and (*c*) a distance. We call the consequent change in position a **displacement**.

DEFINITION: *A displacement is a change in position.*

Although we are not concerned with forces at the moment, it should be noticed that a displacement has the same characteristics as a force (see page 2).

(1) Speed and Velocity

Returning to Fig. 32, let us suppose the field lies on the side of a hill sloping downwards from East to West. Walking from *A* to *E* will probably take longer than walking from *A* to *W* since the first direction is uphill and the second direction is downhill. Possibly we may take the same time to walk from *A* to *N* or *A* to *S*, but any difference in time depends on the rate at which we travel and even this rate may vary. For example, a brisk start uphill may quickly change to a tired climb, depending on the slope and state of the field. We are therefore comparing rates of travel independent of the direction in which the motion takes place.

DEFINITION: *Speed is the rate of change of distance with respect to time.*

If equal distances are travelled in every equal interval of time during the motion, the speed is said to be **uniform.** The standard units of speed are the metre per second (abbreviated to m s^{-1}) and the kilometre per hour (abbreviated to km h^{-1}). Thus to say that a body has a uniform speed of 17 m s^{-1} means that during each second of the motion the body travels a distance of 17 m. In reality it is almost impossible to ensure that a body travels with a uniform speed, but we can approximate to it very closely and it is certainly mathematically convenient to consider that a uniform speed is possible.

When a body travels with a variable speed, i.e. non-uniform speed, we quote its speed at a particular time. For example, if the speedometer of a car registers 36 km h^{-1} it means that, provided we maintain this speed, we shall travel 36 km in the next hour, or 10 m in the next second. In other words, stating the speed of a body at a particular time merely suggests the distance which will be covered in the next interval of time if that speed remains unchanged. For the greatest accuracy we consider the next interval of time to be as short as possible, because it is more realistic to suggest what might happen in the next second rather than the next hour.

Very often we consider the **average speed** of a body, since this is the easiest to measure. If the 50 m journey from *A* to *E* in Fig. 32 had taken 25 s, our average speed would have been 2 metres per second or 2 m s^{-1}. Thus average speed is the total distance divided by the total time for travelling that distance. Even if we had run the first 30 m, then stopped for a rest before continuing, provided the total time taken was 25 s the average speed would have been the same.

When we associate a direction with the speed of a body, we speak of the **velocity** of the body.

DEFINITION: *Velocity is the rate of change of displacement with respect to time.*

Thus the speed of the body is the magnitude of its velocity. Clearly, *uniform* velocity may only take place in a straight line: this is because the condition of uniformity refers not only to the magnitude but also to the sense and direction of the velocity. The distinction between speed and velocity becomes clearer if we return to motion on a circle. It is possible to run in circles at uniform speed but *not* uniform velocity, since the direction of motion is always changing.

Consider a wheel spinning on an axis through its centre of mass, e.g. the wheel of a jacked-up car or an inverted bicycle. Since the centre of mass lies at the axis and is stationary we cannot say that the wheel has a linear velocity. At the same time we cannot say that the wheel is at rest, because its separate parts are moving. We describe the motion by introducing the idea of **angular velocity**.

DEFINITION: *The angular velocity of a body about an axis is measured by the rate of angular displacement of any line in the body which passes through the axis.*

In Fig. 33 we have a wheel or disc rotating anticlockwise about an axis through *C* at right-angles to the plane of the wheel. *CP* is a radius or line fixed in the wheel, e.g. a spoke of the wheel, and *CX* is a fixed line in space, e.g. the horizontal line through *C*. The rate of change of angle *PCX* with respect to time measures the angular velocity of the wheel.

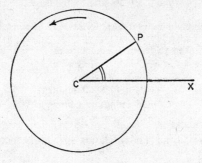

Fig. 33

The angle *PCX* is measured in radians, and the angular velocity is measured in radians per second (abbreviated to rad s^{-1}). Consider, for example, the hands of a clock. The seconds hand makes one complete revolution per minute. Since one revolution is 2π radians, it follows that the angular velocity of the seconds hand is

$$\frac{2\pi}{60} = \frac{\pi}{30} \text{ rad s}^{-1}.$$

Thus the motion of a wheel on a moving car is twofold: it has a linear velocity and an angular velocity.

We now consider some typical calculations related to speed, time, and distance.

Example: Starting at *A*, a body travels 500 m in 20 s to a point *B*, and then travels a further 400 m in 80 s to a point *C*. Find the average speed in travelling from (*a*) *A* to *B*, (*b*) *B* to *C*, (*c*) *A* to *C*, if *ABC* is a straight line.

SOLUTION: (*a*) The average speed from *A* to *B* is

$$\frac{500 \text{ m}}{20 \text{ s}} = 25 \text{ m s}^{-1} \quad (Answer).$$

(*b*) The average speed from *B* to *C* is

$$\frac{400 \text{ m}}{80 \text{ s}} = 5 \text{ m s}^{-1} \quad (Answer).$$

(*c*) The average speed from *A* to *C* is

$$\frac{900 \text{ m}}{100 \text{ s}} = 9 \text{ m s}^{-1} \quad (Answer).$$

Example: A body moves in a straight line with a uniform velocity of 10 m s^{-1}. How far will the body travel in (*a*) 30 s, (*b*) *t* seconds?
SOLUTION: In 1 seconds the body moves 10 metres. Therefore:
(*a*) In 30 s it moves 10 × 30 m = 300 m.
(*b*) In *t* seconds it moves 10*t* metres.

Example: Convert a speed of 10 metres per second to kilometres per hour.
SOLUTION: The distance travelled in 1 minute at 10 m s^{-1} is 10 × 60 metres.
The distance travelled in 1 hour at 10 m s^{-1} is 10 × 60 × 60 metres.
Since 1000 metres = 1 kilometre, the distance travelled in 1 h is

$$\frac{10 \times 60 \times 60}{1000} = 36 \text{ kilometres}$$

Therefore   10 m s^{-1} = 36 km h^{-1} (*Answer*).

Example: Find how many minutes it takes to travel 10 kilometres at 36 kilometres per hour.
SOLUTION: The time taken to travel 1 kilometre is $\frac{1}{36}$ hour. The time taken to travel 10 kilometres is

$$\frac{1}{36} \times 10 \text{ hour}$$
$$= \frac{1}{36} \times 10 \times 60 \text{ minutes}$$
$$= 16\frac{2}{3} \text{ minutes} \quad (Answer).$$

(2) Speed–Time Formulae

We can generalize all the results relating speed, distance and time in the following formulae:
if s = distance
t = time taken
v = average speed

then $v = \dfrac{s}{t}$ (i)

We can multiply both sides of equation (i) by *t* to get

$$vt = s \tag{ii}$$

We can divide both sides of equation (ii) by *v* to get

$$t = \frac{s}{v} \tag{iii}$$

If we look back to the four worked examples above, we see that the first uses equation (i), the second and third use equation (ii), and the fourth uses equation (iii). (Note the difference in the letters *s* for distance and s for the time abbreviation for second.)

Exercise 14

1. Using the result $10 \text{ m s}^{-1} = 36 \text{ km h}^{-1}$, convert the following speeds to kilometres per hour.

(a) 5 m s^{-1}, (b) 20 m s^{-1}, (c) $2 \cdot 5 \text{ m s}^{-1}$, (d) 25 m s^{-1}, (e) 15 m s^{-1}, (f) 1 m s^{-1}, (g) $x \text{ m s}^{-1}$.

2. Convert the following speeds to metres per second:

(a) 72 km h^{-1}, (b) 18 km h^{-1}, (c) 12 km h^{-1}, (d) 30 km h^{-1}, (e) 90 km h^{-1}, (f) 1 km h^{-1}, (g) $x \text{ km } h^{-1}$.

3. It takes a snail about 6 min to travel a distance of 1 m. Calculate its average speed in m s^{-1}.

4. A car takes a half an hour to travel 18 km, and the next half an hour to travel a further 12 km. Find the average speed for (a) the first 18 km, (b) the final 12 km, (c) the complete journey.

5. A car travels 40 km in 45 min, stops for 15 min, and then travels a further 100 km in 1 h. Find the average speed for the complete journey.

6. A man walks uphill with an average speed of 4 km h^{-1} and downhill at 12 km h^{-1}. What is his average speed on a journey to the top of a hill and back to his starting-point by the same route? (Clearly this problem is somewhat artificial, since a walker is more likely to work from a knowledge of the distance and the time he took. However, the problem is posed to warn the reader against the obvious mistake of calculating that the result is $[4 + 12] \div 2$.)

7. The speed of sound in still air is approximately 332 m s^{-1}. If a clap of thunder is heard 5 s after the lightning flash is seen, how far away is the observer from the lightning flash (assuming that the sight of the lightning is instantaneous)?

8. Calculate the angular velocity of the minute and hour hands of a clock.

(3) Distance–Time Graphs

It is helpful to express the relations of Section 2 in graphical form by plotting a graph of distance against time. Before dealing with a particular problem we need to examine the meaning which can be given to the **gradient** of the distance–time graph. Fig. 34 gives three examples of straight-line graphs. In Figs. 34 (a) and (b), y has been plotted against x; in Fig. 34 (c), y has been plotted against t; in Fig. 34 (d), s has been plotted against t.

In Fig. 34 (a) we observe that y increases as x increases: a fact which we feel is intuitively obvious since the line 'slopes upwards'. However, we would like a more precise statement relating the increase in y to the increase in x. Clearly the gradient of the line depends on the scales we choose for y and x. Examination of the graph reveals that y increases by 10 whenever x increases by 1, so we say that the gradient of the line is 10. Alternatively we say that y increases with respect to x at the rate of 10 to 1. We could have described this rate as 20 to 2, or 100 to 10, but it is more convenient to use the unit for the quantity which is plotted along the x axis.

Examination of Fig. 34 (c) reveals, first, that y increases as t increases and, secondly, that the rate of increase is 5 to 2, or $2 \cdot 5$ to 1. If t is the time in seconds and y is measured in metres, then y is increasing at the rate of $2 \cdot 5$ metres per second. Finally, in Fig. 34 (d) it is intuitively obvious that, since the graph 'slopes downwards', s decreases as t increases. Closer examination shows that s decreases with respect to time at the rate of 10 metres per second. Notice that the rate of change is uniform, i.e. the same decrease in s takes place in any period of 1 second during the motion. In other words, the graph represents the motion of a body moving with uniform speed.

Having seen that the gradient of the line at any point on a distance–time graph gives the speed at that point, let us examine a particular problem in detail.

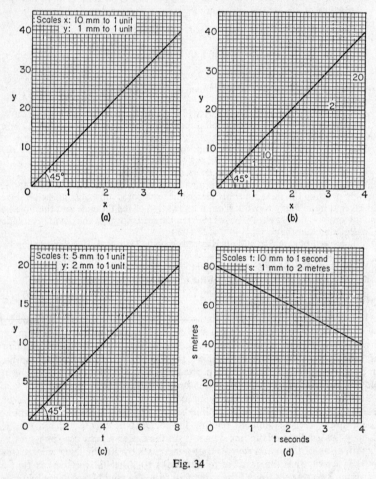

Fig. 34

Example: A patrol car leaves base and travels 25 kilometres along a highway in 30 minutes before parking in a lay-by for a further 15 minutes. Acting on instructions received, the patrol car then returns to base in 20 minutes. Draw a distance–time graph for the motion, assuming that the speeds are uniform.

SOLUTION: Although we think of speed in terms of metres per second or kilometres per hour, it will be more convenient to keep to the units in the question.

Point (i): the first thing to realize is that s measures the distance along the route of the patrol car from its starting position at base. The beginning of the problem takes place with $t = 0$, and $s = 0$. (Compare this with Fig. 34 (*d*) where the start takes place with $t = 0$, $s = 80$ metres.)

Point (ii): since the journey out from base is travelled with uniform speed, the graph is a straight line, *OP* in Fig. 35.

Point (iii): during the next 15 minutes the car is still 25 kilometres from base, the graph is the straight line *PQ*.

Point (iv): on the return journey the distance from base is decreasing at a uniform rate, hence the graph is a straight line *QR*.

The graph of the complete journey consists of the three line segments *OP*, *PQ*, and *QR*. The uniform speed on the outward journey is 25 kilometres per 30 minutes = 50 km h⁻¹. The uniform speed on the return journey is 25 kilometres per 20 minutes = 75 km h⁻¹.

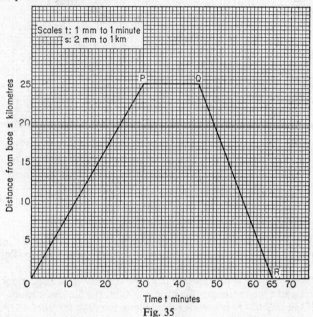

Fig. 35

Example: The distance–time graph for the motion of a car is given in Fig. 36. Interpret the various line segments of the graph. (Try to do this before looking at the solution below.)

SOLUTION:

(i) At point *P*, $t = 0$. Hence the timing of the car's journey starts when it is 40 km from base.

(ii) From $t = 0$ to $t = 1$ h (the line *PQ*), the car travels 20 kilometres in 1 hour with a uniform speed of 20 km h⁻¹.

(iii) From $t = 1$ h to $t = 2$ h (the line *QR*), the car remains at rest (parked) for 1 hour. Its distance from base is 60 kilometres.

(iv) From $t = 2$ h to $t = 3$ h (the line *RS*), the car travels a further 40 kilometres away from base. In this period of 1 h the car maintains a uniform speed of 40 km h⁻¹.

(v) From $t = 3$ h to $t = 6$ h (the line *ST*), the car remains at rest for 3 h. Its distance from base is 100 km.

(vi) From $t = 6$ h to $t = 7$ h (the line *TU*), the car is returning to base. In this hour it maintains a uniform speed of 100 km h⁻¹.

In both the worked examples above *s* represented the distance of the body from a fixed point (called the base or base point) measured along the path taken. For example, the point R on the graph of Fig. 35 represents a return to base ($s = 0$) at time $t = 65$ minutes, but the total distance travelled in this time is 50 kilometres. Similarly the point U on the graph of Fig. 36 represents a return to base ($s = 0$) at time $t = 7$ hours, but the total distance travelled in this time is 160 kilometres.

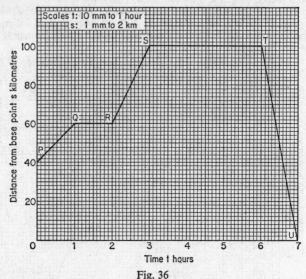

Fig. 36

To illustrate this further, consider walking along the perimeter of the square $ABCD$ in Fig. 37, starting from the point A at $t = 0$ and proceeding to B, C, and D. Each side of the square is 100 m long.

(i) When we reach D the distance from the base point A is 300 m measured along the path; we represent this by $s = 300$ m. (In fact we are only 100 m from A, but *this* 100 m is *not* measured along the path travelled.)

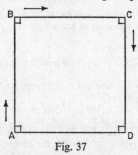

Fig. 37

(ii) If we continue from D to A, then $s = 400$ m because this is the distance from the base point A measured along the path travelled.

(iii) If we now turn and walk back to C, then $s = 200$ m, even though we have actually walked 600 m.

(iv) If we continue back to A, then $s = 0$, even though we have now walked 800 m.

Exercise 15

1. A hiker walks uphill with uniform speed through a distance of 2 km in half an hour, and 4 km downhill with uniform speed on the other side in a further half-hour. Draw a distance–time graph for the walk.

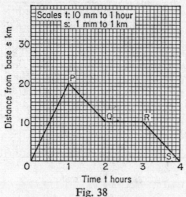

Fig. 38

2. Using the graph of Question 1, draw the distance–time graph which represents walking the total 6 km with uniform speed in one hour.

3. Interpret the distance–time graphs illustrated in Figs. 38 and 39.

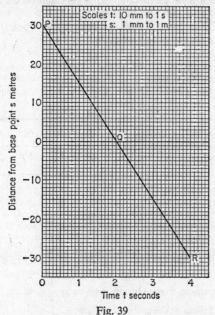

Fig. 39

4. Draw a sketch of the distance–time graph you would expect to get for a body which:

    (*a*) starts from rest and moves with increasing speed;

    (*b*) starts with a speed which decreases until the body comes to rest.

5. Fig. 40 illustrates the distance–time graphs of two men travelling on the same road between towns A and B. Arthur starts from town A to walk to town B; half an hour later Bill rides from town B to town A. Interpret the graphs and find the time and position when Arthur and Bill meet.

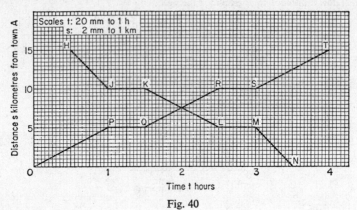

Fig. 40

(4) Non-uniform Speeds

We have just seen how uniform speed is represented on a distance–time graph by a straight line with a fixed gradient, but how can we find the speed at any particular time from a distance–time graph when the speed is not uniform? Consider the following example.

Example: The following table gives the distance–time observations on a rally car at a trial start of a race. Plot a distance–time graph for the motion and find the speed of the car at the time $t = 2 \cdot 8$ seconds; find also the average speed for the first 5 seconds.

Distance from start, s (metres)	0	20	40	60	80	100
Time, t (seconds)	0	2·8	3·8	4·7	5·4	5·9

SOLUTION: We know from page 36 that the speed at any time is measured by that distance which would be travelled in the next unit of time, assuming that the speed remains unchanged. In other words, the distance–time graph is considered to be 'straightened out' or continued in a straight line from the time at which the speed is required. Consider finding the speed at time $t = 2 \cdot 8$ seconds represented on the curve in Fig. 41 by the point P. We 'straighten out' the curve from the point P by drawing the tangent at P; then we find the gradient of this tangent line by choosing any point Q on the tangent and completing the triangle PQR with PR perpendicular to RQ. The gradient of the tangent is given by $\frac{RQ}{PR}$, where both RQ and PR are measured according to the scales of the graph.

(a) From the graph we see that

$$\frac{RQ}{PR} = \frac{16 \text{ metres}}{1 \cdot 2 \text{ seconds}}$$

Since $16 \div 1 \cdot 2 = 13\frac{1}{3}$, the speed at time $t = 2 \cdot 8$ seconds is $13\frac{1}{3}$ m s^{-1}.

(*b*) The average speed for the first 5 seconds is given by the gradient of the line *OT*. The gradient of line *OT* is given by

$$\frac{69 \text{ metres}}{5 \text{ seconds}} = 13\cdot8 \text{ m s}^{-1}.$$

The average speed for the first 5 seconds is therefore $13\cdot8$ m s^{-1}.

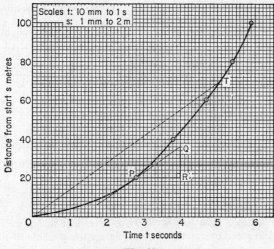

Fig. 41

The main source of error in the above type of problem is due to inaccurately drawn tangents. A reasonably accurate tangent may be obtained by the following method. We shall construct a tangent at the point *A* on the curve in Fig. 42. Select two pieces of glass rod (solid or hollow) of diameter approxi-

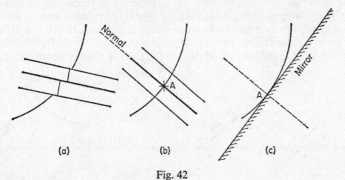

Fig. 42

mately 3 mm and length approximately 60 mm. Place the two rods so that their line of contact passes over the point *A*. Looking at the curve through the rods we find that it seems to be in two broken sections. Rotate the rods until the two sections are joined. Now roll one of the rods away and use the other to draw the normal line through *A* as in Fig. 42 (*b*). The required tangent is at right-angles to this normal line and can be obtained by using a mirror as shown in Fig. 42 (*c*). When the reflecting surface of the mirror lies exactly over the tangent, the normal and its reflection appear to be in one straight line.

Exercise 16

1. Using the graph of Fig. 41, obtain (*a*) the average speed in the first 4 seconds, (*b*) the speed at time *t* = 4·7 seconds.

2. Plot a distance–time graph from the following table of observations:

Distances from starting-point (metres)	0	3	12	27	48	75	108
Time, *t* (seconds)		0	1	2	3	4	5	6

(*a*) Find the speed when (i) *t* = 2 seconds, (ii) *t* = 4 seconds.
(*b*) Find the average speed for the first 4 seconds.

3. Sketch a distance–time graph of the motion of a bus travelling from one bus stop to the next.

4. Sketch a distance–time graph of the motion of a ferry-boat making a complete journey across a river and back.

(5) Speed–Time Graphs

From the distance–time graphs we have examined so far we have been able to find the speed at any time by calculating the gradient of the tangent at the point on the graph which corresponds to that time. In this section we shall consider speed–time graphs. From them we shall deduce not only the distance travelled but also a new measure called **acceleration.**

We have seen (page 36) that when we give the speed of a body at any time during its motion, we are forecasting the distance which would be travelled in the next instant of time if the speed were to remain uniform. Similarly we speak of the velocity of a body at any time during its motion as forecasting what would happen in the next instant of time if the velocity were to remain uniform; this involves moving in an unchanging direction, which is the direction of the tangent to the path at the point concerned. To clarify these ideas let us consider Fig. 43, which illustrates the path of a body starting from *O*

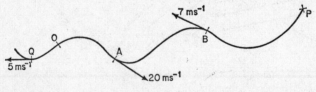

Fig. 43

travelling to *P* and back again through *O* to *Q*. The distance *s* from the base point *O* is measured along the path. If we say that the magnitude of the velocity at *A* is 20 m s⁻¹, we mean that (provided no forces are applied to it) the body will continue to move in the direction of the tangent at *A* and will travel 20 m in the next second. If the speed at *B* is given as −7 m s⁻¹, we know that the body is on its way back to *O* because of the negative sign;

the direction of the velocity is that of the tangent as shown. The illustration for the motion at Q shows a velocity of -5 m s^{-1}.

Fig. 44 illustrates three basic speed–time graphs. As in Section 2 of this chapter, we use the symbols v for speed, s for distance, and t for time. We note from Fig. 44 (a) that the body starts from rest, i.e. $v = 0$ when $t = 0$.

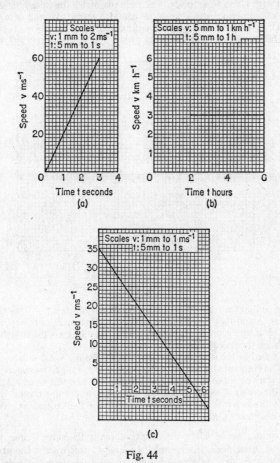

Fig. 44

The speed v increases as t increases. Furthermore, v increases uniformly with t, since v increases by 20 m s^{-1} every second. In Fig. 44 (b) we note that the body has only been timed from $t = 2$ seconds, and that the speed v is constant at 3 km h^{-1}. If the body is travelling in a straight line, we can say that the velocity is also constant.

In Fig. 44 (c) we note that the body already has a speed of 35 m s^{-1} when the timing starts, i.e. $v = 35$ m s^{-1} when $t = 0$. The speed v decreases as t increases. Furthermore, v decreases uniformly with t, since v decreases by 7 m s^{-1} every second. We also note that $v = 0$ when $t = 5$ seconds and that

$v = -7$ m s^{-1} when $t = 6$ seconds, which corresponds to a situation such as B in Fig. 43.

We combine all the above ideas into a single problem as follows:

Example: A train starting from rest travels with uniformly increasing speed to reach a speed of 10 m s^{-1} in 60 seconds. The train maintains this speed for 20 seconds and then uniformly decreases speed to come to rest at the end of a further 40 seconds. Draw a speed–time graph for the complete motion.

SOLUTION: The speed–time graph is given by the line segments OP, PQ, and QR in Fig. 45.

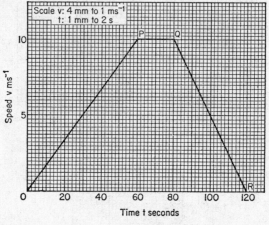

Fig. 45

We note (i) that the train starts from rest, i.e. $v = 0$ when $t = 0$, as represented by the point O, and that $v = 10$ m s^{-1} when $t = 60$ seconds, as represented by the point P. Therefore the graph of the first part of the motion is the straight-line segment OP since the speed is increasing uniformly.

(ii) The speed of the train remains at 10 m s^{-1} from $t = 60$ seconds to $t = 80$ seconds. The graph of this part of the motion is the straight-line segment PQ.

(iii) The train comes to rest when $t = 120$ seconds, and this is represented on the graph by the point R. Since the speed decreases uniformly during the last 40 seconds, it follows that the graph of this part of the motion is given by the straight-line segment QR.

We can obtain some extra information from the speed–time graph by examining the 'area under the graph'. In Fig. 45 this means the area of the figure $OPQR$. Any area on a speed–time graph is made up of small squares or triangles whose area depends on height $\times$ base calculations where heights correspond to speeds and bases correspond to times. But we know that speed v multiplied by time t gives distance s travelled in that time (i.e. $s = vt$). Hence any area under a speed–time graph represents a distance travelled.

For example, if we examine the area under the line segment PQ of the speed–time graph of Fig. 45 we see at once that, since the speed represented by any point on PQ is 10 m s^{-1} and the time interval from P to Q is 20 seconds, the section of the graph from P to Q represents travelling at a speed of 10 m s^{-1} for 20 seconds. This clearly gives a distance travelled of 200 metres, which is also given by the area under the graph: (In Fig. 45 the reader should now draw the perpendicular PN from P to OR and the perpendicular QM from Q to OR).

$$PN \times PQ = 10 \text{ m s}^{-1} \times 20 \text{ s} = 200 \text{ m}.$$

It follows that the area of *OPQR* in Fig. 45 will give the distance travelled in the 120 seconds of the motion.

Example: Obtain the distances travelled in each part of the motion given by the speed–time graph in Fig. 45.

SOLUTION:

(i) Area *OPN* represents a distance given by

$$\tfrac{1}{2}(NP \times ON) = \tfrac{1}{2}(10 \text{ m s}^{-1} \times 60 \text{ s}) = 300 \text{ m}.$$

(ii) Area *PQMN* represents a distance given by

$$(NP \times NM) = (10 \text{ m s}^{-1} \times 20 \text{ s}) = 200 \text{ m}.$$

(iii) Area *MQR* represents a distance given by

$$\tfrac{1}{2}(MQ \times MR) = \tfrac{1}{2}(10 \text{ m s}^{-1} \times 40 \text{ s}) = 200 \text{ m}.$$

Hence the total distance travelled during the motion is 700 m.

N.B. This last result means that the train is 700 m from its starting-point measured along the path or track it followed.

An alternative method of finding the distance would have been to use the formula for the area of a trapezium:

$$\text{Area } OPQR = \tfrac{1}{2}PN(PQ + OR) = 5(20 + 120) = 700 \text{ m}.$$

Exercise 17

1. Using each of the speed–time graphs of Fig. 44, find (*a*) the distance represented by 100 mm^2; (*b*) the distance travelled during the motion.

2. A train starts from rest and increases speed uniformly to 15 m s^{-1} in 100 seconds. The train maintains this speed for 120 seconds, then decreases speed uniformly to come to rest after a further 300 seconds. Draw a speed–time graph and find the distance travelled during the motion.

3. A car being driven along the motorway had its speedometer read every 30 seconds. The readings were tabulated as follows:

Time, t seconds	0	30	60	90	120	150	180	210	240	270	300	330	360
Speed, km h^{-1}	30	37	43	47	49	51	50	48	44	37	29	18	0

(*a*) Draw a speed–time graph using scales of 1 mm to 1 km h^{-1} and 1 mm to 3 seconds.

(*b*) Calculate the distance represented by an area of 100 mm^2.

(*c*) Estimate the distance travelled during the motion by counting the squares under the graph.

4. A cage descends a mine shaft which is 750 metres deep. In the first 250 metres the speed of descent increases uniformly from rest to 20 m s^{-1}; this speed is maintained for the next 300 metres. The final part of the descent is made with the speed decreasing uniformly. Draw a speed–time graph for the motion of the cage and from it find:

(*a*) the time taken to descend,

(*b*) the average speed for the descent,

(*c*) the rate at which the speed is increasing over the first 250 metres of the descent,

(*d*) the rate at which the speed is decreasing in the last 20 seconds of the descent.

CHAPTER FOUR

MOTION WITH UNIFORM ACCELERATION IN A STRAIGHT LINE

Restricting the motion to a straight line has the advantage that any changes in displacement and velocity must take place along the line. For example, if we are told that a body moves along a straight line with a speed of 27 m s⁻¹, all we need to do in order to speak of the velocity of the body is to relate the direction of motion to a positive or negative sign.

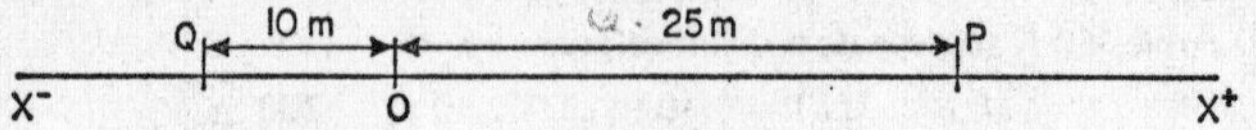

Fig. 46

Consider a body moving along the line X^-OX^+ shown in Fig. 46. The position of the body at any time is given by the point P. Taking O as the base point or origin, we adopt the usual convention of measuring displacements to one side (in this case to the right) as positive and displacements to the other side (in this case to the left) as negative. Thus when a point Q is said to be -10 m from O, we know that it lies to the left of O as illustrated. The position of P would be described as $+25$ m from O; more usually we would just say P is 25 m from O, i.e. all numbers are taken as positive unless otherwise stated. A similar relation applies when we speak of the velocity of a point moving along the line X^-OX^+. For example, a velocity of -13 m s⁻¹ means that the point is moving from right to left.

(1) Acceleration

We have already met in Chapter Three motions in which the velocity does not remain uniform: such changes in the velocity of a body are related to what we call **acceleration**.

DEFINITION: *Acceleration is the rate of change of velocity with respect to time*.

If the increase or decrease in the velocity is the same for every second of the motion, then the acceleration is said to be uniform.

It should be noted that acceleration has magnitude, direction, and sense. When the velocity of a body is increasing, the motion of the body is said to be 'accelerating'; when the velocity is decreasing, the motion is said to be 'retarding'. Again we may associate a positive or negative sign with each of these situations, in fact we refer to a **retardation** as being a negative acceleration.

If a velocity is measured in metres per second, then acceleration (i.e. the change in velocity per second) may be measured in metres per second *per second*. Thus, an acceleration of 3 metres per second per second means that the magnitude of the velocity of the body is being increased by 3 metres per second every second.

Since the motion takes place in a straight line, the acceleration, velocity, and displacement of a body are completely specified by three positive or negative numbers. We shall therefore speak of velocity–time and displacement–time graphs instead of speed–time and distance–time graphs.

Example: A body starting from rest moves in a straight line with a uniform acceleration of 5 metres per second per second for 6 seconds. Draw a velocity–time graph for the motion; from it find (*a*) the velocity at $t = 6$ seconds, and (*b*) the distance travelled in 6 seconds.

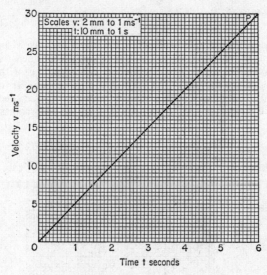

Fig. 47

SOLUTION: Since the velocity increases uniformly we know that the graph will be a straight line. Examination of Fig. 47 shows that the velocity has increased by 5 m s^{-1} for each 1-second increase in time.

(*a*) The point P corresponds to a velocity of 30 m s^{-1} when $t = 6$ seconds. We could have arrived at this result without the graph by reasoning that, since the increase in velocity is 5 m s^{-1} in every second, the total increase in the velocity in 6 seconds must be $(5 \text{ m s}^{-1}) \times 6 = 30 \text{ m s}^{-1}$ (*Answer*).

(*b*) We have already seen that the distance travelled is represented by the area under the graph. From the scales we see that 100 mm^2 represents 5 metres, so the area under the line OP represents a distance of

$$\tfrac{1}{2}(6 \times 30) = 90 \text{ m } (Answer).$$

Example: Interpret the velocity–time graph shown in Fig. 48.

SOLUTION: Since the graph is a straight line the acceleration must be uniform.

(i) The velocity is decreasing at the rate of 20 metres per second per second. We may say, therefore, that the body is moving with a uniform acceleration of -20 metres per second per second, or a retardation of 20 metres per second per second.

(ii) The body starts (i.e. when $t = 0$) with a velocity of 50 m s^{-1}, represented by the point A on the graph.

(iii) The body is at rest (i.e. $v = 0$) when $t = 2\tfrac{1}{2}$ seconds, represented by the point R on the graph.

(iv) The body has a velocity of -50 m s^{-1} when $t = 5$ seconds, represented by the point C on the graph.

(v) The total distance travelled is represented by the sum of the areas of triangles *OAR* and *NCR* with due regard to the scales chosen. Therefore the distance travelled is

$$\tfrac{1}{2}(2\tfrac{1}{2} \times 50) + \tfrac{1}{2}(2\tfrac{1}{2} \times 50) = 62\cdot5 + 62\cdot5 = 125 \text{ m.}$$

We shall use this example to illustrate the important idea of positive and negative areas. If we refer back to the idea of positive and negative displacements and velocities which was discussed at the beginning of this chapter, we can imagine the motion of the body in the above example to consist of starting from a base point P with a velocity of 50 m s^{-1} which decreases uniformly (by 20 metres per second per second) until the body comes to rest $2\cdot5$

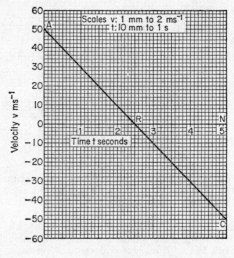

Fig. 48

seconds later and $62\cdot5$ metres away. The velocity of the body continues to decrease at the same rate as before, reaching -50 m s^{-1} when $t = 5$ seconds. In other words, the body travels away from P for $2\cdot5$ seconds and travels back to P in the next $2\cdot5$ seconds. Thus, while the total distance travelled is 125 m, the final displacement is zero since the body has returned to the point from which it first started. We can accommodate this result by giving areas which are below the axis (like triangle *NCR*) a negative measure. In this case the area of triangle *OAR* is $62\cdot5$ units, the area of triangle *NCR* is $-62\cdot5$ units, and the distance from base point P after 5 seconds is $62\cdot5 - 62\cdot5 = 0$. But the total distance travelled is still 125 m as already discussed.

The abbreviation for velocity or speed measured in metres per second is m s^{-1}. The abbreviation for rate per second is s^{-1}. Thus an acceleration measured in (metres per second) per second becomes (m s^{-1})s^{-1}, which is finally abbreviated to m s^{-2}. Thus an acceleration of 5 metres per second per second is written as 5 m s^{-2}. An acceleration of -5 metres per second per second, i.e. a retardation of 5 m s^{-2}, is written -5 m s^{-2}.

The relation between the displacement–time graph and the velocity–time graph will be seen from the following example.

Example: A body travelling in a straight line passes through a point 0 at $t = 0$ with a velocity $v = 40$ m s^{-1} and a uniform acceleration of -10 m s^{-2}. Draw a velocity-time graph for the first 9 seconds of the motion, and use this graph to draw a displacement–time graph for the same time.

SOLUTION: Since the retardation is uniform, the velocity–time graph will be a straight line as shown in Fig. 49. Since the velocity is decreasing at the rate of 10 m s^{-1} every second, it follows that the body is at rest when $t = 4$ seconds; this is represented by the point R on the graph.

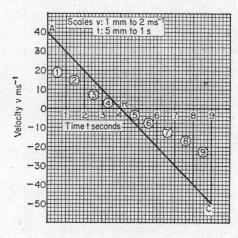

Fig. 49

We must produce AR to C to obtain the velocity–time graph for the required 9 seconds.

The overall motion consists of the body travelling with decreasing velocity until it comes to rest at the end of 4 seconds and then returning to the starting-point with increasing velocity. It passes through the starting-point when $t = 8$ seconds and continues on for a further 1 second.

The distance travelled during

  (i) the 1st or 8th second is represented by the area (1) or (8) $= 35$ m;
  (ii) the 2nd or 7th second is represented by the area (2) or (7) $= 25$ m;
  (iii) the 3rd or 6th second is represented by the area (3) or (6) $= 15$ m;
  (iv) the 4th or 5th second is represented by the area (4) or (5) $= 5$ m;
  (v) the 9th second is represented by the area (9) $= 45$ m.

The values necessary for plotting the displacement–time graph are therefore as follows:

Displacement from starting-point, s metres	0	35	60	75	80	75	60	35	0	−45
Time, t seconds	0	1	2	3	4	5	6	7	8	9

The required displacement–time graph is shown in Fig. 50.

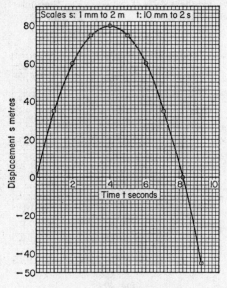

Fig. 50

Exercise 18

1. Starting from rest, a body travels in a straight line with a uniform acceleration of 2 m s^{-2} until it reaches a speed of 20 m s^{-1}. The body maintains this speed for 30 seconds and then comes to rest with a uniform retardation in a further 100 seconds. Draw a velocity–time graph and find (a) the uniform retardation in the last 100 seconds of the motion, (b) the total distance travelled.

2. A body moving in a straight line starts with a velocity of 20 m s^{-1} and has a uniform acceleration of -10 m s^{-2}. Draw a velocity–time graph for the first 5 seconds of the motion, and from this graph obtain a displacement–time graph (see the worked example above). Find the position of the body when $t = 5$ seconds.

3. A body has an initial velocity of -30 m s^{-1} and moves in a straight line with a uniform acceleration of 6 m s^{-2} for 7 seconds. Draw a velocity–time graph and find the position of the body when $t = 7$ seconds. At what time was the body at rest?

4. A body moves in a straight line with an acceleration of 2 m s^{-2}. With the aid of a velocity–time graph find how long the body takes to travel 100 m from its position of rest.

(2) Equations of Motion

We have so far considered motion in a straight line in which displacement, velocity, and acceleration of different signs have been combined to yield considerable groundwork for graphical presentation of various problems. However, some mathematicians find graphical work tedious and look for a more general presentation of ideas in the form of equations.

We have already generalized the relation between velocity (v) distance (s) and time (t) in the equations set out on page 38. Here we shall obtain formulae relating v, s, t with the acceleration (a), and for simplicity the quantities involved will be restricted to units based on the metre and second.

If a body moves in a straight line with a uniform acceleration of 7 m s^{-2}, then the velocity will increase by 7 m s^{-1} every second. At the end of 1 second the velocity has increased by 7 m s^{-1}; at the end of 2 seconds it has increased by 14 m s^{-1}; at the end of 3 seconds it has increased by 21 m s^{-1}; and at the end of t seconds it has increased by $7t$ m s^{-1}. It follows therefore that, if we let the uniform acceleration be a m s^{-2}, at the end of t seconds the velocity will have changed by at m s^{-1}. Note that we say *changed*, because the velocity will increase if a is positive and will decrease if a is negative. If the body concerned already had a velocity of u m s^{-1}, the final velocity v after time t is given by the relation

$$v = u + at.$$

Let us examine this equation with the aid of the velocity–time graph (Fig. 51) drawn for a body moving in a straight line with uniform acceleration a, for time t.

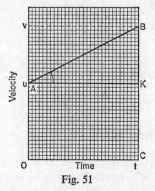

Fig. 51

We relate the graph to the standard notation as follows:

u = initial velocity, represented by point A on the graph
v = final velocity, represented by point B on the graph
t = time, represented by point C on the graph
a = acceleration, represented by the gradient of the line AB.

Note that the upward slope of line AB suggests that a is positive and the velocity is increasing. A negative value for a would be indicated by AB sloping downwards. Whatever the sign of a the reasoning which follows remains unchanged.

The rate of change of velocity with respect to time is measured by $KB \div AK = a$, bearing in mind the scales of the graph. But $KB = v - u$ and $AK = t$. Therefore

$$\frac{v - u}{t} = a$$

Multiplying this equation by t we obtain

$$v - u = at$$

or, $$v = u + at \qquad (i)$$

Example: A body moves in a straight line with an initial velocity of 50 m s^{-1} and a retardation of 20 m s^{-2}. Determine (*a*) the velocity when $t = 5$ seconds, and (*b*) the time at which $v = 0$.

SOLUTION:

(*a*) We are given that

$$u = 50 \text{ m s}^{-1}$$
$$a = -20 \text{ m s}^{-2}$$
$$t = 5 \text{ seconds}$$
$$v = ?$$

Substitution in the equation $v = u + at$ yields

$$v = 50 - 100$$
$$= -50$$

So when $t = 5$ seconds, $v = -50$ m s^{-1} (*Answer*).

(*b*) We are given that

$$u = 50 \text{ m s}^{-1}$$
$$v = 0$$
$$a = -20 \text{ m s}^{-2}$$
$$t = ?$$

Substitution in the equation $v = u + at$ yields

$$0 = 50 - 20t$$

Hence $t = 2 \cdot 5$ seconds (*Answer*).

Readers may like to refer back to Fig. 48 for a graphical interpretation of this example.

Recalling that the distance from the starting- or base-point is represented by the area under the graph, we return to Fig. 51 in order to obtain a relation involving s. The figure $OABC$ is a trapezium with OA and CB as parallel sides. Therefore the area under the graph is $\frac{1}{2}(OA + CB) \times OC$. Bearing in mind the scales used on the graph this area represents the distance travelled. Employing the same symbols as before, we have

$$s = \frac{1}{2}(u + v)t \qquad (ii)$$

But $\frac{1}{2}(u + v)$ is the average velocity throughout the time t, so we now have a general formula for motion in a straight line under uniform acceleration. It may be stated as follows:

Distance from starting-point = average velocity × time.

We can combine equations (i) and (ii) by substituting $(u + at)$ for v in $s = \frac{1}{2}(u + v)t$ to yield

$$s = \frac{1}{2}(u + u + at)t$$
$$= \frac{1}{2}(2u + at)t$$
$$s = ut + \frac{1}{2}at^2 \tag{iii}$$

In equation (iii) we have obtained a relation between s and t not involving v, the final velocity.

Example: A body moves in a straight line with an initial velocity of 60 m s^{-1} under a retardation of 12 m s^{-2}. Calculate (a) the time which elapses before the body returns to its starting point, and (b) the distance travelled in this time.
SOLUTION:

(a) We are given that

$$u = 60 \text{ m s}^{-1}$$
$$a = -12 \text{ m s}^{-2}$$
$$s = 0$$
$$t = ?$$

Substitution in the equation $s = ut + \frac{1}{2}at^2$ yields

$$0 = 60t - 6t^2$$
$$0 = 6t(10 - t)$$

from which we find that $t = 0$ or $t = 10$. Plainly the second of these two solutions is the one required, so $t = 10$ seconds *(Answer)*.

(b) To find the total distance travelled we have to remember that the body moved away from its starting point, came to rest, and then returned. As the whole journey took 10 seconds, intuition tells us that the outward part took 5 seconds and the return took another 5 seconds. We can confirm this by using the equation $v = u + at$ to find the time for which $v = 0$, i.e. the instant in which the body was at rest before starting its return journey. Substituting $v = 0$, $u = 60$, $a = -12$, we obtain

$$0 = 60 - 12t$$

Hence
$$t = 5 \text{ seconds}$$

Now we need to find the distance travelled in the first 5 seconds, so we substitute $t = 5$, $u = 60$, $a = -12$, in the equation

$$s = ut + \frac{1}{2}at^2$$

and obtain
$$s = (60 \times 5) + \frac{1}{2}(-12 \times 25)$$
$$= 150 \text{ metres}$$

So the total distance travelled (150 m out and 150 m back) is 300 metres *(Answer)*.

Example: A body accelerates uniformly from rest to travel in a straight line for a distance of 1000 metres in 10 seconds. Calculate (a) the acceleration, (b) the speed of arrival.
SOLUTION:

(a) We are given that

$$u = 0$$
$$s = 1000 \text{ metres}$$
$$t = 10 \text{ seconds}$$
$$a = ?$$

Substitution in the equation $s = ut + \frac{1}{2}at^2$ yields

$$1000 = 0 + 50a$$

Hence $a = 20$ m s^{-2} *(Answer)*.

(*b*) We now know that

$$u = 0$$
$$a = 20 \text{ m s}^{-2}$$
$$t = 10 \text{ seconds}$$
$$v = ?$$

Substitution in the equation $v = u + at$ yields

$$v = 0 + 200$$

Hence $v = 200 \text{ m s}^{-1}$ (*Answer*).

By combining equations (i) and (ii) in a different way we can obtain a formula which does not involve *t*. From equation (i)

$$v = u + at, \text{ therefore } v - u = at$$

From equation (ii)

$$s = \tfrac{1}{2}(u + v)t, \text{ therefore } v + u = \frac{2s}{t}$$

Multiplication of these two results yields

$$(v - u)(v + u) = 2as$$

Hence

$$v^2 - u^2 = 2as$$

or

$$v^2 = u^2 + 2as \qquad\qquad \text{(iv)}$$

Example: A body moves in a straight line with an initial velocity of 25 m s^{-1} and a uniform retardation of 10 m s^{-2}. Calculate the velocity when the body is 30 metres from its starting-point.

SOLUTION: We are given that

$$u = 25 \text{ m s}^{-1}$$
$$s = 30 \text{ m}$$
$$a = -10 \text{ m s}^{-2}$$
$$v = ?$$

Substitution in the equation $v^2 = u^2 + 2as$ yields

$$v^2 = 625 - 600$$
$$= 25$$

Hence $v = 5 \text{ m s}^{-1}$ or $v = -5 \text{ m s}^{-1}$ (*Answer*).

(The first solution is the velocity on the outward journey; the negative solution is the velocity of the body as it passes the same point on the return journey.)

Exercise 19

1. Explain what is meant when a body is said to have 'uniform acceleration'.
2. Calculate each of the unknown quantities below:

(i) $u = 0$	$v = 10 \text{ m s}^{-1}$	$a = 2 \text{ m s}^{-2}$	$s = ?$	$t = ?$
(ii) $u = 0$	$v = -50 \text{ m s}^{-1}$	$a = -10 \text{ m s}^{-2}$	$s = ?$	$t = ?$
(iii) $u = 50 \text{ m s}^{-1}$	$v = 0$	$a = ?$	$s = 1000 \text{ m}$	$t = ?$
(iv) $u = 100 \text{ m s}^{-1}$	$v = ?$	$a = ?$	$s = 1000 \text{ m}$	$t = 10 \text{ s}$
(v) $u = 96 \text{ m s}^{-1}$	$v = ?$	$a = -16 \text{ m s}^{-2}$	$s = ?$	$t = 4 \text{ s}$
(vi) $u = ?$	$v = 90 \text{ m s}^{-1}$	$a = 18 \text{ m s}^{-2}$	$s = ?$	$t = 5 \text{ s}$
(vii) $u = ?$	$v = 0$	$a = ?$	$s = 500 \text{ m}$	$t = 10 \text{ s}$

3. A body moving in a straight line with a velocity of 15 m s⁻¹ is brought to rest by a uniform retardation in 40 seconds. Find the retardation and the distance travelled in this time.

4. A car travelling at 48 km h⁻¹ brakes to a stop in a distance of 10 metres. Assuming the retardation to be uniform, calculate the time taken and the retardation. (The 48 km h⁻¹ should be converted to m s⁻¹.)

5. During the launching of a spacecraft the following observations were made:

Speed, m s⁻¹	1055	1253	1449	1674
Seconds after lift-off	130	140	150	160

Calculate the distance travelled during each 10-second period between observations, assuming the acceleration to be uniform during each separate period.

(3) Acceleration due to Gravity

The easiest way to give a body an acceleration is to drop it. The acceleration which the body acquires as a result of falling to the ground is called the acceleration due to gravity; with one or two reservations, all bodies fall with the same acceleration. The following experiments go some way to suggest what those reservations might be.

EXPERIMENT 1: Take a circular piece of paper which is small enough to be covered by a coin. Hold the coin and paper horizontally, side by side, and then release together. It will be found that the coin hits the ground before the piece of paper. Clearly these two bodies did not fall with the same acceleration.

Now place the paper on top of the coin. Hold the coin horizontally and allow it and the paper to fall (about 1·5 m is far enough). Coin and paper will reach the ground together. The same result will be obtained with the paper placed underneath the coin. We see therefore that it *is* possible for the two bodies to fall with the same acceleration.

EXPERIMENT 2: Obtain two equal coins and tie one of them with cotton to an inflated balloon. Allow both coins to fall from rest at the same height. The coin attached to the balloon will fall more slowly. Indeed, if the balloon is inflated with the right gas, this coin may well rise instead of fall. Clearly the coin and balloon plus the air inside is heavier than the coin by itself. So we have seen that a heavier body may descend more slowly.

EXPERIMENT 3: Put some coins, a feather, a piece of paper and some pins into a tin without a lid. If the tin is allowed to fall, all the articles inside it will reach the ground at the same time. In this case the acceleration of each body is the same.

If we had screwed the paper in Experiment 1 into a ball, it would have fallen to the ground with the same acceleration as the coin. Now the resistance of the air might account for the different accelerations of the coin, balloon, paper, and feather; so under what conditions is it possible to say that the acceleration due to gravity is the same for all bodies? A first step towards realizing the truth of this statement is to eliminate the resistance of the air by dropping the bodies in a vacuum. We find that if we drop a feather and a coin in a vacuum they both fall with the same acceleration. We can therefore say that in the *unresisted* motion of falling bodies, the acceleration is (*a*) uniform, and (*b*) the same for all bodies. When the motion of a falling body is unresisted, we shall describe it as a 'falling freely'.

The acceleration of a freely falling body is due to the attraction of the earth. The attraction is 'vertically downwards' towards the centre of the earth: the direction is that of a plumb-line (or any string which supports a mass) at the place concerned on the earth's surface. We summarize these ideas by saying that the acceleration is 'due to gravity'.

Since the acceleration at the same place is the same for all bodies, we use the letter g to represent this acceleration due to gravity. The direction of 'vertically downwards' is taken to be towards the centre of the earth, as shown in Fig. 52.

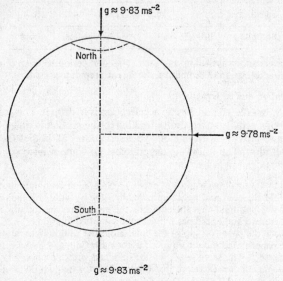

Fig. 52

It is not difficult to find an approximate value for g by using a simple pendulum. This consists of a heavy mass, such as the lead bob of a builder's plumb-line (even a pocket knife will do), suspended by a piece of cotton from a fixed point. The idea is to pull the bob to one side R (Fig. 53), keeping the string taut at an angle of approximately 15° to the vertical, then to release the bob and time 20 swings of the pendulum. Each swing is a complete excursion from R to L and back to R again. By considering the time of each swing to be the same, division by 20 yields the time of one swing: a time which is called the **period** of the pendulum.

We then use the formula

$$T = 2\pi\sqrt{\frac{l}{g}}$$

where T is the period measured in seconds, and l is the length of the pendulum measured from the fixed point A to the centre of gravity of the penknife. (The midpoint of the penknife will usually be good enough.) Rearranging the formula we get

$$T^2 = 4\pi^2\frac{l}{g}$$

$$g = \frac{4\pi^2 l}{T^2}$$

We can see from this formula that our work would be simplified by making $l = 1$ metre.

To check how easy this experiment is, the author used a penknife pendulum of length 1 metre suspended from a fixed point provided by thumb and fore-

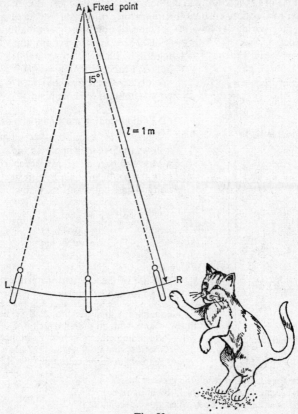

Fig. 53

finger and an arm resting on a wardrobe. After five trials the period averaged 30 swings in 60 seconds (timed by the seconds hand of a watch). These rough experimental results gave a value for g as follows:

$$T = 2 \text{ seconds}, \ l = 1 \text{ metre}, \ \pi = 3 \cdot 14$$

Substituting in the equation $g = \dfrac{4\pi^2 l}{T^2}$ yields

$$g = \frac{4 \times 3 \cdot 14 \times 3 \cdot 14 \times 1}{2 \times 2}$$
$$= 9 \cdot 86 \text{ m s}^{-2}$$

The crudeness of the experiment does not justify giving the result to two decimal places, but the accuracy which the reader can obtain with a little care is always a pleasing experience. Try it but keep the cat out of the room.

Unfortunately, the magnitude of g varies from about 9·83 m s^{-2} at the poles to 9·78 m s^{-2} at the Equator. The internationally accepted value is 9·806 65 m s^{-2}. The value to be used in practical work will clearly depend upon the degree of accuracy required; in this book we shall use either 9·80 m s^{-2} or 9·81 m s^{-2} according to mathematical convenience. The International Associaton of Geodesy ensures the uniformity of measurements of g over all parts of the world. A few of the more notable determinations were: Washington, U.S.A. (1936), $g = 9·800\ 816$ m s^{-2}; Teddington, England (1939), $g = 9·811\ 832$ m s^{-2}; Leningrad, U.S.S.R. (1956), $g = 9·819\ 2$ m s^{-2}; Sèvres, France (1958), $g = 9·809\ 277$ m s^{-2}. We observe that $g = 9·81$ m s^{-2} is a fair approximation to these results.

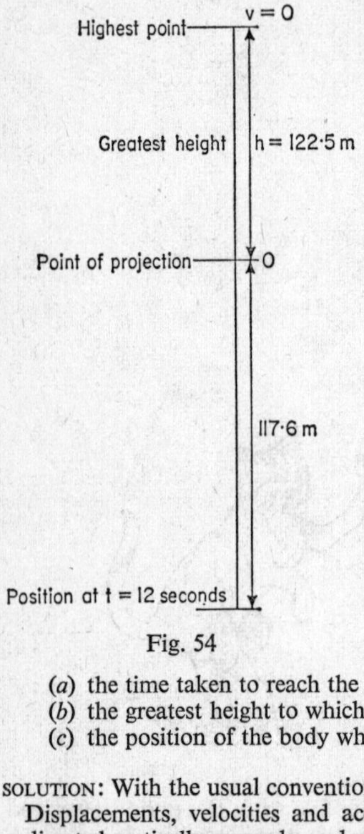

(4) Falling Bodies and Equations of Motion

Since g is a uniform acceleration, it follows that we may apply all the equations we have obtained so far in order to investigate any problem concerning a freely falling body. These equations may now be written as:

$$v = u + gt \qquad \text{(i)}$$
$$s = \tfrac{1}{2}(u + v)t \qquad \text{(ii)}$$
$$s = ut + \tfrac{1}{2}gt^2 \qquad \text{(iii)}$$
$$v^2 = u^2 + 2gs \qquad \text{(iv)}$$

The use of these equations is illustrated by the following examples.

Example: A body is projected vertically upwards with an initial velocity of 49 m s^{-1}. Assuming there is no resistance to the motion, and using 9·80 m s^{-2} as an approximate value for g, calculate:

Highest point — $v = 0$

Greatest height | $h = 122·5$ m

Point of projection — 0

117·6 m

Position at $t = 12$ seconds

Fig. 54

(a) the time taken to reach the greatest height;
(b) the greatest height to which the body rises;
(c) the position of the body when $t = 12$ seconds.

SOLUTION: With the usual convention we shall choose 0 as our base point (Fig. 54). Displacements, velocities and accelerations will be assigned positive values if directed vertically upwards, and negative values if directed vertically downwards.

(a) When the body stops rising it will have reached its greatest height. At the greatest height, therefore, the velocity v will be zero.

We are given that

$$u = 49 \text{ m s}^{-1}$$
$$v = 0$$
$$g = -9·80 \text{ m s}^{-2}$$
$$t = ?$$

Substitution in the equation $v = u + gt$ yields

$$0 = 49 - 9·80t$$

Hence $\qquad\qquad t = 5$ seconds (*Answer*).

(b) In this part of the problem we need to find s when $t = 5$ seconds. Substitution in $s = \frac{1}{2}(u + v)t$ yields

$$s = \frac{1}{2}(49 + 0)5$$
$$s = 122 \cdot 5 \text{ m}$$

Hence the greatest height above the point 0 is 122·5 m   (*Answer*).

(c) We are given that

$$u = 49 \text{ m s}^{-1}$$
$$g = -9 \cdot 80 \text{ m s}^{-2}$$
$$t = 12 \text{ seconds}$$
$$s = ?$$

Substitution in the equation $s = ut + \frac{1}{2}gt^2$ yields

$$s = (49 \times 12) - (4 \cdot 9 \times 144)$$
$$= -117 \cdot 6$$

The negative sign indicates that the body has fallen *below* the base point; hence when $t = 12$ seconds the body is 117·6 m below the point of projection   (*Answer*).

Two important facts emerge from this example:

(i) The equations of motion cater for the complete motion, both upwards and downwards. There is no need to deal with the upward motion separately from the downward motion.

(ii) It is intuitively obvious that, since no effort has to be wasted in overcoming a resistance, the body will return to the point of projection with the same speed with which it was projected. Furthermore, the time of rise will be the same as the time to return.

Let us confirm the observations in (ii) above for any body projected upwards. First, the body is back at the point of projection when $s = 0$. Using the equation $s = \frac{1}{2}(u + v)t$ we see that if $s = 0$, then either $t = 0$ or $u + v = 0$. Now $t = 0$ corresponds to the start of the motion, and $u + v = 0$ (or $v = -u$) corresponds to the time at which it returns. We have confirmed therefore that the speed on return is the same as the speed of projection.

The greatest height is reached when $v = 0$. Using the equation $v = u + gt$ we see that the time to the greatest height is given by $0 = u + gt$ or $t = -\dfrac{u}{g}$ (remember that the numerical value for g is negative).

The return to the point of projection is given by $v = -u$. Using the equation $v = u + gt$, we see that the time of return is given by

$$-u = u + gt$$
$$-2u = gt$$
$$t = \frac{-2u}{g}$$

which is twice the time to the greatest height.

Example: A ball is projected vertically upwards with a speed of 49 m s^{-1}. Assuming that there is no resistance to the motion, and that $g = 9 \cdot 80$ m s^{-2}, calculate the times at which the body is 44·1 m above the point of projection. (The choice of 44·1 m is arithmetically convenient.)

SOLUTION: There will be two answers to this problem since the ball is 44·1 m from the point of projection on two occasions: once when it is rising and once when it is falling. We are given that

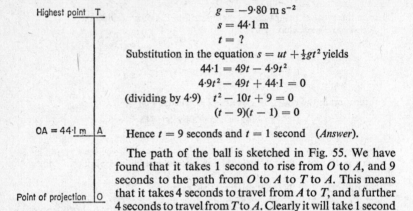

Highest point T

OA = 44·1 m A

Point of projection O

Fig. 55

$$u = 49 \text{ m s}^{-1}$$
$$g = -9 \cdot 80 \text{ m s}^{-2}$$
$$s = 44 \cdot 1 \text{ m}$$
$$t = ?$$

Substitution in the equation $s = ut + \frac{1}{2}gt^2$ yields
$$44 \cdot 1 = 49t - 4 \cdot 9t^2$$
$$4 \cdot 9t^2 - 49t + 44 \cdot 1 = 0$$
(dividing by 4·9) $t^2 - 10t + 9 = 0$
$$(t - 9)(t - 1) = 0$$

Hence $t = 9$ seconds and $t = 1$ second *(Answer)*.

The path of the ball is sketched in Fig. 55. We have found that it takes 1 second to rise from O to A, and 9 seconds to the path from O to A to T to A. This means that it takes 4 seconds to travel from A to T, and a further 4 seconds to travel from T to A. Clearly it will take 1 second more to travel from A to O, so we deduce that the total time of travel is 10 s.

It is easy to show that the speed of the ball as it passes A is the same on both occasions. Using the equation $v^2 = u^2 + 2gs$, we can see at once that, since u, g, and s are the same in each case, the value of v must also be the same. Note that the *velocity* is not the same, since the motion is upwards in the first case and downwards in the second.

Exercise 20

Throughout all these problems assume that $g = 9 \cdot 80 \text{ m s}^{-2}$ and that there is no resistance to the motion.

1. A ball is projected vertically upwards with a velocity of 19·6 m s^{-1}. Find:

   (*a*) the time taken to reach the greatest height;
   (*b*) the greatest height;
   (*c*) the position of the ball when $t = 5$ seconds;
   (*d*) the times taken to reach a height of 14·7 metres.

2. A stone is projected vertically upwards with a velocity of 98 m s^{-1}. Find:

   (*a*) the position of the stone when $t = 20$ seconds;
   (*b*) how far the stone travels during the eleventh second of motion;
   (*c*) the times at which the stone is at a height of 93·1 metres.

3. Draw the velocity–time graph for the motion of a body projected vertically upwards with a velocity of 73·5 m s^{-1}. Indicate on the graph those points which correspond to the greatest height and the return to the point of projection. Find the time of flight and the greatest height reached (to the nearest metre).

When we are concerned with a body which merely falls from a point it is more convenient to consider displacement, velocity, and acceleration as positive when they are measured *downwards*. This means that we should use $g = 9 \cdot 80 \text{ m s}^{-2}$ instead of $-9 \cdot 80 \text{ m s}^{-2}$. The advantage of this system can be seen from a study of the following examples.

Example: A body falls freely from rest at the top of a tower 98 metres high. If $g = 9.80$ m s^{-2}, calculate (a) the time taken to reach the ground, and (b) the velocity of impact.

SOLUTION: We are given that

$$s = 98 \text{ m}$$
$$g = 9.80 \text{ m s}^{-2}$$
$$u = 0$$
$$t = ?$$
$$v = ?$$

(a) Substitution in the equation $s = ut + \frac{1}{2}gt^2$ yields

$$98 = 0 + 4.9t^2$$
$$4.9t^2 = 98$$
$$t^2 = 20$$
$$t = \pm\sqrt{20}$$

Hence, to the nearest half second, $t = 4.5$ seconds   (*Answer*).

(b) Substitution in the equation $v^2 = u^2 + 2gs$ yields

$$v^2 = 0 + 2 \times 9.80 \times 98$$
$$= 1920.8$$
$$v = \pm\sqrt{1920.8}$$
$$= \pm 43.8 \text{ (to one decimal place).}$$

Since the body is moving *downwards*, the velocity is *positive*; hence the required solution is 43.8 m s^{-1} (*Answer*).

Note that s and g were both positive since they were both measured downwards. In part (a) the possible solution $t = -4.5$ seconds is rejected for obvious reasons.

Example: A v.t.o.l. jet plane is rising vertically with a velocity of 10 m s^{-1}. It has reached a certain height when the pilot drops a coin, which takes 4 seconds to hit the ground. Assuming that there is no resistance to the motion of the coin, and taking the acceleration due to gravity as 9.80 m s^{-2}, calculate (a) the height of the plane, and (b) the velocity of the coin on impact with the ground.

SOLUTION: Since the plane is rising with a velocity of 10 m s^{-1} it follows that, at the moment of its release, the coin is also rising with a velocity of 10 m s^{-1}. Applying the 'downwards positive' convention *to the coin*, we are given that

$$u = -10 \text{ m s}^{-1}$$
$$g = +9.80 \text{ m s}^{-2}$$
$$t = 4 \text{ seconds}$$

(a) Substitution in the equation $s = ut + \frac{1}{2}gt^2$ yields

$$s = -40 + 4.9 \times 16$$
$$= 78.4 - 40$$
$$= 38.4$$

Hence the distance through which the coin falls is 38.4 m. Clearly the height of the plane above the ground when the coin was released must also be 38.4 m (*Answer*).

(b) Substitution in the equation $v = u + gt$ yields

$$v = -10 + 9.80 \times 4$$
$$= 39.2 - 10$$
$$= 29.2$$

Hence the coin hits the ground with a velocity of 29.2 m s^{-1} (*Answer*).

Exercise 21

In the following problems assume that $g = 9.80$ m s^{-2} and that there is no resistance to the motion.

1. A stone is dropped down a well and hits the water after 3 seconds. Calculate the depth of the well and the velocity of impact.
2. A cage in a mine shaft is rising with a velocity of 15 m s^{-1}. A piece of coal which falls from the cage takes 8 seconds to reach the bottom of the shaft. Find how high the cage was above the base of the shaft when the piece of coal fell out, and also the velocity of impact.
3. A ball falls from a height of 9.8 m and rebounds to a height of 4.9 m. Calculate the time from the moment of release to the second impact with the ground.
4. A stone is thrown vertically downwards with an initial velocity of 40 m s^{-1}. If it takes 2 seconds to reach the ground, from what height was the stone projected? Find also the velocity of impact.
5. A body drops from rest. Find the distance fallen during the first, second, and third seconds.

CHAPTER FIVE

NEWTON'S LAWS OF MOTION

In the previous two chapters we have discussed the motion of a body without any reference to the forces which we know are necessary to cause the motion. We now relate the forces to the motion by using Newton's Laws, on which the whole of classical mechanics is founded.

(1) Momentum

Newton's Laws involve a quantity called **momentum** which we have not met so far.

DEFINITION: *The momentum of a particle is the product of its mass and velocity.*

Thus the momentum p of a body of mass m moving with a velocity v is given by

$$p = mv.$$

The definition is quite clear if we are dealing with small bodies or particles. But if we try to find the momentum of a rigid body such as a wheel rolling along the ground, difficulties arise because, as a little thought shows, not all parts of the wheel have the same velocity. Consider a wheel of a car which is moving with a velocity v. The axle of the wheel is moving with velocity v; at the same time, the wheel is rotating on the axle. This means that any particle of the wheel will have two velocities: one due to the axle being carried forward with the speed v, and one due to the wheel rotating on the axle. The complete motion of the wheel therefore consists of a linear velocity v, together with an additional angular velocity about the axle.

Now we have just defined the linear momentum of a particle, and we extend this definition to a rigid body, by using the velocity of its centre of mass (which, for our purposes, can be regarded as being the same as the centre of gravity). Whenever the linear velocity of a rigid body is stated, it means the velocity of the centre of mass of that body.

DEFINITION: *The linear momentum of a rigid body is the product of its mass and the velocity of its centre of mass.*

Thus the linear momentum of a car is the product of its mass and the velocity shown accurately on the speedometer. The linear momentum of a horse of mass m which is galloping with a velocity v is simply mv, regardless of the varied movement of its legs. If a boy runs forward with a velocity v carrying a plastic windmill of mass m, the linear momentum of the windmill is mv regardless of the direction in which the windmill is rotating about its centre.

Since linear momentum is the product of a mass (kilogrammes) and a velocity (metres per second), it follows that the unit of momentum is kilogramme metre per second (abbreviated to $kg\,m\,s^{-1}$). For example, if each wheel of a car has a mass of 10 kg, the momentum of each wheel when the car is travelling with a velocity of 15 m s^{-1} is 150 kg m s^{-1}. If the total mass of the car and occupants is 1000 kg, the momentum of the car and occupants is 15 000 kg m s^{-1}.

Newton's Laws of Motion may be stated as follows:

1. Every body will continue in a state of rest or uniform motion in a straight line unless acted upon by an external applied force.
2. The rate of change of motion is proportional to the applied force and takes place in the direction of that force.
3. To each action there is an equal and opposite reaction.

The third law has already been discussed in Chapter One (page 9).

It is not possible to give a proof of these laws, but there is a great deal of experimental evidence for assuming their truth. The assumption that the laws are true is borne out by the fact that the motions of the stars and planets have been, and are, computed with a high degree of accuracy which is constantly confirmed by astronomical observations. The time and place of eclipses and tides throughout the world are all officially noted and predicted in the *Nautical Almanac*; again the overall accuracy of the forecasts convinces us that the laws are valid for all such practical purposes.

(2) The First Law

The first law provides the definition of force which we have already quoted in Chapter One: 'force is that which tends to change the state of rest or uniform motion of a body'. It should be noted that it is the *resultant* force on the body which is being discussed in each case.

Consider for example a body resting on a table. Since the body is in contact with the table, the weight W_1 of the body is being balanced by an equal and opposite reaction force of the table on the body. The resultant force is $R - W_1 = 0$.

If we apply the force of our weight W_2 by standing on the body, and if the table is able to withstand the increased force, then $R - (W_1 + W_2) = 0$. That is, the magnitude of the action force has increased from W_1 to $W_1 + W_2$, but so has the magnitude of the reaction. So the resultant force on the body remains zero and the body continues in its state of rest—unless, of course, the table collapses as in Fig. 56 (*b*), in which case the resultant force on the body is only its weight W_1. We therefore agree that a body will remain at rest unless an external force is used to set it in motion.

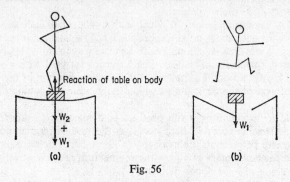

Fig. 56

However, the idea that a body would continue to move with uniform velocity in a straight line if left to itself without any applied forces is not so easy to accept: we have no personal experience of bodies which manage to do this. Nevertheless, there is sufficient evidence to lead us to consider that the truth of this law is a most reasonable assumption.

Here on earth the main difficulty in putting the first law to the test is that when a body moves there is always a resistance to its motion, either by the atmosphere or the surface of contact or both. When a body slides across a surface there is always a force of friction which opposes the continuation of the motion. We attempt to minimize this frictional force by moving the body across what are called 'smoother' surfaces, i.e. surfaces which offer little resistance to the motion across them. Ice is a particularly good surface to attempt an approximation to the first law since its resistance to motion is so small. A puck used in ice hockey will travel a considerable distance across ice, and noticeably it travels in a straight line. A steel ball-bearing will travel even farther, and again in a straight line. Each time we lessen the resistance to the motion, the closer we get to a continuation of the state of uniform motion in a straight line. We infer, therefore, that if we could only remove all resistance to the motion, i.e. have a zero-resisting force, Newton's First Law would be verified.

(3) The Second Law

The original statement of Newton's Second Law referred to a 'change in motion', but Newton interpreted this as being dependent on the mass of the body and its velocity. We measure the change in 'motion' by the change in *momentum*, so the second law is often stated as:

The rate of change of momentum is directly proportional to the applied force and takes place in the direction of this applied force.

If we look further into the influence of this applied force, we realize that the change of momentum will also depend on the length of time the force is applied. For example, if the force is applied for 10 seconds, the change of momentum will be double that achieved by applying the same force for only 5 seconds. Consequently, the second law is sometimes stated as:

The change of momentum is directly proportional to the applied force and the time for which it acts.

To obtain a more concise statement of the relation between momentum and force, let us suppose that the application of the force lasts for time t, and that during this time the velocity of the mass m changes from u to v in the same straight line as the force. Now the change in momentum is $mv - mu$ $= m(v - u)$ in time t. Therefore the rate of change of momentum is $\dfrac{m(v - u)}{t}$ per unit of time. If m is measured in kilogrammes, v and u in metres per second, and t in seconds, then the rate of change of momentum is in kilogramme metres per second per second (abbreviated to $kg\,m\,s^{-2}$).

But, from the equations of motion discussed in Chapter Four, we know that $v = u + at$ where a is the acceleration of the body concerned, and from this equation

$$a = \frac{v - u}{t}$$

Using this result we now see that

$$\frac{m(v - u)}{t} = ma$$

hence the rate of change of momentum is $ma\,\,kg\,m\,s^{-2}$.

Newton's second law enables us to say that the applied force F is directly proportional to ma, which means that we may write

$$F = bma$$

where b is a constant.

At this stage we realize we have yet to define the unit of force; we made use of the unit of force in Chapters One and Two solely for descriptive purposes. Clearly it would be an advantage if we could arrange for b in the equation to have a numerical value of 1, so that $F = ma$. With the standard units being 1 kg for m and $1\,m\,s^{-2}$ for a, this means the standard unit for F is $1\,kg\,m\,s^{-2}$. Therefore the unit force is that force which will give a mass of 1 kg an acceleration of $1\,m\,s^{-2}$. We call this unit of force a **newton** and use N for its abbreviation.

It has been told that the sight of a falling apple gave Newton the first ideas which led to his discovery of universal gravitation, the attraction of each body to every other body in the universe. This is the force which makes bodies fall towards the ground. Whether the story is true or not there is a certain charm in the fact that the S.1. unit of force is such that the weight of the average apple is approximately one newton.

The newton is called an **absolute** unit of force because its value is fixed and independent of where it is applied; consequently it is used throughout all scientific work. The other standard force we have already met is weight, which suggests that we could have used the weight of a mass of 1 kg as a standard of force. Unfortunately the weight of a body varies from place to place, due to the variation in the pull of the earth on the body. A unit dependent on weight, called a **gravitational** unit, is therefore unsatisfactory for scientific work even if it has an on-the-spot convenience.

If we imagine a body to be falling freely, the only force on the body is its own weight W. Since the acceleration due to gravity is g, it follows from the relation $F = ma$ that the weight of any body is related to its mass by $W = mg$. Thus if $m = 1$ kg and the approximation to g is taken as $9\cdot81$ m s^{-2}, the weight of a mass of 1 kg is given by

$$W = 1 \times 9\cdot81 \text{ kg m s}^{-2}$$
$$W = 9\cdot81 \text{ N}$$

A mass of 1 kg therefore has a weight of $9\cdot81$ N.

We may pursue the comparison of mass and weight as in Section 4 of Chapter One. With the obvious notation we have,

$$W_1 = m_1g \text{ and } W_2 = m_2g$$

The value of g is the same in both cases since the weights are obtained at the same place.

Hence $$\frac{W_1}{W_2} = \frac{m_1}{m_2}$$

as before.

These last equations do give a clearer indication of the difference in weight obtained by carrying out the inquiry in two different places. Suppose we take $g = 9\cdot80$ m s^{-2} and $9\cdot81$ m s^{-2} for two different results; then

$$W_1 = m_1 \, 9\cdot80, \, W_2 = m_1 \, 9\cdot81$$

for the same mass in two different places.

Therefore $$\frac{W_1}{W_2} = \frac{9\cdot80}{9\cdot81}$$

from which we can see why weight does not supply us with a satisfactory unit for accurate scientific purposes. With $g = 9\cdot78$ m s^{-2} at the Equator and $g = 9\cdot83$ m s^{-2} at the poles (approximately), the weight of a body could vary by as much as $0\cdot5$ per cent.

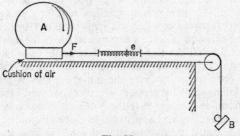

Fig. 57

We still need to clarify the relation between the definition of the newton and the measurement of force by the extension of a spring as discussed in Chapter One. Consider the following experiment:

A body A is able to ride on an almost frictionless cushion of air by hovering on jets on its underside. It is floated over a horizontal table by the pull of a spring attached to a falling body B, as shown in Fig. 57. The experiment consists of measuring the acceleration of body A, and measuring the extension

e of the spring in order to find the tension F. The results indicate that when the body A is in motion, the tension F in the spring is such that

$$F = (\text{acceleration of } A) \times (\text{a constant number } C)$$

Or to put it another way, the acceleration of A is directly proportional to the force in the spring. When the mass of body A is doubled, it is found that the extension of the spring is related to the acceleration of A by

$$F = (\text{acceleration of } A) \times (\text{twice the same constant number } C)$$

From these results we deduce that the constant C is a property of the body A, and this is the property we call **mass**. So we may write

$$F = ma$$

where F is the resultant force on the body A of mass m, and a is the acceleration. Therefore the measurement of force by the extension of a spring is the same as its measurement, in newtons, by the product of mass and acceleration.

Example: A constant force acts for 5 seconds on a mass of 2 kg and changes its velocity from 3 m s^{-1} to 18 m s^{-1} in the same straight line as the force. Calculate the magnitude of the force.
SOLUTION: The change of momentum of the body is 2 kg $\times$ (18 $-$ 3) m s^{-1} = 30 kg m s^{-1}. Since this takes place in 5 s, the rate of change of the momentum is 6 kg m s^{-1} per second = 6 kg m s^{-2} = 6 N   (*Answer*).

Example: A force of 6 N acts for 2 seconds on a mass of 3 kg which is originally moving with a velocity of 10 m s^{-1} in the direction of the force. Calculate the final velocity and the change in momentum of the body.
SOLUTION: Let the velocity be v at the end of the 2-seconds application of the force, so that the change in velocity is $(v - 10)$ m s^{-1}. Therefore the change of momentum is $3(v - 10)$ kg m s^{-1}.
Hence the rate of change of momentum is

$$\frac{3(v - 10)}{2} \text{ kg m s}^{-1} \text{ per second} = \tfrac{3}{2}(v - 10) \text{ N}$$

But this measures the applied force of 6 N so that

$$\tfrac{3}{2}(v - 10) = 6$$
$$v - 10 = 4$$
$$v = 14 \text{ m s}^{-1}   (\textit{Answer}).$$

The change of momentum is $3(14 - 10) = 12$ kg m s^{-1} (*Answer*).

Example: A mass of 10 kg is travelling with a velocity of 3 m s^{-1} when a constant force is applied for 4 s in the opposite direction. At the end of this time the velocity of the body is -2 m s^{-1}. Calculate the magnitude of the applied force.
SOLUTION: The velocity has been changed from 3 m s^{-1} to -2 m s^{-1} during the 4 seconds. This is a change of -5 m s^{-1}. The change of momentum is therefore 10 $\times$ $(-5) = -50$ kg m s^{-1}.
The rate of change of momentum is

$$\frac{-50 \text{ kg m s}^{-1}}{4 \text{ s}} = -12\cdot5 \text{ kg m s}^{-2}$$

The force is $-12\cdot5$ N (*Answer*). The negative sign simply confirms that the force was applied in the opposite direction to the original velocity.

Example: A mass of 10 kg has a velocity of 20 m s^{-1}. A force of 100 N is applied to the mass in the direction of motion and increases the velocity to 35 m s^{-1}. Find the time for which the force is applied.

SOLUTION: The change in velocity of the body is 15 m s^{-1}. The change in momentum of the body is $10 \times 15 = 150$ kg m s^{-1}. If the force is applied for t s, the rate of change of the momentum is $150 \div t$; this is equal to the magnitude of the applied force measured in newtons. Therefore

$$\frac{150}{t} = 100$$

$$t = 1 \cdot 5 \text{ seconds} \quad (Answer).$$

Example: A mass of 6 kg travels in a straight line with a uniform acceleration of 4 m s^{-2}. Find the constant force which is acting on the body, and the change in momentum of the body during 5 s.

SOLUTION: Since $F = ma$ the force on the body is

$$6 \times 4 = 24 \text{ kg m s}^{-2}$$
$$= 24 \text{ N} \quad (Answer).$$

Now $F =$ rate of change of momentum

$$= 24 \text{ kg m s}^{-1} \text{ per second}$$

Hence in 5 s the change in momentum is $24 \times 5 = 120$ kg m s^{-1} *(Answer)*.

Exercise 22

1. Find the momentum of a mass of 5 kg which is moving with a uniform velocity of 10 m s^{-1}.

2. A car with a mass of 560 kg is travelling with a velocity of 20 m s^{-1}. If the driver has a mass of 60 kg, calculate the sum of their momenta.

3. A mass of 4 kg has a momentum of 18 kg m s^{-1}. Find the velocity of the mass.

4. A car which has a total mass (including passengers) of 700 kg increases velocity from 14 km h^{-1} to 50 km h^{-1} in 35 s. Calculate the constant force which has been applied. (Change km h^{-1} to m s^{-1}.)

5. A mass of 8 kg moves under the action of a force of 100 N. Find the acceleration of the body in the direction of the force.

6. A mass of 4 kg falls freely from rest at a height of 4·9 m. Using 9·80 m s^{-2} as an approximation to the value of g, calculate the momentum of the body when it hits the ground.

7. A mass of 2 kg hits the ground with a velocity of 10 m s^{-1} and rebounds in the same vertical line with a velocity of 5 m s^{-1}. Calculate the change in momentum of the body.

8. A constant force acting for 6 s on a mass of 10 kg changes its velocity from 2 m s^{-1} to 20 m s^{-1} in the same straight line. Calculate the magnitude of the force.

9. If a force of 40 N acts for 5 s on a mass, what is the rate of change of momentum of the body?

10. A force of 60 N acts for 4 s and changes the velocity of a body from 15 m s^{-1} to 20 m s^{-1} in the same straight line. Calculate the mass of the body.

(4) Impulse of a Force

So far we have only been concerned with gradual changes in momentum which take place under the action of a constant applied force. We have seen that the time of application of the force has some influence on the change of momentum; for example, doubling the time of application results in a doubling of the change in momentum. However, some changes in momentum are produced in a very short period of time. Bouncing a ball on the ground changes its momentum very quickly from being vertically downwards to vertically upwards. The force is applied only during the time in which the ball is in contact with the ground, a time which is clearly very small. We are therefore confronted by two difficulties: that of measuring the length of time of contact, and that of measuring the magnitude of the force during the contact. To help us overcome these difficulties we introduce the idea of the **impulse** of a force.

DEFINITION: *The change in momentum produced by a force is called the impulse of the force.*

Employing the usual symbols, the initial momentum is mu and the final momentum is mv; the change in momentum is therefore

$$mv - mu = m(v - u)$$

But we know from our equations of motion that when the force is constant,

$$v = u + at \text{ or } v - u = at$$

Hence the change in momentum is

$$m(v - u) = m(at)$$
$$= (ma)t$$
$$= Ft \text{ since } F = ma$$

Change of momentum is therefore measured by the product of the constant force F and the time t during which the force is applied.

From the definition above it follows that Ft is the impulse of the force. For example, the impulse of a force of 10 N acting for 4 s is given by

$$10 \text{ N} \times 4 \text{ s} = (10 \text{ kg m s}^{-2}) \times (4s) = 40 \text{ kg m s}^{-1}$$

By introducing the idea of the impulse of a force, and measuring its effect by the change of momentum it produces, we are able to deal with large unknown forces which act for a short period of time. A ball bouncing on the ground, a bullet hitting a brick wall: these are just two cases where it is practically impossible to measure either the magnitude of the force or the length of time during which it acts. So we measure the effects of the force by comparing the motion before and after the force is applied, and we discover that in the case of a constant force F the product Ft is a means of expressing this effect.

Example: The velocity of a body of mass 2 kg is changed from 10 m s^{-1} to 4 m s^{-1} by a constant force acting for 3 seconds in the same straight line. Find the magnitude of the force and its impulse.

SOLUTION: The change in momentum is $2(4 - 10) = -12$ kg m s^{-1}. Hence the impulse of the force is -12 kg m s^{-1}. (*Answer*).

(The negative sign merely indicates that the force is resisting the motion or is in the opposite direction to that in which the velocity was being measured as positive.)

If we let the force be F, then $Ft = -12$ kg m s^{-1}

Hence
$$F = (-12 \text{ kg m s}^{-1}) \div (3 \text{ s})$$
$$= -4 \text{ kg m s}^{-2}$$
$$= -4 \text{ N} \quad (\textit{Answer}).$$

Again the negative sign indicates that the direction of F is opposite to that in which the velocity has been taken as positive.

Example: A bullet of mass 15 g enters a fixed block of wood with a velocity of 35 m s^{-1}. It is calculated that the bullet comes to rest in 0·01 seconds. Assuming the resisting force is constant, calculate the magnitude of this force and its impulse.

SOLUTION: The change in momentum of the bullet is

$$0·015 \text{ kg} \times (0-35) \text{ m s}^{-1} = -0·525 \text{ kg m s}^{-1}.$$

Therefore the impulse of the force is $-0·525$ kg m s^{-1} (*Answer*).
The rate of change of momentum is

$$\frac{-0·525 \text{ kg m s}^{-1}}{0·01 \text{ s}} = -52·5 \text{ kg m s}^{-2} = -52·5 \text{ N}$$

The magnitude of the resisting force is 52·5 N   (*Answer*).

Exercise 23

1. A force of 12 N acts for 3 s on a mass of 2 kg. Calculate the impulse of the force and the change in velocity of the mass.

2. A hammer-head of mass 3·5 kg strikes a rock with a velocity of 10 m s^{-1}. Calculate the impulse of the force on the rock if the hammer does not rebound.

3. A ball of mass 60 g falling vertically strikes the ground with a velocity of 15 m s^{-1} and rebounds vertically with a velocity of 5 m s^{-1}. Calculate the impulse which the ball receives from the ground.

4. During the launching of a lunar spacecraft, the velocity of the rocket changed from 1674 m s^{-1} to 1771 m s^{-1} in 10 s. Assuming that the propulsive force is constant during this time, calculate its magnitude if the average mass moving is 2500 t (1 t = 1000 kg).

5. A tennis ball of mass 56·5 g has a velocity of 20 m s^{-1} reversed to 30 m s^{-1} by a forehand volley sending the ball back in the same line of flight. Calculate the impulse in kg m s^{-1} the ball has received.

6. A bomb explosion sends pieces of shrapnel in all directions, each piece having a momentum of the same magnitude. If a piece of mass 4 g has an initial velocity of 300 m s^{-1}, what is the initial velocity of (*a*) a piece of mass 3 g, (*b*) a piece of mass 5 g?

Above we considered an example concerning a bullet fired into a block of wood. Clearly one could hardly time the bullet's motion inside the block, but we could measure the depth of penetration and work from this. Such a situation would of course need an equation of motion not involving t, and the obvious choice of $v^2 = u^2 + 2as$. Rearrangement of this equation yields

$$2as = v^2 - u^2$$
$$as = \tfrac{1}{2}v^2 - \tfrac{1}{2}u^2$$

Multiplying by m, $mas = \tfrac{1}{2}mv^2 - \tfrac{1}{2}mu^2$

But we know that $F = ma$, therefore

$$Fs = \tfrac{1}{2}mv^2 - \tfrac{1}{2}mu^2 \qquad\qquad\text{(i)}$$

Equation (i) is of similar form to our equation for the impulse of a force (page 73), namely

$$Ft = mv - mu \qquad\qquad\text{(ii)}$$

Equations (i) and (ii) may be regarded as measuring the distance effect and time effect, respectively, of the constant force F.

Since the quantity mv in equation (ii) has been defined as momentum, we now need to define the corresponding quantity $\tfrac{1}{2}mv^2$ in equation (i). It is in fact called **kinetic energy**.

DEFINITION: *The kinetic energy of a body is half the product of its mass and the square of its velocity.*

We write this as $E_k = \frac{1}{2}mv^2$.

As an example to help in understanding the units involved, let us consider the kinetic energy of a body of mass 12 kg moving with a velocity of 10 m s^{-1}.

$$E_k = \frac{1}{2}mv^2$$
$$= \frac{1}{2} \times 12 \text{ kg} \times 10 \text{ m s}^{-1} \times 10 \text{ m s}^{-1}$$
$$= 600 \text{ kg m}^2 \text{ s}^{-2}$$
$$= 600 \text{ (kg m s}^{-2}) \text{ m}$$

But we have seen that the unit kg m s^{-2} is called the newton. Therefore

$$E_k = 600 \text{ N m}$$

Thus the unit of kinetic energy is the newton metre. However, we have used the newton metre in Chapter Two as a unit for measuring the moment of a force about an axis. So, to avoid ambiguity, we shall use the term **joule** (abbreviated to J) for the unit of energy. Thus the result above is written as $E_k = 600$ J.

DEFINITION: *The work done by a force of 1 N which displaces its point of application through a distance of 1 m in the direction of the force is called a joule.*

We shall have much more to say about work and energy in Chapter Twelve.

We may summarize the results of equations (i) and (ii) by saying that the time effect of the force F is measured by the change in momentum of the body, while the distance effect of the force F is measured by the change in kinetic energy of the body.

Example: A bullet of mass 15 g is fired into a fixed block of wood with a horizontal velocity of 400 m s^{-1}. If the bullet comes to rest after penetrating 80 mm, calculate the resistance of the block, assuming the resisting force to be constant.

SOLUTION: We are given that

$$v = 0$$
$$u = 400 \text{ m s}^{-1}$$
$$s = 0.08 \text{ m}$$
$$m = 0.015 \text{ kg}$$
$$F = ?$$

Substitution in the equation $Fs = \frac{1}{2}mv^2 - \frac{1}{2}mu^2$ yields

$$F \times 0.08 = 0 - \frac{1}{2} \times 0.015 \times 400 \times 400$$
$$F = -\frac{1200}{0.08}$$
$$= -15\,000 \text{ N} \quad (Answer).$$

We may also write the answer as $F = -15$ kN (i.e. 15 kilonewtons). The negative sign indicates that the direction of the force is opposite to the direction of the velocity, as we would expect for a force-resisting motion.

Example: The shortest braking distance for a car travelling horizontally at 80 km h^{-1} is approximately 38 m. Assuming that the braking force remains constant, calculate its magnitude and the time taken to stop a car whose total mass (including passengers) is 450 kg. Assume also that there is no air resistance or similar forces acting.

SOLUTION: We are given that

$$u = 80 \text{ km h}^{-1} = \tfrac{200}{9} \text{ m s}^{-1}$$
$$v = 0$$
$$s = 38 \text{ m}$$
$$m = 450 \text{ kg}$$
$$F = ?$$

Substitution in the equation $Fs = \tfrac{1}{2}mv^2 - \tfrac{1}{2}mu^2$ yields

$$38F = 0 - \tfrac{1}{2} \times 450 \times \tfrac{200}{9} \times \tfrac{200}{9}$$
$$F = -\frac{1\,000\,000}{9 \times 38}$$
$$= -2924 \text{ N}$$
$$= -2 \cdot 9 \text{ kN (to 1 decimal place)} \quad (Answer).$$

Some idea of the size of this force is gained from the fact that it is approximately equivalent to three-quarters of the weight of the car and its passengers. Since the force and mass are constant, so is the acceleration: in this case it is negative.

As we have not worked out the acceleration (although we could), we need an equation of motion which involves only s, t, u, v in order to find t.

We use $s = \tfrac{1}{2}(u + v)t$, which yields

$$t = \frac{2s}{u + v}$$
$$= \frac{2 \times 38 \times 9}{200}$$
$$= 3 \cdot 42 \text{ s} \quad (Answer).$$

Exercise 24

1. Find the kinetic energy of a body of mass 10 kg moving with a velocity of

　(a) 10 m s^{-1}
　(b) 72 km h^{-1}
　(c) −15 m s^{-1}

2. A force of 100 N acting on a body for 3 seconds increases its velocity from 8 m s^{-1} to 12 m s^{-1} in the direction of the force. Calculate

　(a) the change of the momentum of the body
　(b) the mass of the body
　(c) the change in kinetic energy of the body
　(d) the distance over which the force acts.

3. A bullet of mass 30 g passes through a 0·05-m thickness of armour plate. The velocity on entry is 500 m s^{-1} and the velocity on exit is 50 m s^{-1}. Assuming that the resisting force is uniform, calculate

　(a) the time of penetration
　(b) the change in momentum of the bullet
　(c) the resisting force
　(d) the loss in kinetic energy of the bullet.

(5) The Third Law

We have already discussed Newton's Third Law in Chapter Two, when we made use of its applications in simple situations involving forces on a body at rest. Just to recall the method of representing forces diagrammatically, consider those forces which arise from holding an egg at rest between thumb and forefinger and pressing as hard as possible (Fig. 58).

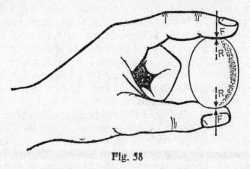

Fig. 58

The egg will not break. Each finger presses on the egg with equal and opposite forces F; the reaction of the egg produces equal and opposite forces R, such that $F = R$. The egg breaks when its shell strength can no longer supply R to equal the magnitude of F. A normal egg will sustain a force of approximately 356 N when applied in the manner of Fig. 58. Few people can exert a force of this magnitude with their fingers.

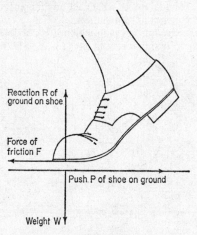

Fig. 59

The act of walking is another everyday example of Newton's Third Law, as shown in Fig. 59. As long as the roughness between the shoe and the ground can supply a frictional force to balance the push force P, the walk will be normal. On ice the frictional force will be very small, so the push force P must also be small to avoid slipping. If no slipping takes place, we simply tip forward and thus momentarily overbalance; we correct this by placing the other foot forward so that walking commences.

Newton's Third Law is particularly useful in its application to the study of changes in momentum which occur when bodies collide.

When two bodies collide, the period of contact may be considered in two parts: (*a*) a period during which a compression of the bodies takes place, and (*b*) a period during which the bodies regain their former shape and, in so doing, rebound apart. The latter period is usually called the **period of restitution,** and we say that the 'force of restitution' makes them rebound apart.

Bodies vary in their ability to regain their former shape. For example, balls of putty retain a permanent distortion and may not rebound at all, whereas steel ball-bearings will regain their original shape very rapidly. This ability to regain shape and rebound after a collision is called the **elasticity** of the body. Bodies which rebound and separate after impact are said to be 'elastic', while bodies unable to recover their shape are said to be 'inelastic'. But knowing whether or not the bodies will separate after impact is not enough: we would like to know something about their velocities after the impact. To simplify our work we shall consider the collision of spheres which are *smooth*: this means that the forces between them during the impact must act along the common normal, i.e. the line joining their centres.

We have seen that the change of momentum caused by a constant force F acting for a time t on a body of mass m is $Ft = m(v - u)$. If we consider the same force acting for the same time t on a different mass m_1, we obtain

$$Ft = m_1(v_1 - u_1)$$

so that $$m(v - u) = m_1(v_1 - u_1).$$

This result means that equal forces which act on different masses for the same time produce equal changes in momentum.

Let us consider the circumstances in which it is possible to ensure that we apply an equal force to two different masses for exactly the same time. Take

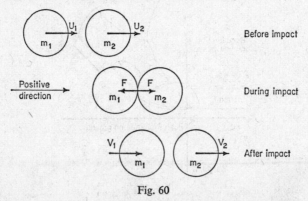

Fig. 60

two spheres of equal size but unequal mass and arrange for them to meet in a head-on collision. The usual 13-mm-diameter steel ball-bearing rolling along a piece of channel iron is excellent for this experiment. Fig. 60 shows the masses of the two spheres and their velocities before and after the impact.

The time of the impact will be very small; but during this time, whatever the magnitude of the force F acting on one sphere, there will be an equal and opposite reaction force F on the other sphere. Since equal forces acting for equal times produce the same change in momentum, it follows that the magnitude of the change in momentum of each sphere will be the same.

In the following calculations we take the positive direction to be from left to right.

For the sphere of mass m, the momentum after impact is m_1v_1, and the momentum before impact is m_1u_1. Therefore the change in momentum is $m_1v_1 - m_1u_1$. But the impulse on the sphere of mass m_1 is $-Ft$. Therefore

$$-Ft = m_1v_1 - m_1u_1 \qquad \text{(i)}$$

For the sphere of mass m_2, the momentum after impact is m_2v_2, and the momentum before impact is m_2u_2. Therefore the change in momentum is $m_2v_2 - m_2u_2$. But the impulse on the sphere of mass m_2 is $+Ft$. Therefore

$$Ft = m_2v_2 - m_2u_2 \qquad \text{(ii)}$$

Comparing equations (i) and (ii) we see that

$$m_1v_1 - m_1u_1 = -m_2v_2 + m_2u_2$$

from which we have

$$m_1v_1 + m_2v_2 = m_1u_1 + m_2u_2 \qquad \text{(iii)}$$

A close examination of equation (iii) reveals a most interesting result. In the direction of the impact the sum of the linear momenta after impact is equal to the sum of the linear momenta before impact. In other words, considering both spheres together, *no momentum has been lost as a result of the impact*.

The above result is a simple example of the **Principle of Conservation of Linear Momentum.** This states that, in any system of particles, the linear momentum in any fixed direction remains constant unless there is an external force acting in that direction. Note that there is no external force on the system shown in Fig. 60. There is an external force F on m_2 at any time during the impact, and likewise an external force $-F$ on m_1 at the same time. But, taking the system of both m_1 and m_2, the total external force is $F - F = 0$. Consequently there will be no change of linear momentum in the direction of F.

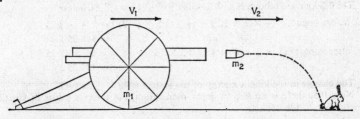

Fig. 61

The recoil of a gun is a simple case of Newton's Third Law. Here the impulse of the force which fires the bullet or shell out of the barrel is equal and opposite to the impulse which recoils on the gun in the direction of the barrel.

As an illustration, let us consider the situation shown in Fig. 61. Using the obvious notation we can apply the principle of conservation of linear momentum (C.O.L.M.) as follows:

Momentum before explosion is 0
Momentum after explosion is $m_2v_2 + m_1v_1$
By C.O.L.M we have $\qquad m_2v_2 + m_1v_1 = 0$

Therefore $\qquad\qquad\qquad m_2v_2 = -m_1v_1$

Clearly the projection of a bullet or shell imparts a considerable kick-back to the gun. This is usually absorbed by the spring mounted in the recoil device, and when the barrel is inclined to the horizontal some of the recoil is absorbed by the ground.

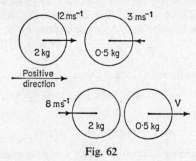

Fig. 62

The following examples illustrate the applications of the principle of conservation of linear momentum.

Example: Two smooth spheres of equal size are moving in the same straight line towards each other. The mass and velocity information is represented by Fig. 62. Calculate the final velocity of the 0·5-kg mass and the loss in kinetic energy of the system.

SOLUTION: By C.O.L.M. we have $(2 \times 12) + (0.5 \times (-3)) = (2 \times 8) + (0.5v)$

$$22.5 = 16 + 0.5v$$
$$6.5 = 0.5v$$
$$13 = v$$

The 0·5-kg mass rebounds with a velocity of $+ 13$ m s^{-1} (*Answer*).

E_k before impact is $(\frac{1}{2} \times 2 \times 12^2) + (\frac{1}{2} \times 0.5 \times (-3)^2) = 144 + 2.25$
$$= 146.25 \text{ J}$$

E_k after impact is $\frac{1}{2} \times 2 \times 8^2 + \frac{1}{2} \times 0.5 \times 13^2 \qquad = 64 + 42.25$
$$= 106.25 \text{ J}$$

The change in the kinetic energy of the system is 40 J (*Answer*).
Note that there is no loss in linear momentum, but there is a loss in kinetic energy due to the impact.

Example: A shell of mass 3 kg is projected horizontally with a velocity of 400 m s^{-1}
If the gun has a mass of 200 kg, calculate the velocity of recoil of the gun.
SOLUTION: Let the velocity of the gun be V. By C.O.L.M. we have,

$$200 V + 3 \times 400 = 0$$

Hence $\qquad\qquad V = -6 \text{ m s}^{-1}$ (*Answer*).

Example: If a gun of mass 200 kg recoils with a velocity of 3 m s^{-1}, calculate (*a*) the velocity with which it projects a 2-kg shell and (*b*) the resisting force necessary to absorb the recoil in a distance of 0·1 m.

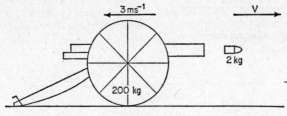

Fig. 63

SOLUTION:

(*a*) Let the velocity of projection of the shell be v m s^{-1}, as shown in Fig. 63. By C.O.L.M. we have

$$2v = 200 \times 3$$

Hence $v = 300$ m s^{-1} (*Answer*).

(*b*) Let the required constant resisting force be F newtons. We are given that

$$v = 0$$
$$u = -3 \text{ m s}^{-1}$$
$$s = -0·1 \text{ m}$$
$$m = 200 \text{ kg}$$

Substitution in the equation $Fs = \frac{1}{2}mv^2 - \frac{1}{2}mu^2$ yields

$$-0·1 \times F = 0 - \frac{1}{2} \times (-3)^2 \times 200$$
$$= -900$$

Hence $F = 9000$ N $= 9$ kN   (*Answer*).

Note that we used $s = -0·1$ m, the negative sign being necessary because the displacement s is in the opposite direction to the resisting force F.

An alternative method of solution is to find the acceleration of the gun, and then find F from the relation $F = ma$.

Substitution in $v^2 = u^2 + 2$ *as* yields

$$0 = (-3)^2 + 2a \times (-0·1)$$
$$0 = 9 - 0·2a$$
$$\therefore a = +45 \text{ m s}^{-2}$$
$$\therefore F = ma = 200 \times 45 \text{ kg m s}^{-2}$$
$$= 9000 \text{ N}$$
$$= 9 \text{ kN}   (\textit{Answer}).$$

Exercise 25

1. *A* and *B* in Fig. 64 are two inelastic spheres, i.e. they will remain together after an impact. *A* has a mass of 2 kg, and *B* has a mass of 3 kg. If both spheres are moving in the same horizontal line with the following velocities, calculate the common velocity after impact.

(*a*) $u_A = 10$ m s^{-1}; $u_B = 4$ m s^{-1}
(*b*) $u_A = 10$ m s^{-1}; $u_B = -4$ m s^{-1}
(*c*) $u_A = 4$ m s^{-1}; $u_B = -10$ m s^{-1}
(*d*) $u_A = -4$ m s^{-1}; $u_B = -10$ m s^{-1}

2. Two inelastic spheres of the same size and of equal mass 4 kg are travelling in the same direction on the same straight line with velocities of 6 m s^{-1} and 10 m s^{-1} respectively. Calculate their common velocity after impact and the loss in kinetic energy due to the impact.

3. Two elastic spheres of equal mass m travel towards one another on the same straight line and collide with velocities of equal magnitude. Show that they separate with velocities of equal magnitude.

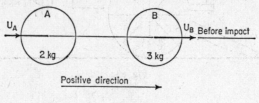

Positive direction

Fig. 64

4. Two bodies are falling freely in the same straight line. If the two bodies collide, is it still possible to apply the principle of the conservation of linear momentum? Try to give a reason for your answer.

5. A man of mass m stands at rest on a small railway truck of mass M which can be assumed to be frictionless and free to move. He picks up a bag of coal of mass m_1 and throws it horizontally out of the truck in the direction of the track with a velocity v. Describe what you think will happen.

The relationship between force and rate of loss of momentum has many commercial applications, including sand blasting and cleaning by water jet.

Recalling once again that $Ft = mv - mu$ is the change in momentum in time t, or that $F = \dfrac{mv - mu}{t}$ is the rate of change of momentum, let us consider a jet of sand or water projected horizontally on to a vertical wall and let us suppose that a mass m of material is ejected each second with a velocity v. If the jet strikes the wall and we consider all the momentum of the jet is destroyed by the impact, then the rate of change in momentum at the face of the wall is mv per second. Therefore the force on the wall is mv. A numerical example will make this clearer.

Example: A jet of water is played at right-angles to a vertical wall. If a mass of 10 kg is ejected with a velocity of 15 m s^{-1} each second, calculate the force on the wall assuming that all the momentum of the water is lost on impact.

SOLUTION: The rate of change of momentum at the wall face is (10 kg × 15 m s^{-1}) per second = 150 kg m s^{-1} per second

$$= 150 \text{ kg m s}^{-2}$$

Therefore the force on the wall is 150 N (*Answer*).

Example: A jet of water is delivered through a hose nozzle of radius 10 mm. Twenty-five litres of water are pumped through each second. If 1 litre of water has a mass of 1 kg, calculate (*a*) the velocity of ejection of the water, and (*b*) the force on a vertical wall from the horizontal jet, assuming that all the momentum of the water is lost at the wall face.

SOLUTION:

(a) The area of the nozzle opening is $\pi(0{\cdot}01)^2$ m². If 25 l of water are ejected, the length of jet passing the nozzle per second is $\dfrac{0{\cdot}025}{\pi(0{\cdot}01)^2}$ m (since 25 l = $0{\cdot}025$ m³).

The velocity of ejection is the length of jet passing the nozzle per second, i.e. $\dfrac{0{\cdot}025}{\pi(0{\cdot}01)^2}$

$$= \frac{250}{\pi} \text{ m s}^{-1} \quad (Answer).$$

(b) The rate of change of momentum at the wall face is therefore (25 kg) $\times$ $\left(\dfrac{250}{\pi} \text{ m s}^{-1}\right)$ per second

$$= \frac{6250}{\pi} \text{ kg m s}^{-1} \text{ per second}$$

$$= 1990 \text{ kg m s}^{-2}$$

Therefore the force on the wall is 1990 N (*Answer*).

Exercise 26

1. A road hammer projects a hammer of mass 2 kg on to the ground at the rate of 200 strikes per minute with an average velocity of 20 m s⁻¹. Calculate the average force on the road due to the hammer.

2. A punch drill projects a small drill bit of mass 1 kg into a brick wall with a strike rate of 500 per minute. A measurement of the mechanism reveals that the bit must be striking with a velocity of 50 m s⁻¹. Calculate the average force exerted on the wall.

3. Water is pumped through a hose at the rate of 30 kg per second with a velocity of 20 m s⁻¹. If the jet is directed horizontally with no loss of velocity on to the side of a frictionless truck of mass 500 kg, with what initial acceleration will the truck move assuming that the jet is in the direction of the track?

4. A jet of water is delivered through a hose nozzle of 10-mm radius, and 30 l of water are pumped through each second. If a litre of water weighs 1 kg, calculate the velocity of ejection of the water and the force on a vertical wall from the horizontal jet assuming that all the momentum of the water is lost at the wall. Use $3{\cdot}14$ as an approximation to the value of π.

CHAPTER SIX

AN INTRODUCTION TO VECTORS

(1) Revision of Trigonometry

In the course of our studies of vectors we shall employ a certain amount of elementary trigonometry, so it is as well to begin with a brief outline of this subject. Readers who are already familiar with the sine and cosine rules for a triangle may omit the next few sections.

We shall confine our attention to the three following **trigonometric functions** of an angle, each of which is derived from a right-angled triangle such as that in Fig. 65.

The trigonometric functions of the angle CAB are defined as follows:

(i) $\dfrac{BC}{AC}$ is called the **sine** of the angle CAB and is written

$$\sin \angle CAB = \frac{BC}{AC}$$

(ii) $\dfrac{AB}{AC}$ is called the **cosine** of the angle CAB and is written

$$\cos \angle CAB = \frac{AB}{AC}$$

(iii) $\dfrac{BC}{AB}$ is called the **tangent** of the angle CAB and is written

$$\tan \angle CAB = \frac{BC}{AB}.$$

Since each of these functions is expressed in terms of a ratio, we often refer to them as **trigonometric ratios.**

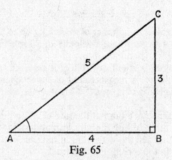

Fig. 65

The first thing to notice is that ratio (iii) may be obtained from ratios (i) and (ii) because

$$\tan \angle CAB = \frac{BC}{AB} = \frac{BC}{AC} \times \frac{AC}{AB} = \frac{BC}{AC} \div \frac{AB}{AC} = \frac{\sin \angle CAB}{\cos \angle CAB}$$

However, the statements made so far are only related to the particular triangle of Fig. 65, and we really need a guide to obtaining these ratios from any triangle whatsoever. We observe that:

(i) AC is the hypotenuse of the triangle.
(ii) BC is the side opposite angle CAB.
(iii) AB is the side adjacent or next to angle CAB.

In (iii) it might be argued that AC is also adjacent to the angle CAB. This is clearly true, but we always agree to reserve the special name of hypotenuse for the longest side of a right-angled triangle.

So we can now express the functions in a manner which is applicable to all right-angled triangles, as follows:

$$\text{Sin angle} = \frac{\text{Opposite side}}{\text{Hypotenuse}}$$

$$\text{Cos angle} = \frac{\text{Adjacent side}}{\text{Hypotenuse}}$$

$$\text{Tan angle} = \frac{\text{Opposite side}}{\text{Adjacent side}}$$

We also notice that, since the hypotenuse is the longest side of any right-angled triangle, the numerical value of the sine and cosine of an angle can never be greater than 1. No such restriction applies to the numerical value of the tangent of an angle because it is possible to have the opposite side longer than the adjacent side.

The most common right-angled triangle has sides of 3, 4, and 5 units respectively, as suggested in Fig. 65. The numerical values are such that:

$$\sin \angle CAB = \tfrac{3}{5} = 0{\cdot}6, \ \cos \angle CAB = \tfrac{4}{5} = 0{\cdot}8, \text{ and } \tan \angle CAB = \tfrac{3}{4} = 0{\cdot}75$$

The problem is to find the measure of the angle CAB, i.e. the number of degrees which give the size of the angle. We could easily draw a scale figure and then measure the angles using a protractor, but this becomes very tedious. However, if we can measure the angle from a figure drawn to scale, the size of the angle will always be the same—no matter which scale we use. That is, an angle will have the same sine, cosine, and tangent regardless of the size of the triangle from which we obtain these ratios. So we may as well catalogue the values of the trigonometric ratios of the angles once and for all in a set of tables like those at the back of the book. (The results in the tables are actually calculated by algebraic methods which are beyond the scope of our discussion.)

Exercise 27

1. Using Fig. 65, write down (i) sin $\angle BCA$, (ii) cos $\angle BCA$, (iii) tan $\angle BCA$. Using the tables on pages 306–311, obtain the following:

2. sin 70° and cos 20°.

3. cos 70° and sin 20°.

4. sin 30° and cos 60°.

5. cos 30° and sin 60°.

6. sin 45° and cos 45° and tan 45°.

7. tan 30° and $\dfrac{1}{\tan 60°}$.

8. tan 21° 48′ and $\dfrac{1}{\tan 68° 12'}$.

9. Obtain the angles $\angle BCA$ and $\angle CAB$ in Fig. 65, giving the result to the nearest degree.

10. In any right-angled triangle ABC where $\angle ABC = 90°$, prove that $\sin^2 \angle CAB + \cos^2 \angle CAB = 1$.

To get some idea of the comparative sizes of sine, cosine, and tangent of any angle, consider Fig. 66 (*a*) in which a circle is drawn with centre A and unit radius. For the moment we shall consider angles in between 0° and 90°.

In Fig. 66 (*b*) angle $CAB = PAX$, and the trigonometric ratios for this angle are given by

$$\sin \angle CAB = \frac{BC}{AC} = \frac{XP}{AP} = XP, \text{ since } AP = 1 \text{ unit}$$

$$\cos \angle CAB = \frac{AB}{AC} = \frac{AX}{AP} = AX, \text{ since } AP = 1 \text{ unit}$$

$$\tan \angle CAB = \frac{BC}{AB} = BC, \text{ since } AB = 1 \text{ unit}$$

Since XP and AX are both less than the hypotenuse $AP = 1$ unit, it follows that both the sine and cosine of the angle are greater than 0 but less than 1.

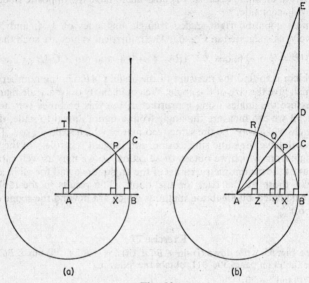

(a)   (b)

Fig. 66

If we make angle CAB smaller and smaller, we see that XP becomes smaller and smaller until eventually (when angle $CAB = 0$) $XP = 0$, therefore

$$\sin 0° = 0$$

At the same time X gets close to B until eventually $AX = AB$, therefore

$$\cos 0° = 1$$

We may use these two results to deduce that $\tan 0 = 0$.

Now let us increase the measure of angle CAB by moving P up to T. As P gets closer to T so XP becomes closer to $AT = 1$ unit and AX comes closer and closer to 0. Therefore

$$\sin 90° = 1 \text{ and } \cos 90° = 0$$

In the meantime $\tan CAB$ has increased beyond all possible bounds as angle CAB has approached 90°. An examination of the tangent tables gives some idea of this rate of increase.

Note that the tangent of an angle is always greater than the sine of the angle. Using Fig. 66 (*b*) we see that

$$\tan \angle CAB > \sin \angle CAB \text{ (because } BC > XP)$$
$$\tan \angle DAB > \sin \angle DAB \text{ (because } BD > YQ), \text{ and so on}$$

Example: *PQR* is an equilateral triangle and *M* is the midpoint of *QR*. If *PQ* = 2 m calculate the length of *MP* and the perpendicular distance of *M* from *PQ*.

SOLUTION: Let us draw a diagram to represent the problem, as in Fig. 67. Since *M* is the midpoint of *QR* we know from the geometry of the equilateral triangle that (i) *PQ* = *QR* = *RP*, (ii) each interior angle of △*PQR* measures 60°, (iii) *PM* ⊥ *QR*, (iv) *MP* bisects angle *QPR*.

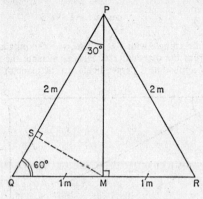

Fig. 67

To find *MP* we observe that *MP* is opposite angle *PQM* in the right-angled triangle *PMQ*, and we know the length of the hypotenuse *QP* is 2 m. Therefore

$$\sin 60° = \frac{MP}{2}, \text{ so that } MP = 2\sin 60°$$

From tables we see that

$$\sin 60° = 0.866, \text{ therefore } MP = 1.732 \text{ m} \quad (Answer).$$

Obviously this is not the only method, for we could use Pythagoras' Theorem to write

$$QP^2 = QM^2 + MP^2$$

Therefore

$$4 = 1 + MP^2$$

and

$$MP = \sqrt{3} \text{ m} \quad (Answer).$$

This second result is more accurate than the first and usually proves to be more useful.

To find *MS*. Let *MS* be the required perpendicular distance of *M* from *PQ*. Examining the right-angled triangle *QMS* we see that *MS* is opposite angle *MQS* and we know the length of the hypotenuse *QM* is 1 m. Therefore

$$\sin 60° = \frac{SM}{QM} = \frac{SM}{1}$$

and since sin 60° = 0.866 it follows that

$$SM = 0.866 \text{ m} \quad (Answer).$$

Looking back over the problem we see that the equilateral triangle provides us with one or two special results which are usually worth remembering.

From triangle *PQM* we have:

$$\tan 60° = \frac{MP}{QM} = \frac{MP}{1} = \sqrt{3}. \quad \therefore \tan 60° = \sqrt{3} \text{ (or 1·732 to three decimal places)}$$

$$\sin 60° = \frac{MP}{QP} = \frac{MP}{2} = \frac{\sqrt{3}}{2}. \quad \therefore \sin 60° = \frac{\sqrt{3}}{2} \text{ (or 0·866 to three decimal places)}$$

$$\cos 60° = \frac{QM}{QP} = \frac{1}{2}. \quad \therefore \cos 60° = \frac{1}{2} = 0·5$$

Example: *ABCD* is a rectangle with $AB = 2BC = 300$ mm. Calculate the distance of the vertex *C* from the diagonal *DB*. Hence calculate the moment about *DB* of a force of 10 N applied at *C* perpendicular to the plane of *ABCD*.

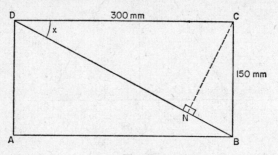

Fig. 68

SOLUTION: Let *CN* be the perpendicular distance from *C* to *DB*. Naming the angle *CDB* = *x* and using the right-angled triangle *CDN* we have,

$$\sin x = \frac{CN}{DC} = \frac{CN}{300}$$

Therefore $\qquad\qquad\qquad CN = 300 \sin x \qquad\qquad\qquad$ (i)

Unfortunately we are unable to look up tables for sin *x* just yet because we do not know *x*. However, if we look at △*CDB* we see that we can find *x* from the relation

$$\tan x = \frac{150}{300} = 0·5$$

and tangent tables, which give *x* = 26° 34′.
Transferring this result to (i) we get

$$CN = 300 \sin 26° 34′ = 300 \times 0·4472$$

Therefore $\qquad CN = 134·16$ mm   (*Answer*).

Finally, the moment about *DB* of the given force of 10 N at *C* is

$$10 \times 134·16 \text{ N m} = 1341·6 \text{ N m} \quad (Answer).$$

Angles of Elevation and Depression

When a point in space is observed, the angle made with the horizontal by the straight line joining the eye to the point is called the angle of elevation when the point is above the eye, and the angle of depression when the point is below the eye.

Example: A man stands 50 m from one of the pillars of a suspension bridge and takes sights at the top of the pillar and at the waterline. The angle of elevation of the top of the pillar is 25° and the angle of depression of the waterline of 10°. Calculate the total height of the pillar above the waterline.

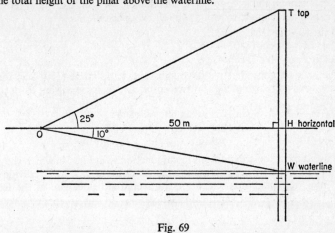

Fig. 69

SOLUTION: Let the problem be represented by Fig. 69. We shall need to work with both the right-angled triangles OHW and OHT and use the relation $WT = WH + HT$.

In $\triangle OHW$, $\dfrac{WH}{OH} = \tan 10°$

∴ $WH = 50 \tan 10° = 50 \times 0 \cdot 1763 = 8 \cdot 815$ m

In $\triangle OHT$, $\dfrac{HT}{OH} = \tan 25°$

∴ $HT = 50 \tan 25° = 50 \times 0 \cdot 4663 = 23 \cdot 315$ m

Hence, $WT = 8 \cdot 815 + 23 \cdot 315$ m

$WT = 32 \cdot 130$ m (*Answer*).

In all probability the accuracy of everyday observations of this kind does not merit the use of four-figure tables, but one of the objects of this problem is an exercise in their use. However, $WT = 32$ m (nearest metre) would be a more realistic answer.

Exercise 28

1. Using Fig. 67 above, calculate (i) sin 30°, (ii) cos 30°, (iii) tan 30°.

2. *ABCD* is a square with $AB = 1$ m. Using Pythagoras' theorem, calculate the length of *AC* and hence find,

(i) sin 45°, (ii) cos 45°, (iii) tan 45°.

3. The Eiffel Tower in Paris is 300 m high. At a point on the horizontal through the base of the tower the angle of elevation of the top of the tower is 15°. How far is this point away from the tower?

4. Using Fig. 69 above, what would be the corresponding angles of elevation and depression from a point which is 100 m from the pillar?

5. A lighthouse on a harbour wall has an observation telescope mounted at a height of 25 m at high tide. A boat is sighted in line for a time of 1 min with an angle of depression which increases from 5° to 10°. Ignoring the curvature of the earth, calculate the distance at which the first sighting took place and the average speed of the boat during the 1 min observation.

So far we have restricted the measure of our angles to lie between 0° and 90°. Since many of our polygons and triangles will contain obtuse angles we shall also need to define the trigonometric functions for obtuse angles.

We can express the trigonometric functions of an obtuse angle in terms of those for an acute angle by using the following definitions.

(i) $\sin x = \sin (180° - x)$.
(ii) $\cos x = -\cos (180° - x)$.
(iii) $\tan x = -\tan (180° - x)$.

We shall see at once that we have substituted the angle $(180 - x)$ for x, together with a change of sign in two cases. An example arising from these definitions is

$$\sin 120° = \sin 60° = 0·866$$
$$\cos 120° = -\cos 60° = -0·5$$
$$\tan 120° = -\tan 60° = -1·732$$

Let us examine these definitions in conjunction with Fig. 70, which shows a circle centre O. Two perpendicular axes X^-OX^+ and Y^-OY^+ divide the circle into four quadrants which are numbered anticlockwise in the figure (this is the standard mathematical practice).

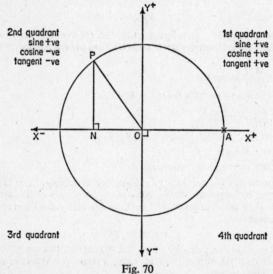

Fig. 70

Any radius OP is drawn from O and thereby creates the angle POA. If OP lies in the first quadrant we say that angle POA is an angle in the first quadrant. Alternatively if the measure of any angle is between 0° and 90°, it is said to be an angle in the first quadrant. Similarly if OP lies in the second quadrant the measure of any angle is between 90° and 180°, and we say that the angle POA is in the second quadrant, and so on.

In order to obtain the relevant trigonometric functions we agree that the required results will be obtained from measurements taken from the triangles ONP formed by drawing a perpendicular from the point P to point N on the line X^-OX^+ wherever P may be on the circle and expressing the trigonometric ratios in terms of the directed line segments ON and NP and the magnitude OP. Thus the numerical values $\sin \angle AOP$, $\cos \angle AOP$ and $\tan \angle AOP$ for the obtuse angle $\angle AOP$ will be obtained from triangle ONP in Fig. 70. Here we have ON expressed as a negative quantity since O to N is in the negative direction along OX^-, while NP will be positive since N to P is in the positive direction parallel to OY^+.

Thus,
$$\sin \angle AOP = \sin \angle NOP = \frac{NP}{OP}, \text{ positive.}$$

$$\cos \angle AOP = \cos \angle NOP = \frac{ON}{OP}, \text{ negative.}$$

$$\tan \angle AOP = \tan \angle NOP = \frac{NP}{ON}, \text{ negative.}$$

However, it must be emphasized that the diagram is only a visual illustration of the statements (i), (ii), and (iii) which are the definition for the sine, cosine, and tangent of an obtuse angle for our elementary work.

Exercise 29

Use the tables at the back of the book to find the numerical values of the following:

1. $\sin 120°$, $\cos 120°$, $\tan 120°$.
2. $\sin 170°$, $\cos 170°$, $\tan 170°$.
3. $\sin 140°$, $\cos 140°$, $\tan 140°$.
4. If $\sin x = 0.5$, find x when x is an obtuse angle.
5. Obtain the obtuse angle solution to $\cos x = -0.866$.
6. Find the solution to $\tan 2x = -1$ if $0 < x < 90°$.
7. Obtain two solutions to the equation $\sin 2x = 0.809$.
8. Given that $\cos x = -0.5$, find x.
9. If $\tan x = -0.3640$, find x.
10. Solve the equation for the acute angle x given by $\sin x = \cos x$.

(2) Sine rule and Cosine rule

The sine and cosine rules for any triangle are best remembered and more systematically obtained by adopting the following simple notation for the typical triangle ABC. We refer to the angle CAB as angle A, the angle ABC

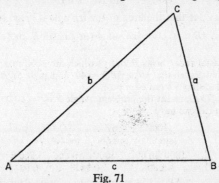

Fig. 71

as angle B, and the angle BCA as angle C. The side of the triangle opposite A is called a, the side opposite B is called b, and the side opposite angle C is called c. Just pause for a moment and see from Fig. 71 how obvious this notation really is.

We now consider two expressions for height CD of the triangle ABC given in Fig. 72.

In the right-angled triangle ADC we have $\sin A = \dfrac{CD}{b}$.

In the right-angled triangle BDC we have $\sin B = \dfrac{CD}{a}$.

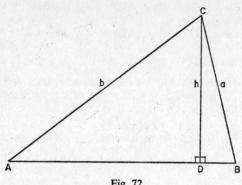

Fig. 72

From these two expressions we get

$$CD = b \sin A = a \sin B$$

Examining the last equation we see that we may write

$$\frac{a}{\sin A} = \frac{b}{\sin B}$$

By choosing another of the three possible heights of the triangle we may complete the relation, which is called the **sine rule**, to read

$$\frac{a}{\sin A} = \frac{b}{\sin B} = \frac{c}{\sin C}$$

Just as an extra we see that the area of the triangle is given by

$$\tfrac{1}{2}AB \times CD = \tfrac{1}{2}cb \sin A \ (\text{or } \tfrac{1}{2}ac \sin B, \text{ or } \tfrac{1}{2}ab \sin C)$$

Example: Two coastal radar posts P and Q which are 1000 m apart receive radio signals from a ship S such that angle $SPQ = 40°$ and angle $SQP = 60°$. Calculate the distance of the ship S from each post.

SOLUTION: Let Fig. 73 represent the problem. Angle $PSQ = 80°$.

Using the sine rule we have

$$\frac{1000}{\sin 80°} = \frac{SP}{\sin 60°} = \frac{SQ}{\sin 40°}$$

$$SP = \frac{1000 \sin 60°}{\sin 80°} = \frac{1000 \times 0\cdot866}{0\cdot9848} = \frac{866\cdot0}{0\cdot9848}$$

$$\therefore \ SP = 879 \text{ m (to nearest metre)} \quad (\textit{Answer}).$$

Similarly

$$SQ = \frac{1000 \sin 40°}{\sin 80°} = \frac{1000 \times 0\cdot6428}{0\cdot9848} = \frac{642\cdot8}{0\cdot9848}$$

$$SQ = 653 \text{ m (to nearest metre)} \quad (\textit{Answer}).$$

Example: Triangle *ABC* in Fig. 74 is drawn to scale so that *AB* represents the magnitude and direction of a force of 200 N. If angle *CAB* = 20° and angle *ABC* = 50°, calculate the forces which are represented by *BC* and *CA*.

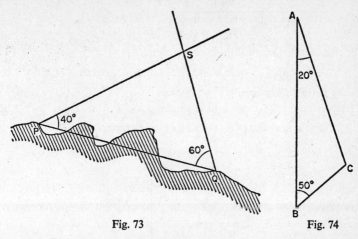

Fig. 73   Fig. 74

SOLUTION: With angle *ACB* = 110° we apply the sine rule to △*ABC* and write

$$\frac{AB}{\sin 110°} = \frac{BC}{\sin 20°} = \frac{CA}{\sin 50°}$$

$$BC = \frac{AB \sin 20°}{\sin 110°} = \frac{AB \sin 20°}{\sin 70°} = 0.3640 \ AB$$

If *AB* represents a force of 200 N, then *BC* represents a force of 0·3640 × 200 = 72·8 N   (*Answer*).

Similarly,

$$CA = \frac{AB \sin 50°}{\sin 110°} = \frac{AB \sin 50°}{\sin 70°} = 0.8153 \ AB$$

CA represents a force of 0·8153 × 200 = 163·06 N   (*Answer*).

Exercise 30

In the usual notation for △*ABC* calculate the remaining unknown sides and angles in the following:

1. *A* = 45°,   *B* = 60°,   *C* = ?,   *a* = 100 m,   *b* = ?,   *c* = ?
2. *A* = 40°,   *B* = ?,   *C* = 20°,   *a* = ?,   *b* = 1000m,   *c* = ?
3. *A* = 20°,   *B* = ?,   *C* = ?,   *a* = 200 m,   *b* = 300 m,   *c* = ?

4. If in Fig. 74 the side *AB* of the given triangle had represented a force of 2000 N, what forces would have been represented by *BC* and *CA*?

5. In *ABC*, angle △*ABC* = 40°, and angle *BCA* = 10°. If *BC* is drawn to scale so as to represent a force of 1000N what forces are represented by the other two sides?

The cosine rule requires a little more skill in manipulation as the following work will show. In Fig. 75 let *CD* be the perpendicular from *C* to *AB* at *D*, and put *CD* = *h*. Now that *AB* = *c* is made up of two segments *AD* and *DB* put *AD* = *x* and *DB* = *c* − *x*.

Using Pythagoras' theorem in the right-angled triangle CDB we have

$$a^2 = h^2 + (c - x)^2$$
$$= h^2 + x^2 + c^2 - 2cx \qquad \text{(i)}$$

Similarly in the right-angled triangle CDA we have

$$b^2 = h^2 + x^2 \qquad \text{(ii)}$$

Substituting the result (ii) into (i) we obtain

$$a^2 = b^2 + c^2 - 2cx, \text{ and since } \cos A = \frac{x}{b}, \, b \cos A = x$$

we may write the final result as

$$a^2 = b^2 + c^2 - 2bc \cos A. \qquad \text{(iii)}$$

This result is known as the **cosine rule** for any triangle ABC.

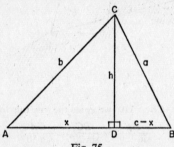

Fig. 75

A closer study of (iii) should convince the reader of the existence of two similar expressions (see end of solutions to Exercise 31). The reader might think that the overall usefulness of the sine rule makes the cosine rule unnecessary. Unfortunately this is not so as the next worked example demonstrates.

Example: The sides of a triangle are 10 m, 8 m, and 4 m respectively. Calculate the measure of the greatest angle in the triangle.

SOLUTION: Let Fig. 76 represent the problem, with angle A the greatest angle being opposite the greatest side. As a first step the sine rule is unhelpful since we do not know the numerical value of any of the sine ratios.

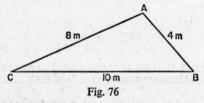

Fig. 76

However the cosine rule yields

$$10^2 = 8^2 + 4^2 - 2 \times 8 \times 4 \times \cos A$$

i.e.

$$100 = 64 + 16 - 64 \times \cos A$$
$$20 = -64 \cos A$$
$$\cos A = \frac{-5}{16} = -0 \cdot 3125$$

The negative sign indicates that A is an obtuse angle.

$$\therefore \cos A = -0.3125 = -\cos 71° \ 47'$$
$$= \cos 108° \ 13'$$

angle $BAC = 108° \ 13'$ (*Answer*).

Now that we have obtained one of the angles we may employ the sine rule to find either of the other angles in the triangle, even though it would be easier to continue using the cosine rule.

Example: Two sides of a parallelogram are 10 m and 4 m respectively, and the included angle measures 70°. Calculate the length of each diagonal.

SOLUTION: Let Fig. 77 represent the problem, with angle $COA = 70°$. Since $COA = 70°$ and $OABC$ is a parallelogram, we know angle $OAB = 110°$, and the reader should now insert points B and C. Also, since opposite sides of a parallelogram are equal in length, $AB = 4$ m and $CB = 10$ m.

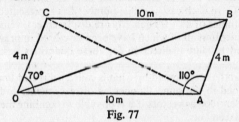

Fig. 77

Applying the cosine rule in $\triangle OAB$ in order to find OB we have,

$$OB^2 = OA^2 + AB^2 - 2 \, OA . \, AB \cos 110°$$
$$OB^2 = 100 + 16 - 80 \cos 110°$$
$$= 116 + 80 \cos 70°$$
$$= 116 + 80 \times 0.342$$
$$= 116 + 27.36$$
$$OB^2 = 143.36$$

Using square root tables

$$OB = 11.98 \text{ m} \quad (\textit{Answer}).$$

Applying the cosine rule in $\triangle OAC$ in order to find CA we have

$$AC^2 = 10^2 + 4^2 - 2 \times 10 \times 4 \cos 70°$$
$$= 116 - 80 \cos 70°$$
$$= 116 - 27.36$$
$$AC^2 = 88.64$$

Using square root tables

$$AC = 9.415 \text{ m} \quad (\textit{Answer})$$

Exercise 31

1. If Fig. 77 above is drawn to a scale such that a line of length 1 m represents a force of 10 N, calculate the forces represented by OC, OA, OB, and AC.

2. The sides of a triangle measure 12 m, 10 m, and 4 m respectively, calculate the measure of the greatest and smallest angles in the triangle.

3. Two sides of a parallelogram measure 100 mm and 80 mm, and include an angle of 80°. Calculate the length of the diagonal passing through their point of

intersection. If the sides of the parallelogram containing the angle 80° represent forces to the scale of 10 mm to 20 N, what is the magnitude of the force represented by the diagonal whose length has just been calculated?

4. *ABCD* is a rectangle with $AB = 120$ mm and $BC = 50$ mm. Calculate the length of *AC* and the measure of each of the angles of $\triangle ACB$ to the nearest 0·1°.

(3) Vector Quantities

In our work we have in general two different types of physical quantities called either scalars or vectors. Examples of a **scalar quantity** are length, area, volume, mass, temperature. Thus we may speak of a rectangle which is 10 m × 4 m at a temperature of 10° Celsius so that the number 10 m represents the length of one side of the rectangle and the number 4 m represents the length of the other side. We may go further by noting that the number 400 m² represents the area of the rectangle at a temperature represented by the number 10°C. In each case a single number has represented the quantity concerned. Examples of a **vector quantity** are displacement, velocity, acceleration, force, momentum all of which have been discussed in previous chapters so that the need to state a direction for these quantities has already been stressed.

Any vector can be represented by what is called a **directed line segment** and they are combined in the same manner as displacements. Indeed, physical quantities are identified as vectors if it is possible to combine them in the same manner as displacements.

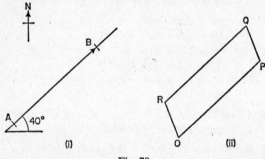

Fig. 78

In Fig. 78 (*i*) we have a line segment *AB* directed by the aid of the arrowhead placed at the end *B* so that the complete arrow represents the vector. The length of *AB* to the chosen scale represents the magnitude of the vector, and the arrow indicates the direction of the vector along the line *AB*. If the length of *AB* is 30 mm and the scale is 10 mm to 1 N, the directed line segment represents a force of 3 N in the direction N 50° E from *A* to *B*.

The vector represented by the directed line segment *AB* is written $\overrightarrow{AB}$ and referred to as 'vector *AB*'.

The magnitude of the vector is written $|\overrightarrow{AB}|$ and referred to as the 'modulus $\overrightarrow{AB}$' or more simply as 'mod $\overrightarrow{AB}$'. Printers also use a bold type such as **AB** for indicating vectors, but this is very difficult for the reader to copy. Very often we only use one letter to represent a vector which has already been given either in a diagram or a previous statement. Thus we might start by saying that $\overrightarrow{AB} = a$ and thereafter use only a when discussing $\overrightarrow{AB}$, since it makes for a neater and more concise-looking statement.

The question of the representation of a vector raises one or two problems. For instance, in Fig. 78 (*i*) shall we take *any* segment of length 30 mm on the line? Can we slide *AB* anywhere along the line and still say that it represents the vector? Again, in Fig. 78 (*ii*) the parallelogram *OPQR* has been drawn with *OP* and *RQ* both of length 30 mm and parallel to *AB*, in which case could either *OP* or *RQ* (with an attached arrowhead) have represented the original vector? Most problems will impose their own special restrictions on the vectors they employ—such as particular points of application of the forces, the starting-points of a displacement, and so on. We do, however, gain the maximum advantage from our geometrical model of a problem by allowing a vector to be represented by any equal and parallel line segment in the same direction as the vector. Thus the line segments *OP* and *RQ* in Fig. 78 (*ii*) may each represent the same vector. We have three different types of vectors as follows:

(*a*) A **free vector** is one whose point of application is not specified.

(*b*) A **sliding vector** is one whose line of action is specified. The vector is said to be localized in that line of action.

(*c*) A **bound vector** is one whose point of application is specified. It is said to be localized at that point. For example, a force is a bound vector since it is always bound to a point of application.

DEFINITION: *Two vectors are said to be equal if they have the same magnitude and direction.*

The equality of the two vectors referred to above in Fig. 78 (*ii*) is stated as $\overrightarrow{OP} = \overrightarrow{RQ}$, and the equality of the other two vectors is stated as $\overrightarrow{OR} = \overrightarrow{PQ}$.

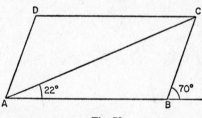

Fig. 79

To see what is meant by the 'sum of two vectors', consider the displacement diagram of Fig. 79 and let it represent a two-stage change in position, first from *A* to *B* and then from *B* to *C*. In vector language this means $\overrightarrow{AB} + \overrightarrow{BC}$. The resultant displacement is equivalent to the sum of these two vectors and is represented by $\overrightarrow{AC}$, the single displacement which would achieve the same final position as the other two combined.

Since we are speaking of the addition of the two vectors, we call the resultant the **vector sum** and we agree to continue using the same plus sign as always to represent the addition of vectors even though the process is different from that of ordinary addition. So we write $\overrightarrow{AC} = \overrightarrow{AB} + \overrightarrow{BC}$, and refer to $\overrightarrow{AC}$ as the vector sum or the resultant of the two vectors $\overrightarrow{AB}$ and $\overrightarrow{BC}$. This is called the **triangle law** for the addition of two vectors.

Now we have already agreed that either BC or AD may represent the same vector quantity. So the equation above may be written as

$$\overrightarrow{AC} = \overrightarrow{AB} + \overrightarrow{AD} \quad \text{or} \quad \overrightarrow{AC} = \overrightarrow{AD} + \overrightarrow{DC}$$

Since a force is a vector we could summarize this example in conjunction with Fig. 79 by saying that the resultant of two forces of magnitude $\overrightarrow{AB}$ and $\overrightarrow{AD}$ acting at the point A and inclined at 70° to each other is $\overrightarrow{AC}$ at 22° to the greater force. A practical confirmation of the general result may be obtained as follows.

Knot three pieces of strong thread together, attach spring balances to each thread and hook them on to three nails at P, Q, and R in a vertical board, as shown in Fig. 80.

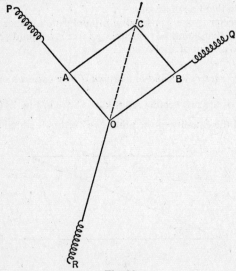

Fig. 80

There are now three forces in equilibrium applied at the point O. The direction of the forces is given by the direction of the strings OP, OQ, OR; their magnitude can be read off from the spring balances. Since the forces are in equilibrium each force is balancing the other two, in which case we call each force the **equilibrant** of the other two. To put it another way, each pair of forces balances the third force. But we know that any force is balanced by an equal and opposite force; so the resultant of each pair of forces must be equal and opposite to their equilibrant, which is the third force.

Making a note of the position of O and the forces recorded by the springs, remove the apparatus and by choosing a suitable scale measure OA and OB to represent the forces in strings OP and OQ. On completion of the parallelogram $OACB$ it will be found that OC will represent the magnitude of the force in the string OR. Thus, $\overrightarrow{OC}$ represents the resultant of $\overrightarrow{OA}$ and $\overrightarrow{OB}$, i.e.

$$\overrightarrow{OC} = \overrightarrow{OA} + \overrightarrow{OB} \quad \text{or} \quad \overrightarrow{OC} = \overrightarrow{OB} + \overrightarrow{OA}$$

since it keeps in equilibrium the force in the string OR.

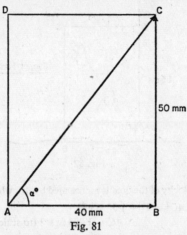

Fig. 81

We see, therefore, that we may also add vectors by using the **parallelogram law** stated as follows: *If two vectors are represented by the adjacent sides of a parallelogram, then the sum of the two vectors will be represented by the diagonal of the parallelogram which passes through their point of intersection.*

Example: Two forces are represented in magnitude and direction by the sides $AB = 40$ mm and $AD = 50$ mm of the rectangle $ABCD$ with the scale 1 mm to 15 N. Calculate the magnitude and direction of the resultant force.

SOLUTION: Let Fig. 81 represent the problem. We are required to find $\overrightarrow{AC} = \overrightarrow{AB} + \overrightarrow{BC}$ (or $\overrightarrow{AC} = \overrightarrow{AD} + \overrightarrow{DC}$).

Using Pythagoras' theorem we get

$$\begin{aligned} AC^2 &= AB^2 + BC^2 \\ &= 40^2 + 50^2 \\ &= 4100 \end{aligned}$$

Therefore $\qquad AC = \sqrt{4100} = 64 \cdot 03$ mm

Therefore AC represents the magnitude of a force of $64 \cdot 03 \times 15 = 960$ N (nearest newton).

The direction of $\overrightarrow{AC}$ will be adequately described by finding the angle $BAC = \alpha$

$$\tan \alpha = \tfrac{50}{40} = 1 \cdot 25, \text{ and from tables } \alpha = 51° \, 20'$$

Hence $\overrightarrow{AC}$ is given by $|\overrightarrow{AC}| = 960$ N (nearest newton) and angle $BAC = 51° \, 20'$ (*Answer*).

Example: Calculate the resultant velocity of a boat which is rowed at 1·5 m s⁻¹ at right-angles to the direction of a current of 1 m s⁻¹.

SOLUTION: Let Fig. 82 represent the problem, with $\overrightarrow{AD}$ representing the forward 1·5 m s⁻¹ velocity of the boat and $\overrightarrow{AB}$ representing the velocity of the boat in the direction of the current.

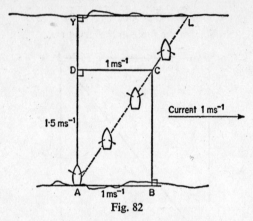

Fig. 82

The resultant velocity of the boat is represented by $\overrightarrow{AC}$, where

$$AC^2 = 1^2 + 1\cdot5^2 = 3\cdot25$$
$$AC = 1\cdot803 \text{ m s}^{-1} \text{ (to scale)}$$

The direction of AC is adequately described by angle DAC

$$\tan \angle DAC = \frac{1}{1\cdot5} = \frac{2}{3}$$
$$\angle DAC = 33° 41'$$

Therefore the boat travels with a velocity of 1·8 m s⁻¹ (to the nearest 0·1 m s⁻¹) at an angle of 33° 41' to the direct crossing, as shown in Fig. 82 (*Answer*).

Note that if these velocities are maintained, the boat will eventually land at the point L on the opposite bank. The distance the boat is carried downstream is given by YL, which is clearly related to the width of the river by $\tan \angle DAC = \dfrac{YL}{AY}$.

Imagine you were in a helicopter hovering over the river: you would then see the boat move in the direction AL. If there had been no current, the boat would have travelled straight across with a velocity of 1·5 m s⁻¹. Hence we often describe this as 'the velocity in still water'.

Exercise 32

1. $ABCD$ is a square of side 50 mm. Using a scale of 10 mm to 25 N, calculate the resultant of the two forces represented by $\overrightarrow{AB}$ and $\overrightarrow{AD}$.

2. Using Fig. 81, and with the same scale, calculate the resultant of the two forces represented in magnitude and direction by BA and BC.

3. Two forces of magnitude 300 N and 400 N act at right-angles to each other at the point O. Calculate the resultant of the two forces.

4. If in Fig. 82 above the river had been 50 m wide, calculate the distance the boat was carried downstream by the current.

5. A boat crosses a stream which is 100 m wide. If the boat moves directly across with a velocity of 5 m s^{-1} but is carried downstream at 2 m s^{-1}, calculate the resultant velocity of the boat and the distance carried downstream.

6. A hiker examines a map of the route he has just taken and sees that it consists of 30 mm in a straight line East followed by 50 mm in a straight line North East. Calculate the resultant displacement on the map and the actual displacement if the map scale is 10 mm to 1 km.

(4) Vector Addition

We now turn our attention to calculating the resultant of two vectors which are not perpendicular to each other. Before going any further let us examine the trigonometric relations within the parallelogram. Fig. 83 shows a parallelo-

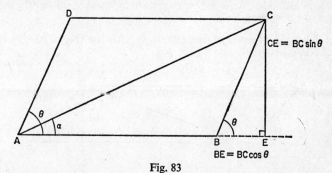

$CE = BC \sin \theta$

$BE = BC \cos \theta$

Fig. 83

gram *ABCD* in which side *AB* has been produced to *E*, the foot of the perpendicular from *C*. Applying the cosine rule to triangle *ABC*, we obtain

$$AC^2 = AB^2 + BC^2 - 2AB \cdot BC \cdot \cos \angle ABC \qquad (1)$$

But we know that $\qquad \angle ABC = 180° - \angle DAB$

so that $\qquad \cos ABC = \cos(180° - DAB)$

$$= -\cos \angle DAB = -\cos \theta \text{ for short (see page 90)}$$

Since $BC = AD$ we may rewrite eqn. (1) as

$$AC^2 = AB^2 + AD^2 + 2AB \cdot AD \cdot \cos \theta \qquad (2)$$

noting that the angle θ is the angle between the two vectors $\overrightarrow{AB}$ and $\overrightarrow{AD}$. In brief, the resultant of the two vectors $\overrightarrow{AD}$ and $\overrightarrow{AB}$ is given by $\overrightarrow{AC}$, where the modulus of $\overrightarrow{AC}$ (i.e. the magnitude of the vector $\overrightarrow{AC}$) is given by eqn. (2). Now all we need to completely specify $\overrightarrow{AC}$ is its direction, and this we obtain by finding angle *CAB*. In the right-angled triangle *CAE* we have

$$\tan \angle CAE = \frac{CE}{AE} = \frac{CE}{AB + BE} = \frac{BC \sin \theta}{AB + BC \cos \theta} \qquad (3)$$

This form of eqn. (3) enables us to express the direction given by angle *CAE* in terms of what we are originally given, i.e. *AB*, *AD*, and the included angle θ.

Let us now suppose that we are to find the resultant R of two forces P and Q acting at a point O inclined to each other at an angle θ. Since the line segments represent the forces such that $P = \overrightarrow{AB}$ and $Q = \overrightarrow{AD}$, we may rewrite eqns. (2) and (3) as

$$R^2 = P^2 + Q^2 + 2PQ \cos \theta; \quad \tan \alpha = \frac{Q \sin \theta}{P + Q \cos \theta}$$

where $P = |P|$, $Q = |Q|$, and $R = |R|$

Example: Two forces of 1 N and 3 N act at a point O and are inclined at an angle of 135° to each other. Calculate the magnitude of the resultant and the angle it makes with the greater force.

SOLUTION: Let Fig. 84 represent the problem, with $|\overrightarrow{OC}| = 3$N and $|\overrightarrow{OA}| = 1$N to the chosen scale.

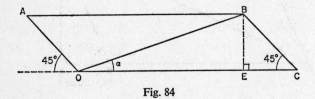

Fig. 84

If the resultant is represented by $\overrightarrow{OB} = F$, then

$$F^2 = 1^2 + 3^2 + 2 \cdot 1 \cdot 3 \cos 135°$$
$$= 10 - 6 \times 0{\cdot}7071$$
$$= 5{\cdot}7574$$
$$F = 2{\cdot}399 \text{ N}$$

i.e. the magnitude of the resultant is 2·4 N (to the nearest 0·1 N). Rather than blindly apply the formula in eqn. (3) above, let us derive the angle from first principles using Fig. 84.

$$\tan \alpha = \frac{BE}{OE} = \frac{BE}{OC - EC} = \frac{BC \sin 45°}{OC - BC \cos 45°} = \frac{1 \sin 45°}{3 - 1 \cos 45°}$$

$$\tan \alpha = \frac{0{\cdot}7071}{3 - 0{\cdot}7071} = \frac{0{\cdot}7071}{2{\cdot}2929}$$

$$= 0{\cdot}3084$$

$$\therefore \alpha = 17° \, 8'$$

The resultant is $\overrightarrow{AC}$ such that $AC = 2{\cdot}4$ N (to the nearest 0·1 N) and making an angle of 17° 8′ with the direction OC (*Answer*).

Note that direct substitution in the formula would yield

$$\tan \alpha = \frac{1 \sin 135°}{3 + 1 \cos 135°} = \frac{\sin 45°}{3 - \cos 45°} \text{ as before.}$$

Example: Calculate the resultant of two equal accelerations of 10 m s⁻² acting at a point A and inclined at an angle of 80° to each other.

SOLUTION: If the resultant acceleration is $\overrightarrow{AC} = a$, then

$$a^2 = 10^2 + 10^2 + 2 \cdot 10 \cdot 10 \cos 80°$$
$$= 200 + (200 \times 0 \cdot 1736)$$
$$a^2 = 234 \cdot 72$$
$$a = 15 \cdot 32 \text{ m s}^{-2}$$

For the direction of $\overrightarrow{AC}$ as given by the angle α we have

$$\tan \alpha = \frac{BC \sin 80°}{AB + BC \cos 80°} = \frac{10 \sin 80°}{10 + 10 \cos 80°} = \frac{\sin 80°}{1 + \cos 80°}$$
$$= \frac{0 \cdot 9848}{1 + 0 \cdot 1736}$$
$$= \frac{0 \cdot 9848}{1 \cdot 1736}$$
$$\tan \alpha = 0 \cdot 8389$$
$$\alpha = 40°.$$

The resultant acceleration is $15 \cdot 32$ m s⁻² at 40° to either of the original accelerations of 10 m s⁻¹ (*Answer*).

Note that since $AB = AD$ the figure $ABCD$ is a rhombus, and the diagonals of a rhombus bisect its opposite angles. It therefore follows that the resultant of two *equal* vectors bisects the angle between them.

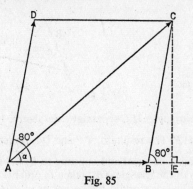

Fig. 85

Exercise 33

1. Calculate the resultant of two forces of 10 N and 6 N acting at a point A and inclined at an angle of 60°.

2. Calculate the resultant of two velocities of 10 m s⁻¹ and 6 m s⁻¹ applied to a body and acting in directions inclined at 60° to each other.

3. A body of mass 2 kg initially at rest receives two impulses simultaneously. One impulse of magnitude 10 kg m s⁻¹ is in the direction due East, and the other impulse of magnitude 20 kg m s⁻¹ is in a direction North East. Calculate the velocity of the body immediately after the blows are struck.

4. Calculate the resultant of two equal forces of magnitude 10 kN acting at a point A and inclined to each other at an angle of 120°.

5. Calculate the resultant of two forces of equal magnitude 3 kN acting at a point A in directions inclined at 60° to each other.

(5) Resultant of Three Vectors

So far we have been concerned with finding the resultant of only two vectors, but, as we saw on page 98, the method really enables us to discuss more fully the relation between three forces (or vectors) which act at a point in equilibrium. If we consider three forces P, Q, and R acting at a point A in equilibrium, as shown in Fig. 86, we know now that R must be equal and opposite to the resultant of P and Q as determined by $\overrightarrow{AC}$ in the parallelogram $ABCD$ in the usual manner.

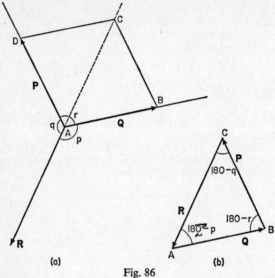

Fig. 86

If we consider the triangle ABC separately, we observe that $\overrightarrow{AB}$ represents Q, and $\overrightarrow{BC}$ represents P. The resultant of P and Q is represented by $\overrightarrow{AC}$ and, since this resultant must be equal and opposite to R, $\overrightarrow{CA}$ represents R. This means that the sides of the triangle ABC taken in order (i.e. with the arrowheads all going anticlockwise round the perimeter of $\triangle ABC$) represent the three forces. These results are summarized as follows:

(a) **The Triangle of Forces:** If three concurrent forces can be represented in magnitude and direction by the sides of a triangle taken in order, the forces will be in equilibrium.

(b) **The Converse of the Triangle of Forces:** If three concurrent forces are in equilibrium, the forces can be represented in magnitude and direction by the sides of a triangle taken in order.

We can express the above results trigonometrically by applying the sine rule to triangle ABC in Fig. 86 (b) to obtain:

$$\frac{R}{\sin ABC} = \frac{Q}{\sin BCA} = \frac{P}{\sin CAB}$$

Returning to the position of the forces at A with the angles notated as $p, q,$ and r, we observe that the above equations may be rewritten as

$$\frac{R}{\sin (180° - r)} = \frac{Q}{\sin (180° - q)} = \frac{P}{\sin (180° - p)}$$

Using the result $\sin (180° - x) = \sin x$ (see page 90) we can amend the above equation to read

$$\frac{R}{\sin r} = \frac{Q}{\sin q} = \frac{P}{\sin p}$$

a result which is known as Lami's Theorem.

Lami's Theorem: *If three forces acting at a point are in equilibrium, each force is proportional to the sine of the angle between the other two forces.*

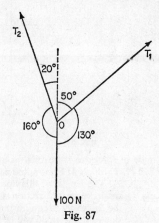

Fig. 87

Example: A body with a weight of 100 N is supported by two attached strings inclined to the vertical at angles of 20° and 50° respectively. Calculate the tension in each string.

SOLUTION: Let Fig. 87 represent the problem. Since the three concurrent forces are in equilibrium we may apply Lami's theorem.

$$\therefore \frac{100}{\sin 70°} = \frac{T_1}{\sin 160°} = \frac{T_2}{\sin 130°}$$

$$\therefore \frac{T_1}{\sin 160°} = \frac{100}{\sin 70°}, \quad \frac{T_2}{\sin 130°} = \frac{100}{\sin 70°}$$

$$T_1 = \frac{100 \sin 160°}{\sin 70°}, \, T_2 = \frac{100 \sin 130°}{\sin 70°}$$

$$T_1 = \frac{100 \sin 20°}{\sin 70°}, \quad T_2 = \frac{100 \sin 50°}{\sin 70°}$$

$$T_1 = 36·4 \text{ N}, \qquad T_2 = 81·53 \text{ N} \quad (Answer).$$

Example: A body of weight 200 N is supported at the point L by two attached strings, one of which is inclined to the vertical at 40°. Calculate the direction and tension in the other string if the tension is to be a minimum.

SOLUTION: Let Fig. 88 represent the problem, with the three forces applied at the point L. The centre of gravity of the body must lie vertically below L as shown.

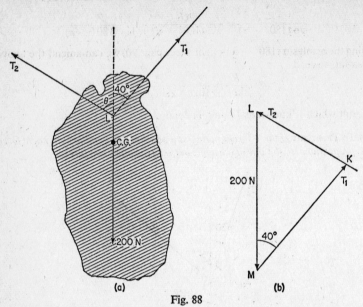

Fig. 88

We shall use the triangle of forces to solve this problem, so consider LM to represent the weight of 200 N. Through M draw a line parallel to the direction of

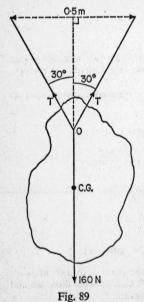

Fig. 89

T_1. The length of this line is as yet unknown, but we do know its direction is 40° to the vertical. Since the forces are in equilibrium we know that the third side of the triangle will be parallel to T_2, so draw a line through L parallel to T_2 to complete the triangle of forces KLM. Now we know that the direction of MK is fixed but we know neither the length nor the direction of KL; however, we do know that its length represents the magnitude of T_2 which we require to be a minimum. The length of KL is minimum when angle $MKL = 90°$; hence the angle between T_1 and T_2 measures 90° and, from the right-angled triangle of forces KLM, we see that

$$T_2 = 200 \sin 40°$$
$$\therefore T_2 = 128 \cdot 56 \text{ N at 50° to the vertical} \quad (Answer).$$

Example: A body of weight 160 N is smoothly threaded on a light string of length 1 m and is suspended in equilibrium with the free ends of the string fixed 0·5 m apart in the same horizontal line. Calculate the tension in the string.

SOLUTION: Let Fig. 89 represent the problem, with A and B the points at which the string has been fixed. Since the string is smoothly threaded through the body the tension is the same throughout the string.

So the body is being supported by two forces of equal magnitude T applied at the point O, say. Since $AO = OB = 0.5$ m, ABO is an equilateral triangle, and each force T is inclined to the vertical at an angle of $30°$.

Since the three concurrent forces are in equilibrium we have

$$\frac{160}{\sin 60°} = \frac{T}{\sin 150°} = \frac{T}{\sin 150°} \quad \text{(by Lami's Theorem)}$$

$$T = \frac{160 \sin 150°}{\sin 60°} = \frac{160 \sin 30°}{\sin 60°}$$

$$= \frac{160 \sin 30°}{\cos 30°} = 160 \tan 30°$$

$$\therefore T = 92.38 \text{ N} \quad (Answer).$$

Exercise 34

1. Three forces of equal magnitude act at a point O in equilibrium. Show that the angle between any two of the forces measures $120°$.

2. A body of weight 3 kN is supported by two strings having equal lengths of 1 m attached to a point O on the body with their free ends fixed at two points 1 m apart in the same horizontal line. Calculate the tension in each string.

3. A body of weight 100 N is supported by two strings attached to a point O of the body. If the direction of one of the strings is horizontal and the tension in the other string is 200 N, calculate the angles between the strings and the tension in the horizontal string.

4. A body of weight 300 N is supported at the point L by two attached strings, one of which is inclined to the vertical at an angle of $60°$. Calculate the direction and tension in the other string when that tension is at a minimum.

5. Calculate the tensions in the strings in Question 2 if the weight of the body is increased to 6 kN, assuming that the strings do not break.

6. AB is a light string of length 1 m. The upper end A is fixed and the lower end B is attached to a weight of 100 N. A horizontal force is applied at B and pulls it aside a distance of 0.5 m. Calculate this horizontal force and the tension in the string.

7. The ends A and D of a light string $ABCD$ are fastened to two fixed points in the same horizontal line. A weight of 100 N is fastened to each of the points B and C such that $AB = BC = CD$. If BC is horizontal and AB and CD are both inclined at an angle of $30°$ to the vertical, calculate the tension in each of the three separate sections of string. (Hint: apply Lami's theorem at both B and C.)

CHAPTER SEVEN

VECTOR ALGEBRA AND COPLANAR FORCES

In the previous chapter we were introduced to the idea of vectors, the use of the parallelogram or triangle laws, and the relevant trigonometry. We saw that we must use the triangle law to combine two vectors in order to find their resultant or vector sum. Furthermore, to avoid the inconvenience of introducing new symbols we have agreed to continue useing the $+$ sign of ordinary addition in order to indicate vector addition. However, we must discover whenever introducing any new system whether the similarity with the 'ordinary' number system is likely to go any deeper than the mere use of the same symbols.

(1) Addition and Subtraction

The statement $a + b = c$ or $c = a + b$ in vector algebra means that c is the resultant of the two vectors a and b, the value of c being obtained by the triangle law. Thus in Fig. 90 $\overrightarrow{OA} = a$, $\overrightarrow{AC} = b$ and $\overrightarrow{OC} = c$ leads us to

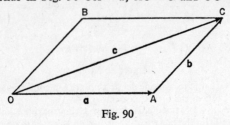

Fig. 90

$a + b = c$. But if we use $\triangle OBC$ instead of $\triangle OAC$, the fact that $\overrightarrow{OB} = b$ and $\overrightarrow{BC} = a$ leads us to $b + a = c$. These two results give us $a + b = b + a$, which means that vector addition is a **commutative** operation just as ordinary addition is commutative for real numbers, e.g. $3 + 4 = 4 + 3$ or $5 + 9 = 9 + 5$, etc.

If $\overrightarrow{OA} = a$ what is represented by $\overrightarrow{AO}$? Clearly the magnitude of this vector is the same as $|\overrightarrow{OA}|$, and the only difference is the direction of the vector. We agree therefore to let $\overrightarrow{AO}$ represent the vector $-a$.

This gives us a suggestion for a solution to $a + b = c = 0$, where the zero vector is understood to be a vector of zero magnitude (i.e. $|c| = 0$) and its direction is irrelevant. Therefore if $a + b = 0$, then $a = -b$ or $b = -a$, a situation which corresponds to O and C being the same point in Fig. 90. Notice that the addition of the zero vector to any other vector leaves that vector unchanged; thus $b + 0 = b$, $0 + a = a$, and so on.

We now have a clue as to how to define the subtraction of two vectors. Since we already know how to add vectors, it follows that if we can interpret subtraction in terms of addition we shall be back on familiar ground. The relation $a - b = a + (-b)$ is a definition which serves our purpose, for we are here defining the subtraction of b to be the same as the addition of the negative vector $-b$. Very often $-b$ is called the **additive inverse** of b because $b + (-b) = 0$.

We can adapt Fig. 90 to become Fig. 91 to give us an idea for obtaining the result $a - b$. Produce BO to D so that OD represents $-b$. Complete the parallelogram $ODEA$ with $DE = a$. From $\triangle ODE$ we see that

$$\overrightarrow{OD} + \overrightarrow{DE} = \overrightarrow{OA} + \overrightarrow{AE} = \overrightarrow{OE}$$

Therefore
$$(-b) + a = a + (-b) = \overrightarrow{OE}$$

i.e.
$$a - b = \overrightarrow{OE}$$

Since the parallelogram $ODEA$ and $BOAC$ are congruent, it also follows that $\overrightarrow{OE} = \overrightarrow{BA}$, therefore $a - b = \overrightarrow{BA}$. This means that each diagonal of a parallelogram has a special significance with respect to the vectors represented by its adjacent sides. Thus one diagonal may represent the sum while the other diagonal may represent the difference of the two vectors represented by the adjacent sides of the parallelograms.

Exercise 35

1. $ABCD$ is a square with $\overrightarrow{AB} = a$ and $\overrightarrow{AD} = b$. Find the magnitude and direction of the two vectors (i) $a + b$, (ii) $a - b$.

2. $ABCD$ is a rectangle with $\overrightarrow{AB} = a$ and $\overrightarrow{AD} = b$. If $|a| = 4$ and $|b| = 3$ calculate the magnitude and direction of the two vectors (i) $a + b$, (ii) $a - b$.

3. Given that O is a fixed point find the path of P such that $|\overrightarrow{OP}| = 3$.

4. Solve the equation $a + x = a$.

5. Solve the equation $a - x = a$.

6. In preparation for the next section try to give a meaning to the following vectors:

$$\text{(i) } 2a, \text{ (ii) } 3b, \text{ (iii) } 2a + 3b.$$

(2) Distributive Law and Associative Law

In Fig. 91 we have $EA = AC$ or $EC = 2AC$. Since $\overrightarrow{AC} = b$ it seems logical to call $\overrightarrow{EC} = b + b = 2b$. We extend this to the multiplication of a vector a by any scalar n in the following definition.

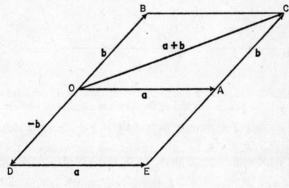

Fig. 91

DEFINITION: *The product na where n is a scalar is represented by a vector whose magnitude is $|na| = |n||a|$ in a direction parallel to a.*

If n is positive, the sense of na is the same as a; if n is negative, the sense of na is opposite to a.

Again from Fig. 91 we see that

$$\overrightarrow{DC} = \overrightarrow{DE} + \overrightarrow{EC}$$
$$= a + 2b$$

and
$$\overrightarrow{BE} = \overrightarrow{BD} + \overrightarrow{DE}$$
$$= (-2b) + a$$
$$= a + (-2b)$$
$$= a - 2b.$$

From the above definition we can accept that

$$(m + n)a = ma + na$$

but what is not so obvious is that

$$n(a + b) = na + nb$$

Here we have to make use of the similar triangles as shown in Fig. 92. Each side of $\triangle PQR$ is obtained by multiplying each corresponding side of $\triangle LMN$ by n. Since $\overrightarrow{PR} = n\overrightarrow{LN}$, $\overrightarrow{PQ} = na$, and $\overrightarrow{QR} = nb$, it follows from the relation $\overrightarrow{PR} = \overrightarrow{PQ} + \overrightarrow{QR}$ that $n(a + b) = na + nb$. This is called the **distributive law** of scalar multiplication over vector addition, and corresponds to $3(x + y) = 3x + 3y$ in ordinary algebra.

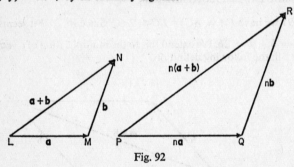

Fig. 92

Vector addition is called a *binary* operation because it combines only two vectors at a time, and so far this is the greatest number of vectors we have dealt with at a time. For example we had $\overrightarrow{DC} = a + 2b$, which suggests that $\overrightarrow{DC} = a + (b + b)$, which in turn suggests that the vectors in the brackets should be dealt with first and the result $2b$ then added to a. But suppose we had started with $\overrightarrow{DC} = a + b + b$, without the bracket, how would we interpret this statement? Which end should we start from? Since $+$ is a binary operation we are as yet unable to sum *three* vectors in a single move, so we must devise a law for collecting or associating the vectors with each other. Such a law will be called the **associative law** for addition of vectors. As an aid to its understanding consider Fig. 93 and the question of obtaining the vector sum $a + b + c$.

We start by associating *a* with *b* and we indicate this by the use of a bracket. Thus

$$(a + b) + c = \overrightarrow{OL} + c$$
$$= \overrightarrow{OM}$$

If we start by associating *b* with *c* first, we have

$$a + (b + c) = a + \overrightarrow{KM}$$
$$= \overrightarrow{OM}$$

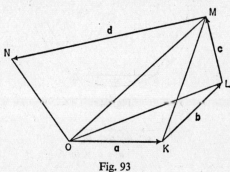

Fig. 93

Both results are the same, so we can write

$$a + (b + c) = (a + b) + c$$

Since it makes no difference which arrangement is chosen we may leave out the brackets and write, without any ambiguity,

$$a + b + c = \overrightarrow{OM}$$

In other words, vectors obey the associative law of ordinary algebra, e.g. $13 + (4 + 7) = 24 = (13 + 4) + 7$, $(11 + 3) + 9 = 23 = 11 + (3 + 9)$, and so on.

If we return to Fig. 93 we see that we may extend the summation to as many vectors as we please, although only four are used in the following result:

$$a + b + c + d = \overrightarrow{ON}$$

The result is more readily appreciated by writing

$$\overrightarrow{OK} + \overrightarrow{KL} + \overrightarrow{LM} + \overrightarrow{MN} = \overrightarrow{ON}$$

and examining the order of the lettering.

Example: *ABCDEF* is a regular hexagon. Using $\overrightarrow{AB} = a$ and $\overrightarrow{BC} = b$ express the following vectors in terms of *a* and *b*.

(i) $\overrightarrow{AC}$, (ii) $\overrightarrow{BE}$, (iii) $\overrightarrow{AF}$.

SOLUTION: Let Fig. 94 represent the problem. The geometry to recall at this stage is that (*a*) the sides of a regular hexagon are each equal in length to the radius of the circle which circumscribes the hexagon, (*b*) diagonals such as AD are diameters of the circle and parallel to two opposite sides such as FE and BC.

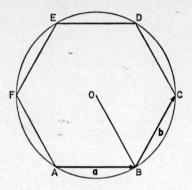

Fig. 94

(i)
$$\overrightarrow{AC} = \overrightarrow{AB} + \overrightarrow{BC}$$
$$= a + b \quad (Answer).$$

(ii)
$$\overrightarrow{BE} = 2\overrightarrow{BO} \qquad \text{or} \quad 2\overrightarrow{BO} = 2(b - a)$$
$$= \overrightarrow{BO} + \overrightarrow{BO} \qquad\qquad = 2b - 2a$$
$$= (b - a) + (b - a)$$
$$= 2b - 2a \quad (Answer).$$

Note that in this result we have used $\overrightarrow{BO} = \overrightarrow{BA} + \overrightarrow{AO} = (-a) + b = b - a$.

(iii)
$$\overrightarrow{AF} = \overrightarrow{BO}$$
$$= b - a \quad (Answer).$$

Example: $ABCD$ is a square whose centre is O; $M, N, P,$ and Q are the midpoints of its sides. If $\overrightarrow{AM} = a$ and $\overrightarrow{AQ} = b$ express the following vectors in terms of a and b.

(i) $\overrightarrow{MQ}$, (ii) $\overrightarrow{MC}$,(iii) $\overrightarrow{QP}$.

SOLUTION: With the problem represented by Fig. 95 we work as follows:

(i)
$$\overrightarrow{MQ} = \overrightarrow{MA} + \overrightarrow{AQ}$$
$$= (-a) + b$$
$$= b - a \quad (Answer).$$

(ii)
$$\overrightarrow{MC} = \overrightarrow{MB} + \overrightarrow{BC}$$
$$= a + 2b \quad (Answer).$$

(iii)
$$\overrightarrow{QP} = \overrightarrow{QO} + \overrightarrow{OP}$$
$$= a + b \quad (Answer).$$

The methods above are not unique as seen by reconsidering (ii):

$$\vec{MC} = \vec{MO} + \vec{ON} + \vec{NC}$$
$$= b + a + b$$
$$= a + b + b$$
$$= a + 2b \quad (Answer).$$

or alternatively,

$$\vec{MC} = \vec{MA} + \vec{AQ} + \vec{QO} + \vec{OP} + \vec{PC}$$
$$= (-a) + b + a + b + a$$
$$= a + 2b \quad (Answer).$$

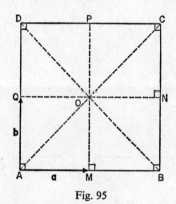

Fig. 95

Exercise 36

1. Using the example in Fig. 94 express the following vectors in terms of *a* and *b*:

(i) $\vec{CE}$, (ii) $\vec{DA}$, (iii) $\vec{FD}$, (iv) $\vec{DC}$, (v) $\vec{BD}$.

2. Using the example in Fig. 95 express the following vectors in terms of *a* and *b*:

(i) $\vec{DO}$, (ii) $\vec{DB}$, (iii) $\vec{PB}$, (iv) $\vec{AP}$, (v) $\vec{DM}$.

3. Using the example in Fig. 93 express the following vectors in terms of any of *a*, *b*, *c*, and *d*:

(i) $\vec{LN}$, (ii) $\vec{KN}$, (iii) $\vec{MK}$, (iv) $\vec{NK}$, (v) $\vec{NO}$, (vi) $\vec{OK} + \vec{KL} + \vec{LO}$

(3) Vectors in Elementary Geometry

It is useful at this stage to relate vector algebra to some of the standard elementary geometry results. The exercise is interesting and especially valuable for reinforcing the simple ideas behind the general application of vectors. Consider the following examples.

Example: Prove that the line joining the midpoint of two sides of a triangle is parallel to the third side and is also half its length.

SOLUTION: In Fig. 96 let A and B be the midpoints of their respective sides and let

$$\vec{AQ} = a, \vec{QB} = b$$

Since $\vec{PA} = \vec{AQ}$, then $\vec{PQ} = 2a$

Similarly $\vec{QR} = 2b$

$$\therefore \vec{AB} = \vec{AQ} + \vec{QB} \quad \text{and} \quad \vec{PR} = \vec{PQ} + \vec{QR}$$
$$= a + b \qquad\qquad = 2a + 2b$$
$$\therefore \vec{PR} = 2(a + b) = 2\vec{AB}$$

This last vector statement means that AB is parallel to PR and also that $AB = \frac{1}{2}PR$.

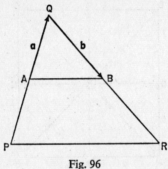

Fig. 96

Example: Prove that the diagonals of a parallelogram bisect each other.

SOLUTION: We start by lettering $\vec{AD} = b$ and $\vec{AB} = a$ in Fig. 97 and assuming that M and N are the midpoints of the diagonals as shown. In other words, we assume that the diagonals do not intersect in their midpoints and then set about proving that M and N are one and the same point so that the diagonals must bisect each other.

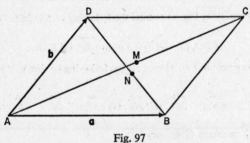

Fig. 97

Now $\qquad \vec{AC} = \vec{AB} + \vec{BC} = a + b$, since $\vec{BC} = b$

But, $\qquad \vec{AC} = 2\vec{AM}$, since M is the midpoint of AC

$$\therefore \vec{AM} = \tfrac{1}{2}(a + b)$$

Also, $\qquad \vec{BD} = \vec{BA} + \vec{AD} = (-a) + b$

But, $\qquad \overrightarrow{BD} = 2\overrightarrow{BN}$, since N is the midpoint of BD

$$\therefore \ \overrightarrow{BN} = \tfrac{1}{2}(b - a)$$

Finally, $\qquad \overrightarrow{AN} = \overrightarrow{AB} + \overrightarrow{BN}$

$$= a + \tfrac{1}{2}(b - a)$$
$$= \tfrac{1}{2}(a + b)$$

$\therefore \ \overrightarrow{AN} = \overrightarrow{AM}$ which means that M and N are the same point.

Exercise 37

1. Using Fig. 96 express the following vectors in terms of a and b:

(i) $\overrightarrow{PB}$, (ii) $\overrightarrow{AR}$.

2. In triangle ABC, $\overrightarrow{AB} = a$, $\overrightarrow{AC} = b$, M is the midpoint of BC and N lies on BC such that $BN = \tfrac{1}{3}BC$. Express the following vectors in terms of a and b:

$\qquad$ (i) $\overrightarrow{BM}$, (ii) $\overrightarrow{BN}$, (iii) $\overrightarrow{NM}$, (iv) $\overrightarrow{AM}$, (v) $\overrightarrow{AN}$

3. In triangle ABC, $\overrightarrow{AB} = b$, $\overrightarrow{AC} = c$. Express $\overrightarrow{AP}$ in terms of b and c where (i) P is the point on BC such that $BP = \tfrac{1}{4}BC$, and (ii) P is the point on BC such that $CP = \tfrac{1}{5}CB$.

(4) The Polygon of Forces

If a system of coplanar forces acts at a point it becomes convenient to know which single force acting at the point will produce the same result as the original system. In other words, we wish to find the resultant of a system of forces acting at the point. We already know that force is a vector quantity and so the resultant force is simply the vector sum of all the forces in the system. If, therefore, we have a system of coplanar forces a, b, c, d, and e acting at a point as shown in Fig. 98 (*a*), the resultant will be given by

$$r = a + b + c + d + e$$

We have already seen (page 111) that all we need do to find r is construct the polygon of Fig. 98 (*b*) such that $\overrightarrow{OP} = a$, $\overrightarrow{PQ} = b$, $\overrightarrow{QR} = c$, $\overrightarrow{RS} = d$, $\overrightarrow{ST} = e$ to the chosen scale and in the appropriate parallel directions.

So far we have $\qquad \overrightarrow{OQ} = a + b$

$$\overrightarrow{OR} = \overrightarrow{OQ} + \overrightarrow{QR} = a + b + c$$
$$\overrightarrow{OS} = \overrightarrow{OR} + \overrightarrow{RS} = a + b + c + d$$
$$\overrightarrow{OT} = \overrightarrow{OS} + \overrightarrow{ST} = a + b + c + d + e$$

$\overrightarrow{OT}$ was the resultant we were hoping to find. We see, therefore, that the resultant force is obtained by the closure of the polygon of forces.

Note that each of these forces acts at O, so it follows that the resultant must act at O.

Finally, if the force polygon is closed (i.e. T coincides with O), then the resultant is zero, which means that the forces acting at the point O must be in equilibrium.

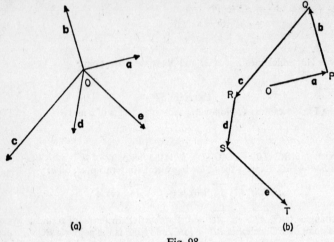

(a)   (b)

Fig. 98

Example: Four forces act at a point O in the following directions: 10 N directed East, 30 N directed North, 20 N directed West, and 25 N directed South. Find the magnitude and direction of the resultant force at O.

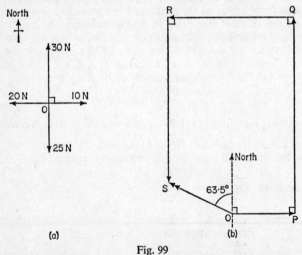

(a)   (b)

Fig. 99

SOLUTION: Let Fig. 99 (a) represent the problem. Choosing a scale of 2 mm to 1 N we take O as the starting-point for the polygon of forces as sketched in in Fig. 99 (b) and draw OP directed East and of length 20 mm, PQ directed North and of length 60 mm, QR directed West and of length 40 mm, RS directed South and of length 50 mm.

The vector sum of the four forces is given by $\overrightarrow{OP} + \overrightarrow{PQ} + \overrightarrow{QR} + \overrightarrow{RS} = \overrightarrow{OS}$.

Closing the force polygon we find by measurement that $|\overrightarrow{OS}| = 11$ N and the direction of $\overrightarrow{OS}$ is North 63·5° West (*Answer*).

Example: Four equal billiard balls *A*, *B*, *C*, and *D* each at rest and of equal mass 0·5 kg are struck simultaneously by the cue ball and set in motion as follows:

A: 1·2 m s⁻¹ North, *B*: 0·6 m s⁻¹ North East, *C*: 0·8 m s⁻¹ West, *D*: 0·6 m s⁻¹ North West.

Find the magnitude and direction of the resultant momentum.

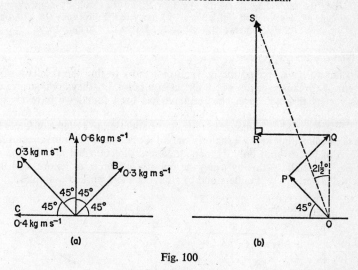

(a)   (b)

Fig. 100

SOLUTION: Let Fig. 100 (*a*) represent the momentum of the four separate billiard balls. The reader will have probably noticed that the vectors in the previous example were taken in a clockwise order round the point *O* when constructing the polygon for the forces in Fig. 99 (*b*). With the vectors all directed out from *O* this procedure is adopted to ensure that we remember to include all the vectors, the order in which they are taken makes no difference to the final result.

In Fig. 100 (*b*) we have drawn the polygon of vectors in the following manner. Choosing a scale of 5 mm to 0·1 kg m s⁻¹ we start with *OP* directed North-West and of length 15 mm, followed by *PQ* directed North-East and of length 15 mm, then *QR* directed West and of length 20 mm, followed finally by *RS* directed North and of length 30 mm. The resultant vector is given by $\overrightarrow{OP} + \overrightarrow{PQ} + \overrightarrow{QR} + \overrightarrow{RS} = \overrightarrow{OS}$; measurement from the diagram shows that $|\overrightarrow{OS}| = 11$ kg m s⁻¹ and the direction is N 21·5 W (*Answer*).

Note that since the balls were originally at rest, $\overrightarrow{OS}$ represents the gain in momentum of the system *A*, *B*, *C*, and *D*. By the principle of the conservation of momentum it follows that the cue ball has *lost* a momentum of 11 kg m s⁻¹ in the direction N 21·5 W. The problem is somewhat artificial since the likelihood of measuring the velocity of the balls after a random impact by the cue ball is very remote. However the example gives a helpful illustration of the principles of the polygon of vectors.

Exercise 38

1. Find the magnitude and direction of the following four forces acting at a point 0:100 N directed East, 200 N directed North, 150 N directed West, 150 N directed South.

2. A, B, C, and D are four points in order on a circle centre O such that $\angle AOB = 90°$, $\angle BOC = 60°$, $\angle COD = 90°$, $\angle DOA = 120°$. Calculate the magnitude and direction of the resultant of forces of 100 N, 200 N, 300 N, 400 N which act along OA, OB, OC, OD, respectively, the directions of the forces being given by the order of lettering.

3. Using the same diagram as for Question 2, find the resultant momentum of four runners who each set off with a speed of 3 m s^{-1} in the directions OA, OB, OC, OD if the masses of the runners are respectively 50 kg, 60 kg, 80 kg, 60 kg.

(5) Non-concurrent Vectors

So far we have applied the polygon of vectors to find the resultant of a system of coplanar vectors which all act at a point, i.e. they are concurrent. This means that whenever the resultant has been found we have always known that it acts through the point of concurrency of the system of forces it can replace. We shall continue to restrict our attention to systems of forces in the same plane, but now we shall try to find the resultant of a non-concurrent system of forces—not only its magnitude and direction but also its line of action. We shall do this by using the polygon of forces as before, and the method is perhaps best understood by studying a few worked examples.

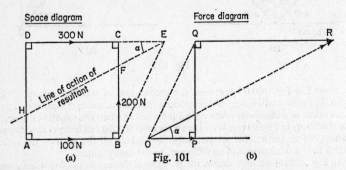

Fig. 101

Example: Forces of 100 N, 200 N and 300 N respectively act along the sides AB, BC, and DC of a square $ABCD$. Find the magnitude and line of action of the resultant of these three forces.

SOLUTION: We shall draw two diagrams, one for the space relations showing the lines of action of the various forces or vectors, and the other for the force or vector polygon.

With the chosen scale we draw $\overrightarrow{OP}$ parallel to AB to represent the force of 100 N along AB, similarly $\overrightarrow{PQ}$ represents the force of 200 N along BC, and $\overrightarrow{QR}$ represents the force 300 N along DC. The resultant force is given by $\overrightarrow{OP} + \overrightarrow{PQ} + \overrightarrow{QR} = \overrightarrow{OR}$.

By measurement $|\overrightarrow{OR}| = 447$ N and the direction of OR is given by $\alpha = 26 \cdot 5°$. Since the forces in Fig. 101 (b) have been drawn parallel to their directions in Fig. 101 (a), it follows that the line of action of the resultant force to be drawn in (a)

must be parallel to $\overrightarrow{OR}$. The difficulty lies in finding any one point on this line of action through which we may draw a line parallel to $\overrightarrow{OR}$ as required.

Let us start again to see how we arrived at $\overrightarrow{OR}$. We first drew $\overrightarrow{OP} + \overrightarrow{PQ} = \overrightarrow{OQ}$, and this meant that $\overrightarrow{OQ}$ was the resultant of 100 N along *AB* and 200 *N* along *BC*, but, since these two forces intersect at *B*, their resultant $\overrightarrow{OQ}$ must also pass through *B*. We have now reduced the system from three forces to two forces: $\overrightarrow{OQ}$ through *B* and 300 N along *DC*. The resultant of these two forces must pass through their point of intersection at *E*. We have now found that the resultant $\overrightarrow{OR}$ passes through *E*, so that since we know the angle α we have also determined the line of action of the resultant.

Note that at no time should we become so inflexible as to overlook the possibility of a profitable use of elementary common sense. In this example we could have noticed that two of the forces were parallel. The resultant of 300 N along *DC* and 100 N along *AB* is a parallel force of 400 N passing through *F* such that *BF* = 3 *FC* (see Chapter Two). By this result we are left with two forces of 400 N and 200 N acting through the point *F*, hence *F* must lie on the line of action of resultant $\overrightarrow{OR}$.

We complete the picture by drawing the line of action through *F* parallel to $\overrightarrow{OR}$.

Example: *ABC* is an equilateral triangle with *AB* = 25 mm. Three forces act along the sides of the triangle as follows: 150 N along *AB*, 200 N along *BC*, 100 N along *CA*, the order of the lettering indicating the direction of the forces. Find the magnitude and line of action of the resultant force, stating its position by finding its intersection with *AB*.

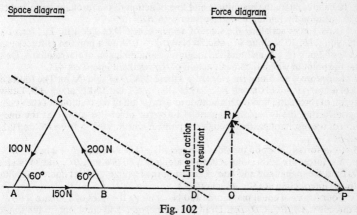

Fig. 102

SOLUTION: Let Fig. 102 represent the space and force diagrams of the problem.
Choosing a scale of 1 mm to 5 N we construct the force polygon in the now familiar manner.

Thus, $\overrightarrow{OP}$ represents 150 N along *AB*

$\overrightarrow{PQ}$ represents 200 N along *BC*

$\overrightarrow{QR}$ represents 100 N along *CA*

The resultant force is given by the vector sum $\overrightarrow{OP} + \overrightarrow{PQ} + \overrightarrow{QR} = \overrightarrow{OR}$. Measurements on the force diagram yield $|\overrightarrow{OR}| = 90$ N (the calculated answer would be 86·6 N) and $\angle POR = 90°$. These results give the magnitude and direction of the resultant we now need to find the line of action.

Examining the force diagram in a little more detail we see that $\overrightarrow{PR} = \overrightarrow{PQ} + \overrightarrow{QR}$, which tells us that the resultant of the two forces intersecting at C is represented by $\overrightarrow{PR}$. We therefore draw DC parallel to PR to become the line of action of the resultant of the forces of 200 N and 100 N acting at C. We are now left with PR along DC and 150 N along AB, which must clearly yield the final resultant through their point of intersection D. Again by measurement, $BD = 25$ mm.

In conclusion, the resultant is a force of 90 N (to the nearest 10 N) acting perpendicular to AB through D on AB produced such that $BD = 25$ mm. The sense of the resultant is indicated on the diagrams.

The tedium of finding the line of action even in simple examples like these suggests that it might be profitable to pursue the possibility of an alternative method. We shall certainly do this in the next section, in the meantime the following exercise contains questions which have reasonable solutions to illustrate the usefulness of the polygon of vectors.

Exercise 39

1. Forces of 200 N, 400 N and 600 N respectively act along the sides AB, BC, DC of a square $ABCD$. Using the example of Fig. 101 above, *state* the magnitude and line of action of the resultant of these three forces.

2. ABC is an equilateral triangle with $AB = 30$ mm. Three forces act along the sides of the triangle as follows: 450 N along AB, 600 N along BC, 300 N along CA, the order of lettering indicating the directions of the forces. Using the example of Fig. 102 above, state the magnitude and line of action of the resultant force, indicating its position by finding its intersection with AB.

3. Three forces act along the sides of a square $ABCD$ of side 1 m. The forces are 40 N along AB, 20 N along BC, and 20 N along AD. Draw a polygon of forces using a scale of 2 mm to 1 N and find the magnitude and direction of the resultant. Determine the point in which the line of action of the resultant intersects AB.

4. Four forces act along the sides of a square $ABCD$ of side 0·5 m. The forces are 20 kN along AB, 40 kN along BC, 30 kN along DC, and 10 kN along AD. Draw a polygon of forces and find the magnitude and direction of the resultant of the system. By considering the two pairs of parallel forces or otherwise show that the line of action of the resultant passes through the point E which is 0·2 m from DC and 0·1 m from BC.

5. Four forces act along the sides of a rectangle $ABCD$ with $AB = 2$ m and $BC = 1$ m. The forces are 50 N along AB, 40 N along BC, 80 N along DC, and 25 N along AD. Find the magnitude and line of action of the resultant. (Hint: find the resultant of forces intersecting at C or A separately.)

6. Four forces of equal magnitude 100 N act along the sides of a square $ABCD$ in the direction AB, BC, CD, DA. Draw a force polygon and comment on the result.

(6) Resolution of Vectors

In our examples so far we have always reduced the number of forces in the system one by one until we obtained the resultant force. With a system containing only a small number of forces the use of the polygon of forces has worked reasonably well, but it must be admitted that graphical work, while reinforcing the principles, is tedious. So we begin to look for a possible alternative method. Strangely enough we now find that replacing a single force or vector by *two* forces or vectors results in a considerable simplification.

Two vectors are combined by using the triangle law or parallelogram law as discussed in Chapter Six. Thus, instead of regarding $a + b = c$ as replacing two vectors a and b by one vector c, let us reverse the notion and think of it as c replaced by $a + b$. Just to identify our thoughts we shall call a and b two **components** of c. The problem now arises as to whether we might be able to replace *any* other vector d not by any components but by components directly related to a and b already given. Fig. 103 should help to make this idea clearer.

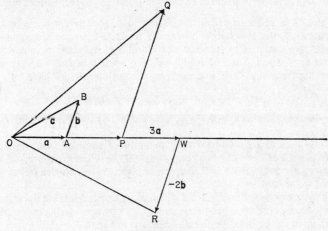

Fig. 103

Triangle OAB illustrates the relation $c = a + b$ with $\overrightarrow{OB} = c$. We may locate the vector d at O by making $\overrightarrow{OQ} = d$. Through Q draw a line parallel to b to intersect OA in P. We now have $\overrightarrow{OQ} = \overrightarrow{OP} + \overrightarrow{PQ}$. Since $\overrightarrow{OP}$ lies along a we may express $\overrightarrow{OP}$ as a multiple of a and write $\overrightarrow{OP} = ma$. Similarly, since $\overrightarrow{PQ}$ is parallel to b, we may express $\overrightarrow{PQ}$ as a multiple of b and write $\overrightarrow{PQ} = nb$.

Therefore $\overrightarrow{OQ} = d = ma + nb$, which shows that any vector d may be expressed as a sum of two components which are multiples of any two non-parallel vectors a and b.

To appreciate the simplification immediately, let us suppose that we have already expressed four vectors $\overrightarrow{OK}$, $\overrightarrow{OL}$, $\overrightarrow{OM}$, $\overrightarrow{ON}$ in terms of a and b as follows:

$$\overrightarrow{OK} = (-a) + 3b$$
$$\overrightarrow{OL} = (-4a) - 2b$$
$$\overrightarrow{OM} = 2a + 2b$$
$$\overrightarrow{ON} = 6a - 5b$$

The resultant is

$$\overrightarrow{OK} + \overrightarrow{OL} + \overrightarrow{OM} + \overrightarrow{ON} = (-a) - 4a + 2a + 6a + 3b - 2b + 2b - 5b$$
$$= 3a - 2b$$

By drawing $\overrightarrow{OW} = 3a$, $\overrightarrow{WR} = (-2b)$ in Fig. 103 we obtain the resultant $\overrightarrow{OR}$ with its direction specified by the measure of angle WOR.

The reader has probably already noticed the obvious disadvantage of first having to construct triangles such as $\triangle OPQ$ in order to express each vector such as $\overrightarrow{OQ}$ in terms of a and b, and secondly, having to reconstruct resultants such as $\overrightarrow{OR}$ in order to obtain their directions. However, we are bound to conclude that, since rectangles are easier to draw than parallelograms and triangles, we should if possible choose a and b at right-angles to each other and we should make a and b unit vectors, such that $|a| = 1$, $|b| = 1$.

Looking back at $\triangle OAB$ and applying the cosine rule, we have

$$c^2 = a^2 + b^2 - 2ab \cos \angle OAB$$

which shows us at once that if we make angle OAB a right-angle all our resultants such as $\overrightarrow{OB}$ will have their magnitudes obtained from simpler expressions such as $c^2 = a^2 + b^2$ because $\cos 90° = 0$. Furthermore, the direction given by angle BOA will likewise be much easier to determine if a is perpendicular to b.

The two perpendicular directions which come to mind first are the directions of the axes of coordinate geometry, so we begin by considering the resolution of vectors in the coordinate plane as outlined in Fig. 104.

The unit vector in the direction OX is represented by i and the unit vector in the direction OY is represented by j. From the diagram, with the position of the point P given by its coordinates (5, 2), we know that $OQ = 5$ and $QP = 2$. In vectors this is represented by the statement

$$\overrightarrow{OP} = 5i + 2j.$$

Similarly,
$$\overrightarrow{OS} = -3i + 3j,$$
$$\overrightarrow{OT} = -2i - 4j,$$
$$\overrightarrow{OU} = 0i - 3j, \text{ (to be inserted by the reader)}$$
$$\overrightarrow{OR} = 4i - 2j,$$

and each vector is said to have been *resolved* in the directions i and j.

If we now require the resultant $\overrightarrow{ON}$ of these five vectors, we merely find the vector sum as before.

$$\overrightarrow{OP} + \overrightarrow{OS} + \overrightarrow{OT} + \overrightarrow{OU} + \overrightarrow{OR} =$$
$$(5 - 3 - 2 + 0 + 4)i + (2 + 3 - 4 - 3 - 2)j$$
$$= 4i - 4j$$

The resultant is $\overrightarrow{ON} = 4i - 4j$

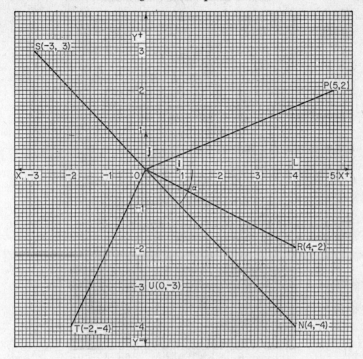

Fig. 104

Hence N is the point $(4, -4)$ i.e. $OL = 4$, $LN = 4$, and the direction of $\overrightarrow{ON}$ is given by

$$\tan \alpha = \frac{LN}{OL} = 1$$

Hence $$\alpha = 45°$$

The magnitude of $\overrightarrow{ON}$ is obtained by using Pythagoras' theorem in the right-angled triangle NOL:

$$ON^2 = 4^2 + 4^2 = 32$$

$$ON = 4\sqrt{2} = 5\cdot7 \quad \text{(one decimal place).}$$

The resultant is $\overrightarrow{ON}$, where $|\overrightarrow{ON}| = 5\cdot7$ in a direction at 45° to the x-axis and below the axis.

Example: A, B, and C are the three points $A\,(6, 2)$, $B\,(-1, 9)$, $C\,(-3, -5)$. Calculate the magnitude and direction of the resultant of the vectors $\overrightarrow{OA}$, $\overrightarrow{OB}$, $\overrightarrow{OC}$.

SOLUTION: Let Fig. 105 represent the problem.

Resolving each vector in the directions of the axes we have

$$\overrightarrow{OA} = 6i + 2j$$
$$\overrightarrow{OB} = -i + 9j$$
$$\overrightarrow{OC} = -3i - 5j$$

The resultant $\quad \overrightarrow{ON} = \overrightarrow{OA} + \overrightarrow{OB} + \overrightarrow{OC}$
$$= 2i + 6j,$$

which means that N is the point (2,6).

The magnitude of $\overrightarrow{ON}$ is given by

$$ON^2 = 2^2 + 6^2 = 40$$
$$ON = 6 \cdot 3 \text{ (one decimal place).}$$

The direction of $\overrightarrow{ON}$ is given by $\tan \alpha = 6/2 = 3$
Therefore: $\alpha = 71 \cdot 6°$ (nearest $\frac{1}{10}$ degree).

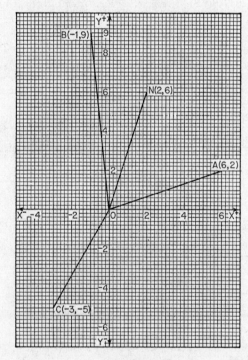

Fig. 105

Example: Four forces act at a point in the following directions: 10 N directed East, 30 N directed North, 20 N directed West, and 25 N directed South. Find the magnitude and direction of the resultant force.

SOLUTION: We consider OY to represent the North direction and OX to represent the East direction as shown in Fig. 106.

The force 10 N, East, is represented by $10\,\boldsymbol{i} + 0\boldsymbol{j}$

30N, North, is represented by $0\boldsymbol{i} + 30\boldsymbol{j}$

20 N, West, is represented by $-20\boldsymbol{i} + 0\boldsymbol{j}$

25 N, South, is represented by $0\boldsymbol{i} - 25\boldsymbol{j}$

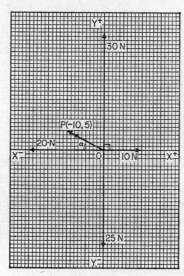

Fig. 106

The resultant of all four forces is the vector $-10\boldsymbol{i} + 5\boldsymbol{j} = \overrightarrow{OP}$, where P is the point $(-10, 5)$.

Since $OP^2 = 10^2 + 5^2 = 125$, $|\overrightarrow{OP}| = 11 \cdot 2$ N (one decimal place).

Referring to Fig. 106 the direction of $\overrightarrow{OP}$ is given by

$$\tan \alpha = 5/10 = 0 \cdot 5000$$

Therefore $\alpha = 26 \cdot 6°$ (nearest $\frac{1}{10}$ degree).

The resultant is a force of $11 \cdot 2$ N in the direction West $26 \cdot 6°$ North   (*Answer*).

Exercise 40

1. Using the example of Fig. 106 above find the resultant when each of the forces is (i) doubled, (ii) halved.

2. A, B, C, and D are the four points $A\,(2, 3)$, $B\,(-1, 4)$, $C\,(4, 5)$, $D\,(-2, -8)$. Calculate the magnitude and direction of the resultant of the vectors $\overrightarrow{OA}$, $\overrightarrow{OB}$, $\overrightarrow{OC}$, $\overrightarrow{OD}$.

3. $ABCD$ is a rectangle with $AB = 12$ units, $BC = 5$ units. Calculate the magnitude and direction of the resultant of the vectors $\overrightarrow{AB}$, $\overrightarrow{AC}$, $\overrightarrow{AD}$. (Hint: take the x axis along AB and the y axis along AD.)

4. $ABCD$ is a square with centre O. Four forces act as follows: 10 kN along OA, 2 kN along OB, 5 kN along OC, and 8 kN along OD. Calculate the magnitude of the resultant of the four forces and the inclination of its direction to OA.

5. With the unit vectors *i* and *j* in the positive direction of the co-ordinate axes *OX* and *OY* respectively, express each of the following forces in terms of *i* and *j* using a scale of 1 unit to 1 N as drawn in Fig. 107.

(i) 500 N in the direction *OA*.
(ii) 500 N in the direction *OB*.

6. Find the resultant of the two forces in Question 5.

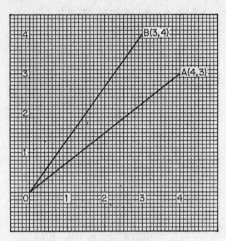

Fig. 107

(7) Resolving Non-concurrent Vectors

We have already used the polygon of vectors to find the resultant of a system of vectors whether they are concurrent or not. Since the method of the last section is essentially the same method as the polygon of vectors we may extend it to find the resultant of a system of coplanar vectors which are not concurrent. The same reservation as before still applies: namely that a zero resultant does not necessarily imply that the system is in equilibrium (as we saw in Question 6 of Exercise 39). We shall still have difficulties in finding the line of action of the resultant but its magnitude and direction will be easier and quicker to obtain than before.

Consider any vector **F** inclined at an angle θ to the line *AB* in Fig. 108 (*a*) and let **F** be represented by $\overrightarrow{OP}$ to some chosen scale. With *PN* perpendicular to *AB* we have a right-angled triangle in which

$$\frac{ON}{OP} = \cos \theta \text{ so that } ON = OP \cos \theta$$

$$\frac{NP}{OP} = \sin \theta \text{ so that } NP = OP \sin \theta.$$

But, since $\overrightarrow{OP} = \overrightarrow{ON} + \overrightarrow{NP}$, we have expressed $\overrightarrow{OP}$ in terms of two components $\overrightarrow{ON}$ and $\overrightarrow{NP}$, one component $\overrightarrow{ON}$ being along or (as in Fig. 108 (*b*)) parallel to $\overrightarrow{AB}$ and the other component $\overrightarrow{NP}$ being perpendicular to $\overrightarrow{AB}$.

In the previous section the direction AB was the positive x axis, here we are seeing how to resolve a vector in any direction whatsoever. If the vector had been a force of magnitude 10 N and the angle $\theta = 27°$, then the components would have been

$$ON = 10 \cos 27° = 8·910 \text{ N}$$
$$NP = 10 \sin 27° = 4·540 \text{ N}$$

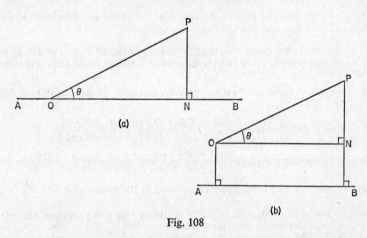

(a)

(b)

Fig. 108

In the special case of $\theta = 90°$, then $ON = 0$, $NP = OP$ which merely confirms what common sense tells us, that a force has a zero component, perpendicular to itself. Being able to resolve forces in any direction enables us to choose x and y axes wherever we please as the next examples will show.

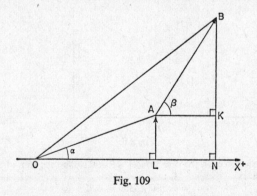

Fig. 109

Before going any further let us confirm that this method of resolution does in fact give the resultant we require. The reader is left to insert the vectors onto Fig. 109 as they arise. Let $\overrightarrow{OA}$, $\overrightarrow{AB}$ be two vectors with resultant $\overrightarrow{OA} + \overrightarrow{AB} = \overrightarrow{OB}$, and let OX be in any direction. Draw AL, BN perpendicular to OX. From Fig. 109 we have $\overrightarrow{OA} = \overrightarrow{OL} + \overrightarrow{LA} = \overrightarrow{OL} + \overrightarrow{NK}$, $(\overrightarrow{LA} = \overrightarrow{NK})$

and $\overrightarrow{AB} = \overrightarrow{AK} + \overrightarrow{KB} = \overrightarrow{LN} + \overrightarrow{KB}$, $(\overrightarrow{AK} = \overrightarrow{LN})$. Therefore the sum of the resolved parts of $\overrightarrow{OA}$ and $\overrightarrow{AB}$ in the direction OX is given by

$$\overrightarrow{OL} + \overrightarrow{LN} = \overrightarrow{ON}$$

But $\overrightarrow{ON}$ is the resolved part of the resultant $\overrightarrow{OB}$ in the direction OX, which means that we have proved the following:

> *The sum of the resolved parts of two forces acting at a point in any given direction is equal to the resolved part of their resultant in the same direction.*

But, even more important, we see that the direction of the resultant $\overrightarrow{OB}$ is given by

$$\text{Tan } \angle BON = \frac{NB}{ON} = \frac{NK + KB}{OL + LN} = \frac{OA \sin \alpha + AB \sin \beta}{OA \cos \alpha + AB \cos \beta}$$

$$= \frac{\text{Sum of the resolved parts perpendicular to } OX}{\text{Sum of the resolved parts along } OX}$$

We shall make use of these important results in the following examples.

Example: Forces of 100 N, 200 N, 300 N act respectively along the sides AB, BC, DC of the square $ABCD$. Find their resultant.

SOLUTION: Let Fig. 110 represent the problem.

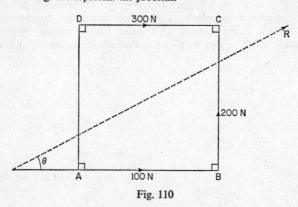

Fig. 110

Resolving the forces in the direction of AB and calling the sum of the resolved parts X we have

$$X = 100 + 300 = 400$$

Resolving the forces in the direction of AD and calling the sum of the resolved parts Y we have

$$Y = 200$$

Now let the resultant be of magnitude R acting at an angle θ to AB as shown in the diagram.

$$\therefore \ R \cos \theta = X = 400$$
$$R \sin \theta = Y = 200$$

Since $\sin^2 \theta + \cos^2 \theta = 1$ we have

$$R^2 = X^2 + Y^2 = 160\,000 + 40\,000$$
$$= 200\,000$$
$$R = 100\,\sqrt{20}$$
$$= 447\cdot2 \text{ N}$$

and
$$\frac{R \sin \theta}{R \cos \theta} = \tan \theta = \frac{Y}{X} = 0\cdot5000$$

Hence
$$\theta = 26° \, 34'$$

The resultant is of magnitude 447·2 N acting at an angle of 26° 34′ to the direction of *AB* (*Answer*).

Compare this method with that used in the corresponding example on page 118. Note that we have not found the line of action by this method.

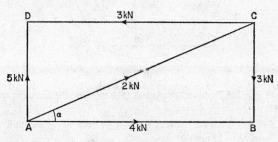

Fig. 111

Example: *ABCD* in Fig. 111 is a rectangle with $AB = 1\cdot2$ m and $BC = 0\cdot5$ m. Forces act along the following lines: 4 kN along *AB*, 3 kN along *CB*, 3 kN along *CD*, 5 kN along *AD*, and 2 kN along *AC*. Calculate the magnitude and direction of the resultant.

SOLUTION: Since $CB = 0\cdot5$ m, $AB = 1\cdot2$ m, then $AC = 1\cdot3$ m by Pythagoras' theorem. Resolving the forces in the direction *AB* we have

$$X = 4 - 3 + 2 \cos \alpha = 1 + 2 \times \frac{12}{13} = \frac{37}{13}$$

Resolving in the direction *AD*

$$Y = 5 - 3 + 2 \sin \alpha = 2 + \frac{2 \times 5}{13} = \frac{36}{13}$$

Let R be the resultant inclined at an angle θ to *AB*.

$$R^2 = X^2 + Y^2 = \frac{37^2 + 36^2}{13^2}$$

$$= \frac{2665}{13^2}$$

$$R = \frac{51\cdot63}{13} = 3\cdot97 \text{ N (two decimal places)}$$

Also
$$\tan \theta = \frac{Y}{X} = \frac{37}{13} \times \frac{13}{36} = 1\cdot0278$$

$$\theta = 45° \, 47'$$

The resultant has a magnitude of 3·97 N and acts at an angle of 45° 47′ to the direction of *AB* (*Answer*).

Example: *ABCDEF* is a regular hexagon and the following forces act along its sides: 3 kN along *AB*, 2 kN along *CB*, 6 kN along *CD*, 4 kN along *DE*, 2 kN along *FE*, and 6 kN along *AF*. Calculate the magnitude of the resultant and its inclination to *AB*.

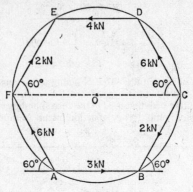

Fig. 112

SOLUTION: Let Fig. 112 represent the problem. Resolving the forces in the direction *AB* we have

$$X = 3 - 2\cos 60° - 6\cos 60° - 4 + 2\cos 60° - 6\cos 60°$$
$$= -1 - 12\cos 60° = -7 \qquad (\cos 60° = \tfrac{1}{2})$$

Resolving the forces in the direction *AE* perpendicular to *AB* we have,

$$Y = -2\sin 60° + 6\sin 60° + 2\sin 60° + 6\sin 60°$$
$$= 12\sin 60°$$
$$Y = 6\sqrt{3} = 10.39 \text{ (two decimal places) } (\sin 60° = \tfrac{1}{2}\sqrt{3})$$
$$\therefore \ R^2 = X^2 + Y^2 = 49 + 108 = 157.$$
$$R = 12.53 \text{ kN (two decimal places)}$$

If *R* acts at an angle θ to *AB*, then

$$\tan θ = \frac{Y}{X} = \frac{10.39}{-7} = -1.4843$$

From tables, tan 56° 2′ = 1·4845

$$\therefore \ θ = 123° 58′ = 124° \text{ (nearest degree).}$$

The resultant has a magnitude of 12·5 kN (3 significant figures) and acts at an angle of 124° to the direction of *AB* (*Answer*).

Note that the negative value for *X* combined with a positive value for *Y* shows that the resultant is directed into the second quadrant. If *X* and *Y* are both negative, the resultant is directed into the third quadrant; finally, if *X* is positive but *Y* is negative, the resultant is directed into the fourth quadrant.

Exercise 41

1. In the last two examples above state the magnitude and direction of the resultant if all the forces are doubled in magnitude.

2. In the last two examples above state the magnitude and direction of the resultant if all the forces are reversed in direction but unaltered in magnitude.

3. In the example of Fig. 111 calculate the magnitude and direction of the resultant if the force in *BC* is reversed.

4. In the example of Fig. 112 calculate the magnitude and direction of the resultant if the force in *ED* is reversed.

5. *ABCD* is a square. Calculate the magnitude and direction of the resultant of the following forces: 15 kN along *AB*, 6 kN along *BC*, 5 kN along *DC*, and 14 kN along *AD*, the direction of each force being indicated by the order of the lettering.

6. *ABCDEF* is a regular hexagon (see Fig. 112). Calculate the magnitude and direction of the resultant of the following forces: 10 kN along *AB*, *BC*, *CD*, *AF*, *FE*, and *ED*, the direction of each force being indicated by the order of the lettering.

7. By considering the pairs of parallel forces in opposite sides of the hexagon, prove that the resultant in Question 6 passes through the centre of the hexagon.

8. Four forces acting at a point are in equilibrium. Calculate the magnitude and direction of the fourth force if three of the forces are as follows: 20 kN directed West, 30 kN directed South, 10 kN directed North.

9. *ABCD* is a square and the following forces act along its sides: 11 kN along *AB*, 2 kN along *BC*, 3 kN along *CD*, 6 kN along *AD*, and 2 kN along *AC*. Calculate the magnitude and direction of the resultant.

10. State the magnitude and direction of the resultant in Question 9 if the force in *BC* is transferred to act along *BA*, the direction of the force again being given by the order of lettering.

(8) Locating the Line of Action

In the last section we have used the method of resolution in order to find the magnitude and direction of the resultant of the system of coplanar forces. By using the method of taking moments for the system we shall now see how to find the line of action of the resultant without some of the difficulties arising from the methods discussed in Section 5.

We have already discussed how the resultant of a system of coplanar forces is that single force, if it exists, which produces the same effect as the original system. Thus we have resolved all the forces in the system in two perpendicular directions and we have said that the sum of the resolved parts, or **resolutes**, in these directions is equal to the resultant resolved in these directions. These results have been expressed by the equations

$$X = R \cos \theta, \text{ and } Y = R \sin \theta$$

in the last few problems. Taking this one stage further, we argue that if we take moments about any point *P* in the plane, for all the forces in the system, then the sum of these moments will be equal to the moment of the resultant about the same point *P*. If *P* is a point on the line of action of the resultant, then its moment about *P* is zero. Conversely, if we take moments about a point *P* and find that the sum of the moments of the forces in the system is zero, we know that *P* lies on the line of action of the resultant.

It follows, therefore, that all we have to do to find the line of action is to take moments about a point *P* and equate the sum to zero to give us the position of *P* and thereby a point on the line of action of the resultant. The reader will understand the method by examining its application in the following examples.

Example: *ABCD* is a square of side 1 m. Five forces act on the figure as follows: 70 N along *AB*, 20 N along *BC*, 30 N along *CD*, 60 N along *DA*, and 30 N along *AC*, the direction of the forces being given by the order of the lettering. Find the magnitude, direction, and line of action of the resultant.

SOLUTION: Let Fig. 113 represent the problem. Resolving parallel to *AB*

$$X = 70 + 30 \cos 45° - 30$$
$$= 40 + 15\sqrt{2} = 61{\cdot}21$$

Resolving parallel to *AD*

$$Y = -60 + 30 \sin 45° + 20$$
$$= -40 + 15\sqrt{2} = -18{\cdot}79$$
$$R^2 = X^2 + Y^2 = 4099$$
$$R = 64 \text{ N (nearest newton)}$$

Let the resultant act at an angle θ to the direction *AB*. Thus

$$\tan \theta = \frac{Y}{X} = \frac{-8 + 3\sqrt{2}}{8 + 3\sqrt{2}} = \frac{-3{\cdot}758}{12{\cdot}242} = -0{\cdot}307$$

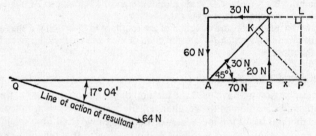

Fig. 113

Now $\tan 17° 04' = 0{\cdot}307$, and from the fact that *X* is positive and *Y* negative we see that the resultant acts into the fourth quadrant at an angle of $17° 04'$ to *AB*.

To find the line of action let us suppose that the resultant passes through the point *P* on *AB* produced such that *BP* = *x*. Since we are assuming that *P* lies on the line of action of the resultant, the sum of the moments of the forces about *P* must be zero. Taking moments about *P* we have:

$$60 \times PA + 30 \times PL - 20 \times PB - 30 \times PK = 0$$

Alternatively, putting the sum of the anticlockwise moments = the sum of the clockwise moments,

$$60 \times PA + 30 \times PL = 20 \times PB + 30 \times PK$$

From the geometry of the figure,

$$PA = 1 + x, PL = 1, PB = x, PK = (1 + x) \sin 45°$$
$$60(1 + x) + 30 = 20x + 30 (1 + x) \, 0{\cdot}7071$$
$$90 + 40 x = 21{\cdot}213 + 21{\cdot}213 \, x$$
$$18{\cdot}787 \, x = -68{\cdot}787$$
$$x = \frac{-68{\cdot}787}{18{\cdot}787} = -3{\cdot}662$$

This result tells us that the point *P* is really 3·662 m on the other side of *B*, i.e. at the point *Q* such that *BQ* = 3·662 m or *AQ* = 2·662 m. The line of action of *R* is thus determined as shown at *Q*   (*Answer*).

Example: *ABCD* is a rectangle with *AB* = 1 m, *BC* = 0·5 m. The following forces act on the rectangle: 40 N along *AB*, 70 N along *BC*, 20 N along *DC*, 30 N along *AD*, and 100 N along *CA*. Find the line of action of the resultant and where it intersects *AB*.

SOLUTION: Let Fig. 114 represent the problem with P the point in which the line of action intersects AB. As in the previous example let $BP = x$.

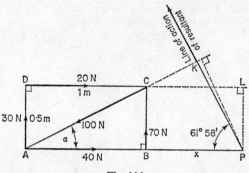

Fig. 114

Taking moments about P we have

$$100 \times PK - 20 \times PL - 70 \times PB - 30 \times PA = 0$$
$$100 (1 + x) \sin \alpha = 20 \times 0 \cdot 5 + 70 x + 30 (1 + x) \qquad \text{(i)}$$

From the geometry of the figure we have

$$\sin \alpha = \frac{1}{\sqrt{5}} = 0 \cdot 4472$$

so that

$$\alpha = 26° \ 34'$$

Substituting this result in eqn. (i) we get

$$44 \cdot 72 + 44 \cdot 72x = 40 + 100x$$
$$55 \cdot 28x = 4 \cdot 72$$
$$x = \frac{4 \cdot 72}{55 \cdot 28} = 0 \cdot 08539$$

We now need the direction of the resultant. Resolving parallel to AB

$$X = 40 + 20 - 100 \cos \alpha$$
$$= 60 - 89 \cdot 44 = -29 \cdot 44$$

Resolving parallel to AD

$$Y = 30 + 70 - 100 \sin \alpha$$
$$= 100 - 44 \cdot 72 = 55 \cdot 28$$

Hence:

$$\tan \theta = \frac{Y}{X} = \frac{+55 \cdot 28}{-29 \cdot 44} = -1 \cdot 878$$

Now $\tan 61° 58' = 1 \cdot 8779$, and the signs of X and Y indicate that the resultant is directed into the second quadrant   (*Answer*).

Example: A regular hexagon *ABCDEF* of side 1 m has the following forces acting. in its plane: 30 N along *AB*, 40 N along *BC*, 20 N along *CD*, 100 N along *AD*, 80 N along *FA*, the direction of the forces being given by the order of the lettering. Find the line of action of the resultant and the point in which it intersects *AB*.

SOLUTION: Let Fig. 115 represent the problem, with *P* the suggested point of intersection of the line of action of the resultant and *AB*.

Let $AP = x$.

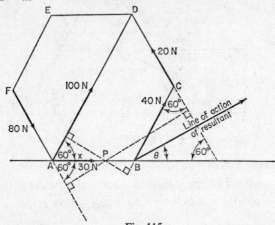

Fig. 115

Taking moments about *P* we have
$$80x \sin 60° + 40(1 - x) \sin 60° + 20(2 - x) \sin 60° = 100x \sin 60°$$
Dividing throughout by $20 \sin 60°$ this simplifies to
$$4x + 2(1 - x) + (2 - x) = 5x$$
$$4 = 4x$$
$$1 = x$$
which indicates that *P* lies at *B*.

We now obtain the direction of the resultant. Resolving parallel to *AB*
$$X = 30 + 40 \cos 60° + 100 \cos 60° - 20 \cos 60° + 80 \cos 60°$$
$$= 30 + 20 + 50 - 10 + 40 = 130$$
Resolving parallel to *AE*
$$Y = 100 \sin 60° + 40 \sin 60° + 20 \sin 60° - 80 \sin 60°$$
$$= 80 \sin 60° = 40\sqrt{3} = 69.28$$
$$\tan \theta = \frac{Y}{X} = \frac{69.28}{130} = 0.5329$$
Hence $$\theta = 28° 03'$$
The line of action of the resultant passes through *B* at an angle of 28° 03′ to *AB* and is directed into the first quadrant.

Exercise 42

1. Using the example of Fig. 113 above, find the line of action of the resultant if each of the forces is doubled in magnitude.

2. Using the example of Fig. 114 above, find the line of action of the resultant and the point in which it intersects *AB* if the force of 100 N in *CA* is removed.

3. Using the example of Fig. 115 above, find the line of action of the resultant and the point in which it intersects *AB* if the force in *AB* is increased to 100 N.

4. Using the example of Fig. 115 above, find a point on the line of action of the resultant of the three forces in *FA*, *AD*, *CD*.

5. *ABCD* is a square of side 1 m. Forces of equal magnitude 10 N act along the lines *AB, BC, AD, AC, BD*. Find the resultant and the point at which it intersects *AB*.

6. Show that if an extra force of 10 N acts along *DC* in Question 5, then the resultant passes through the centre of the square.

7. *ABCDEF* is a regular hexagon of side 0·1 m. Forces of equal magnitude 100 N act along the lines *AF, AE, BE, AB*. Find the line of action of the resultant, giving the point in which this line intersects *AB*.

8. Find the line of action of the resultant in Question 7 if all the forces are reversed.

CHAPTER EIGHT

RELATIVE MOTION

In the last two chapters we used vectors to present in detail the relation between a set of forces and its resultant. A force, however, is only one example of a vector and we could have replaced forces in the problems by any other vectors such as displacements, velocities, or momenta without altering the method of obtaining the final result. It is this convenience of applying the same set of rules to problems in general which makes the study of vectors so attractive to the mathematician. We shall now apply the language of vectors to the study of motion in a plane, and we start by considering the subject of relative velocity.

It is often said that 'all things are relative', to which one replies 'relative to what?'. Consider, for example, the meaning we gave to the ideas of being in a state of rest or motion in Chapter One. In both cases rest and motion were defined relative to the earth. That is, we consider the earth to be at rest and we call the earth a frame of reference in the sense that all our usual measurements are referred to or relative to its surface. There are, of course, many situations in which relative measurements are taken which have nothing to do with the earth at all, as we shall now suggest.

Two sisters are on a diet. Bertha loses 3 kg during the first week but only 2 kg during the second week. Her sister Nelly loses only 1 kg during the first week but 3·5 kg during the second week. These data enable us to say that, compared with Nelly, during the first week Bertha is losing mass at the rate of 2 kg per week. In mathematics we prefer to use the term 'relative to' rather than 'compared with', so that looking at the two sisters again we can say that over the two-week period Bertha is losing 0·5 kg per fortnight relative to Nelly. If a third sister, Lizzie, is eating to gain weight at the rate of 1 kg per week, we can say that during the second week she is gaining weight at the rate of 4·5 kg per week relative to Nelly. Since the rates concerned are only scalar quantities, we either add or subtract the rates in order to achieve the relation 'relative to'. Let us take the idea a stage further with examples in which the rates are vectors like velocity.

Suppose that you are travelling along a straight road in a car *A* which is towing a caravan at a speed of 20 km h^{-1} when another car *B* passes you, travelling in the same direction with a speed of 30 km h^{-1}. Each speed just quoted is taken to be a speed relative to the earth, i.e. the speed which would be registered on a radar screen at the roadside. Your first reaction is to notice how much faster than you the other car is travelling so that you can adjust your speed to avoid the awkward moment of catching up the other car if it should slow down too quickly after overtaking. In other words, what is the other car's speed relative to yours, or at what rate is the gap between you increasing or decreasing? Clearly the other car is travelling in the same direction at 10 km h^{-1} relative to your car. A more concise statement would be: '*B* is travelling at 10 km h^{-1} relative to *A* in the same direction as *A*'.

Let us make a slight change in the problem by considering another car *C* to be travelling towards you at a speed of 30 km h^{-1}. If you were at rest, then *C* would be approaching at a speed of 30 km h^{-1}, but since you are travelling towards *C* with your own speed of 20 km h^{-1}, then the rate of approach is increased to $(20 + 30)$ km h^{-1}. Again we can put this more concisely by saying that the speed of *C* relative to *A* is 50 km h^{-1} in the opposite direction to *A*.

To put our own difficulties with relative motion in perspective let us consider for a moment the problems facing navigators from the fourteenth to seventeenth centuries who were busy sailing in all directions at unknown speeds without much idea of when or where they would arrive. Their biggest problem arose out of not having an accurate clock to help them obtain their longitude. The sand-glass (like the kitchen hour-glass or egg-timer) was not good enough for precise calculations. Even the method of obtaining latitude was more appropriate to standing on dry land, and this accounts for that strange ritual described in captains' log-books as 'putting ashore to find our position'. Knowing how far they travelled each day might have been useful, but even this was obtained with considerable misgivings.

With so many guesses for obtaining a ship's position it is small wonder that the ancient mariner trusted only that which he knew best, finding latitude, and from bitter experience, the seasons and directions of the prevailing winds. His method of voyaging became known as 'sailing the latitudes': he sailed North or South using his magnetic compass and as soon as he knew the latitude he sailed due East or West until he sighted land. A more colourful description of this method of sailing from England to the West Indies was known as 'South till the butter melts and then sail due West'. With such ideas did Columbus mistake the West Indies for Asia.

Let us take a closer look at man's travels relative to the earth. The old mariner used to estimate the speed of his ship by timing the flotsam as it floated by the length of the ship. Suppose, for example, that on the first voyage of Columbus in the *Santa Maria* in 1492, a ship of length approximately 23 metres, it took 8 seconds to pass the flotsam. The calculation of the ship's speed would have to be as follows:

$$23 \text{ metres in 8 seconds is } \frac{23}{8} \text{ m s}^{-1}$$

$$= \frac{23}{8} \times 60 \times 60 \text{ m h}^{-1}$$

$$= \frac{23}{8} \times \frac{60 \times 60}{1000} \text{ km h}^{-1}$$

$$= 10 \cdot 35 \text{ km h}^{-1}$$

$$= \frac{23 \times 60 \times 60}{8 \times 1852} \text{ knots (international nautical miles per hour,}$$
$$1 \text{ knot} = 1852 \text{ m h}^{-1})$$

$$\Rightarrow \text{ approximately } 5\tfrac{1}{2} \text{ knots}$$

Naturally the old sailors would not have carried out this calculation, they would probably have used a rule of thumb which amounted to the following:

$$\frac{23}{8} \times \frac{60 \times 60}{1852} \text{ is approximately } \frac{23}{8} \times 2$$

$$= \frac{\text{length of ship (metres)} \times 2}{\text{time for flotsam (seconds)}}$$

or, double the length of the boat in metres, then divide by the time in seconds and the result is the speed in knots. A speed of 7–8 knots would be very good for the ships of those days.

Example: A ship of length 100 metres takes 20 seconds to pass flotsam. Use the rough approximation above to calculate the speed of the ship.
SOLUTION:

$$\text{Speed} = \frac{\text{Double length of the boat (metres)}}{\text{Time for flotsam (seconds)}}$$

$$= \frac{100 \times 2}{20} = 10 \text{ knots} \quad (Answer).$$

Note the following simplifications which have been used in putting forward the above calculations:

(*a*) The metric system was not adopted until 1799.
(*b*) The international nautical mile (Kn) has only recently been accepted in Britain.
(*c*) The best timer the sailors had was a 30-second sand-glass, so in all probability seconds were 'counted' out.
(*d*) Allowance was seldom made for the influence of currents or winds on the speed of the flotsam.

It is the last point which is of interest because it means that the calculations of speed were made relative to the flotsam and not relative to the earth.

When there was no flotsam to be seen, wood or other debris was thrown overboard and timed as it floated by. With a shortage of wood it wouldn't take long before some one invented a more economical device, and so it was that a line was attached to a piece of wood which was then thrown overboard. The length of line which passed through the hand was timed, and from this the speed of the boat was calculated. In order to get a quicker result for measuring the line, knots were tied in the line at regular intervals (Fig. 116) and counted as they slipped through the hand. It was from this device that the name for nautical measurement of speed was first derived.

For a closer examination of relative velocity in a straight line consider an observer standing on a railway bridge watching two trains *A* and *B* travelling in opposite directions on parallel lines directed East to West. This is similar to the mariner's problem but this time the observer on the bridge is at rest and only the train is moving. We shall try to answer two questions:

(*a*) How can we find the approximate speed of the train?
(*b*) What is the speed of each train relative to the other?

To answer the first question requires some knowledge of railway stock, so let us assume that the observer knows that the length of the standard railway carriage and diesel car is 20 metres each, including buffers. If train A has nine carriages and one diesel car and is timed at 30 seconds under the bridge,

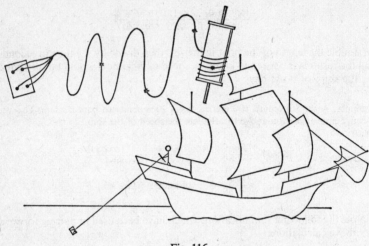

Fig. 116

then the speed of train A relative to the stationary observer can be found as follows:

Length of train A is $20 \times 10 = 200$ m

Distance travelled by train A in 30 seconds is 200 m

Speed of train A is $\dfrac{200}{30}$ m s^{-1} = $\dfrac{200 \times 60 \times 60}{30 \times 1000}$ km h^{-1} = 24 km h^{-1}

Similarly, train B has twelve carriages and one diesel car and is timed at 32 seconds under the bridge. Therefore

Length of train B is $20 \times 13 = 260$ m

Distance travelled by train B in 32 s is 260 m

Speed of train B is $\dfrac{260}{32}$ m s^{-1} = $\dfrac{260 \times 60 \times 60}{32 \times 1000}$ km h^{-1}

$$= \frac{117}{4} = 29 \cdot 25 \text{ km h}^{-1}$$

(On the basis of one timing only we are not justified in taking the result to four figures, so we approximate the result to 29 km h^{-1}.)

Now let us consider the velocity of A relative to B. The driver of train B sees the driver of train A coming towards him at a speed of $(29 + 24) = 53$ km h^{-1}, so the velocity of A relative to B is 53 km h^{-1} in the direction due West (Fig. 117).

Another way of deriving this result is to consider imposing on everything in the problem a velocity which brings B to rest. Thus a velocity of 29 km h^{-1}

West on everything will bring *B* to rest, increase the velocity of *A* to 53 km h⁻¹ due West, and give the observer (and the bridge) a velocity of 29 km h⁻¹ due West, and all these velocities are relative to *B*. Of course the bridge doesn't really move, but relative to *B* it *appears* to move towards *B* with a velocity of 29 km h⁻¹. We have all experienced those moments in a train or bus during which we have had the feeling that it was the station or bus stop which was moving while we remained at rest in the seat.

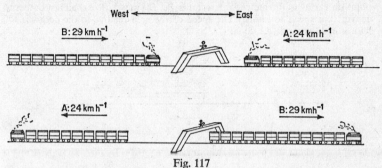

Fig. 117

Examining the situation after the trains have passed each other, what is the velocity of *A* relative to *B* now? Again we are asking ourselves how is everything travelling compared with *B*? Or, to put it another way, considering *B* to be the frame of reference which is at rest, how does everything else appear to travel? So again let us impose a velocity of 29 km h⁻¹ due West on to the whole problem in order to bring *B* to rest. We see that train *A* is still travelling West as before with a velocity of 53 km h⁻¹ relative to *B*. In conclusion, the relative velocities are the same before and after the trains pass each other.

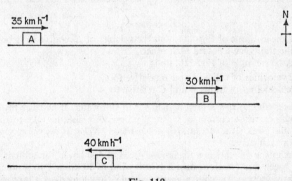

Fig. 118

Example: Three trains *A*, *B*, and *C* are travelling on parallel lines. The velocities of the trains at the moment of observation are *A*, 35 km h⁻¹ East; *B*, 30 km h⁻¹ East; *C*, 40 km h⁻¹ West. Find the velocity of the trains relative to *B*.

SOLUTION: Let Fig. 118 represent the problem.

Since we require the velocities relative to *B* we impose a velocity of 30 km h⁻¹ West on the whole problem in order to calculate as though *B* were at rest. It follows at once that the velocity of *A* relative to *B* is $(35 - 30) = 5$ km h⁻¹ East, and the velocity of *C* relative to *B* is $(40 + 30) = 70$ km h⁻¹ West.

Thus to an observer on train B the train A appears to be overtaking him with a velocity of 5 km h^{-1} due East and train C appears to be moving away from him with a velocity of 70 km h^{-1} due West.

Example: Debris floats by the 100-m length of a ship in 12 seconds when the ship takes only 8 seconds to pass a fixed buoy. Calculate the true speed of the ship and the speed of the current, assuming that ship and current are travelling in the same line.
SOLUTION: Since the buoy is fixed, the true speed of the ship (i.e. relative to a fixed point on earth) is 100 metres in 8 seconds, i.e. $12\frac{1}{2}$ m s^{-1}. (We shall not bother to convert this speed to knots.) The speed of the ship relative to the debris is 100 metres in 12 seconds, i.e. $8\frac{1}{2}$ m s^{-1}.

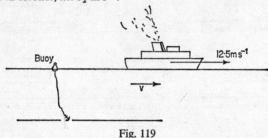

Fig. 119

With this information let us now examine Fig. 119, where the speed of the debris is shown, as v m s^{-1}. If we require the speed of the ship relative to the debris we obtain the result by considering the debris to be at rest by imposing a velocity of $-v$ on the whole problem. This means that the speed of the ship relative to the debris becomes $(12\frac{1}{2} - v)$ m s^{-1}, a result we already know to be $8\frac{1}{2}$ m s^{-1}. Therefore

$$8\tfrac{1}{2} = 12\tfrac{1}{2} - v$$

i.e. $$v = 4 \text{ m s}^{-1}$$

Thus the speed of current is 4 m s^{-1} in the direction we suggested in Fig. 119.

Exercise 43

1. Using the example of Fig. 117, find the velocity of B relative to A both before and after the trains have passed each other.
2. Using the example of Fig. 118, find:
   (i) the velocities of trains A and B relative to C;
   (ii) the velocities of trains B and C relative to A.

3. An oarsman can row a boat at 1 m s^{-1} in still water. If he rows his boat in a stream which carries a current of 0·5 m s^{-1}, how long does he take to row 100 m relative to the bank (i) upstream, (ii) downstream? What is his velocity relative to the water in each case?
4. An aeroplane which has a maximum speed of 500 km h^{-1} in still air flies with maximum speed due North against a North wind of 50 km h^{-1} to a point 100 km away. What is its velocity relative to the ground? How long does it take to arrive at the point?
5. If the aeroplane of Question 4 flies the return journey of 100 km under the same conditions, find the velocity of the plane relative to the ground and the time taken for the return journey.
6. Using the example of Fig. 119, calculate the speed of the current if the flotsam had passed by in (i) 10 s, (ii) 5 s.
7. A passenger in a train travelling at 20 m s^{-1} walks with uniform speed a distance of 40 m along the corridor to the buffet car in 60 s. Find the passenger's velocity (i) relative to the train, (ii) relative to the track.

(1) Latitude and Longitude

As we have already noted, the old sailors kept to their own ways of estimating their position. Their favourite method was called 'dead reckoning', a name which is believed to be an abbreviation of 'deduced reckoning'. The idea was to estimate a position by knowing the speed of the ship throughout the time it sailed on a fixed compass bearing, the magnetic compass being the only instrument they regarded as trustworthy. It was for this reason that estimates of the speed were taken every half-hour. We have already seen that serious errors could be introduced by the presence of unsuspected currents, and clearly the longer the voyage the greater the overall error.

Since we are speaking of positions relative to the earth let us examine briefly how these positions are given. The earth may be regarded as a sphere covered by a grid consisting of circles of latitude, which are parallel to the Equator, and circles of longitude, also called meridians, which are at right-angles to the Equator and pass through the North and South poles. As always we must start measuring from somewhere, so circles of latitude are measured from the Equator, as 0° or zero latitude, into either the northern hemisphere or the southern hemisphere.

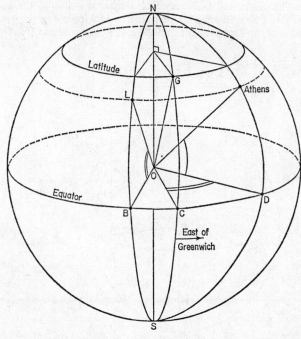

Fig. 120

In 1884 the meridian passing through Greenwich Observatory in South-East London was internationally accepted as the zero meridian from which all other longitudes were measured up to 180° either East or West. Thus Greenwich (G in Fig. 120) has the position given by Latitude 51° 29′ N, Longitude 0°.

This means that $\triangle COG$ in Fig. 120 is 51° 29' on the arc given by *NGCS* as the zero meridian. The position of Lisbon (*L* in Fig. 120) is given by Latitude 38° 43' N, Longitude 9° 10' W. This means that $\angle LOB$ in Fig. 120 is 38° 43' and $\angle COB$ is 9° 10'. Athens is on Latitude 37° 59' N and Longitude 23° 42' E. This means that $\angle AOD = 37° 59'$ and $\angle COD = 23° 42'$.

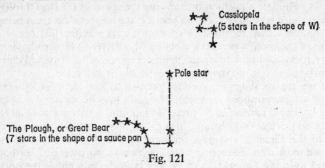

Fig. 121

Now the Pole Star always appears to be in the same place in the sky relative to the earth, and with very little variation it lies on the line *SON* produced. The Pole Star may be identified by looking along the line of the two end stars of the Plough as shown in Fig. 121. Since the star is so far away,

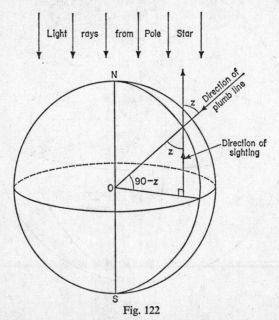

Fig. 122

all its rays of light reaching the earth are considered to be parallel to each other. Therefore, on looking through a telescope mounted on a vertical scale carrying a plumb line (Fig. 122), we can read the latitude of the point of observation directly, as shown in Fig. 123.

(2) Finding Latitude

Using stiff card cut out two circles of diameters 0·15 m and 0·13 m respectively, and mark two perpendicular diameters on each circle. On the larger of the two circles mark out one quadrant from 0° to 90° as a protractor; use a drawing pin to make holes through the centre and at either end of the zero diameter, as shown in Fig 123 (*a*). Using a small piece of wood or a small

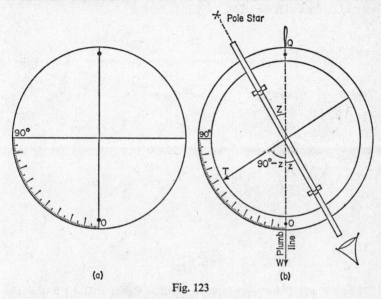

Fig. 123

india rubber as backing, fasten the two circles together, centre to centre, with a drawing pin. Now take a sheet of paper about the same size as this page and roll it round a pencil to make a tube. Use gummed paper to prevent the tube unrolling, and then stick it along one of the diameters of the smaller circle as in Fig. 123 (*b*). Hang a heavy body from *O* (the zero mark), and hang the complete apparatus from *Q* so that the diameter *OQ* is now vertical.

To use the apparatus we sight the Pole Star with the tube, and our latitude is given by the marker *T* which is at the end of the diameter perpendicular to the tube. To correspond with Fig. 122 the angle *z* has been inserted. Anyone who tries this when standing on the deck of a small boat will soon realize why the early maps were so inaccurate: a mistake of 1° of latitude means a difference of about 110 km.

(3) Finding Longitude

The calculation of *longitude* is essentially the calculation of a difference in time. The earth rotates on its axis once every 24 h and moves round the sun approximately once every 365¼ days. Since the calendar we use has only 365 days it is necessary to make a periodic adjustment in order to keep the calendar in step with the seasons: the correction is made by employing the leap-year of 366 days.

Let us suppose that a celestial observer near the Pole Star is watching the earth move round the sun, as in Fig. 124. He sees the earth move along the arc *AB* with the sun as centre, and at the same time the earth spins on its own North–South axis. In other words, this is the motion of the earth relative to someone in the vicinity of the Pole Star. For further simplicity imagine the sun as a football placed on a table and the earth as a tennis ball moving along the arc *AB* round the football and spinning on an axis not quite through the point of contact with the table.

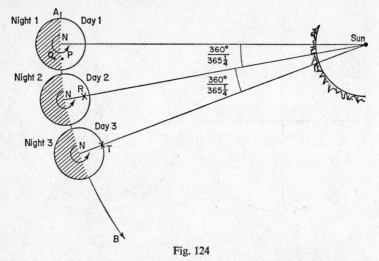

Fig. 124

Now if *P* and *Q* are two people on earth as shown at Day 1 in Fig. 124, then *P* is to the East of *Q*. Since *P* is in the daytime region and *Q* is in the night-time region, it follows that *P* will have seen the sun first. This means that the sun appears to rise in the East so that relative to *P* the sun moves across the sky from East to West. Let us apply our ideas about relative velocity again. To obtain the motion relative to *P* we impose a velocity on the whole problem (universe) which brings *P* to rest. Fig. 125 illustrates the result of this suggestion. We now have *P* at rest and the sun moving at this particular instant in the direction *SK* so that, relative to an observer such as *P* on the earth, the sun appears to travel across the sky from East to West.

Re-examining Fig. 124 we notice that anyone on the meridian through *R* or *T* will describe the sun as being 'directly overhead', a time which is described as noon or midday for obvious reasons on examining the geometry of the figure. It is always possible to discover the midday time by observing the length of the shadows thrown by a sundial. A sundial can be made by fixing any vertical object in a horizontal plane. A tall wine bottle filled with water (to prevent it being blown over) and placed upright on horizontal ground which takes chalk marks will do. When the bottle throws the shortest shadow the sun is 'directly overhead'.

To obtain the shortest shadow one can take readings every 5 min; alternatively one can note the times when two shadows of equal length are thrown, midday is then halfway between. For example, if equal shadows were thrown at times of 11.35 a.m. and 12.15 a.m. according to the observer's watch, then midday by the sun occurs at 11.55 a.m. on the same watch. There is no need to take readings every 5 minutes all day long because we can always anticipate when midday occurs by observing that at midday the sun is due South in

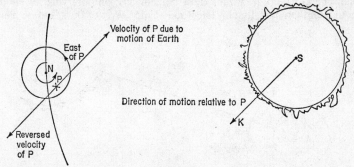

Fig. 125

the northern hemisphere and due North in the southern hemisphere. Thus the use of a magnetic compass will decrease the number of observations required (unless the sundial is being used to determine the direction of South).

We know that on average the earth spins through 360° in 24 hours, which

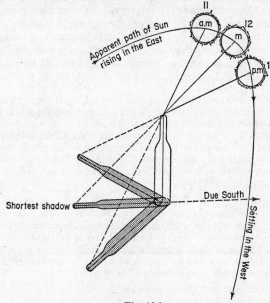

Fig. 126

means that the earth has an angular velocity of 15° per hour. It is this rotation of the earth about its own axis which makes the finding of longitude dependent upon time. Examining Fig. 127 we see that it is midday at the point *R* and that *S* lies on the meridian which is 15° West of *R*. This means that in one hour's time it will be midday at S, i.e. the sun will be overhead and the sundial will throw the shortest shadow.

Let us now reverse the situation and start at R on land at midday, setting our watches to the local time of midday. If someone at S sets his watch at the same time as we do, he would know immediately, on observing his midday 1 hour later than ours, that his longitude was 15° West of ours; but he would

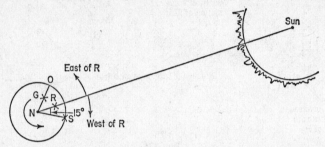

East of R

West of R

Fig. 127

have the inconvenience of his midday taking place at one o'clock instead of twelve o'clock. Clearly it would be most convenient if everyone in the world could measure their time from a basic zero time and zero meridian so that everyone could arrange to have the midday when their time was twelve o'clock. It has been internationally agreed that this basic time will be called **Greenwich Mean Time** (G.M.T.). Everyone in the world can keep up to date with Greenwich Mean Time by the use of radio signals and calculate their longitude simply by finding their own local midday time relevant to Greenwich. The B.B.C. broadcasts the Greenwich time signal of six pips, the last of which gives the exact moment of G.M.T.

Example: Two observers A and B obtain their local midday times and find from the radio that Greenwich Mean Time for A was 2.30 p.m. and for B was 11.00 a.m. Find the longitude of the positions of A and B.

SOLUTION: G.M.T. for A being 2·30 p.m. means that at noon local time (as measured by the sundial at A) it was 2.30 p.m. at Greenwich. In other words, the earth has rotated for $2\frac{1}{2}$ hours since the sun was overhead or crossed the zero meridian through Greenwich. It follows therefore that A lies on a longitude of $15° \times 2\frac{1}{2}$ West of Greenwich, i.e. $37\frac{1}{2}°$ W.

G.M.T. for B being 11.00 a.m. means that at noon local time (by the sundial at B) it was only 11.00 a.m. at Greenwich. So the earth has still to rotate for another hour before the sun crosses the zero meridian through Greenwich. In other words, B is leading Greenwich by 1 hour—which is equivalent to a rotation of the earth through 15°. The longitude of B is therefore 15° East.

The abbreviations a.m. and p.m. which we write after the numerical statement of the time have the following meanings:

a.m. = *ante meridian* (Latin: before noon)
p.m. = *post meridian* (Latin: after noon)

Hence, to quote the time at any place at '4 p.m.' means 4 hours after local midday.

Example: The position of New York is 40° 45′ N, 74° W. If the world heavyweight championship fight starts at 9 p.m. Monday New York time, at what time must I tune in to the 'live' broadcast if I live in Greenwich?

SOLUTION: Since the longitude of New York is 74° W we know that New York is $\frac{74}{15}$ h behind Greenwich, i.e. 4 h 56 min behind (or, if we give the time relative to New York, Greenwich is 4 h 56 min ahead). Therefore 9 p.m. Monday in New York corresponds to a p.m. time of 9 h *plus* 4 h 56 min in Greenwich. Thus the time in Greenwich is 1.56 *a.m.* on *Tuesday*.

We have been discussing how places to the East of Greenwich are ahead of G.M.T. and places to the West of Greenwich are behind G.M.T. If we make a journey from Greenwich and travel due East round the earth, there must come a time when we change from being ahead of G.M.T. to being behind G.M.T. It should be fairly clear that this is bound to coincide with the crossing of the meridian 180° E—which coincides with the meridian 180° W— so that travelling East round the earth we shall find ourselves travelling back from Tuesday to Monday. Anyone living on or near this meridian would find life intolerable when a short walk down the road might start on Tuesday and end the previous Monday, as for example in the East Cape of Siberia. In order to avoid these difficulties we have what is called the **International Date Line**: an imaginary line on the earth's surface which fixes the beginning of every new day. We observe a change of 24 hours relative to Greenwich time whenever the date line is crossed.

The line coincides approximately with the 180° meridian and is so chosen because it touches very little land. It deviates slightly from the meridian in some places to enable people living on islands close together to have the same day. For example, it is turned slightly westward so as to include Wrangel Island with Alaska. It is turned westward to have the Aleutian Islands on the same side of the date line as the United States, and it is turned eastward to include Fiji and Tonga islands with New Zealand.

Exercise 44

1. Find the local time to the nearest 5 min at the following places when it is Tuesday 10 p.m. G.M.T.:

New York:	40° 45′ N, 74° W
Tokyo:	35° 48′ N, 139° 45′ E
Sydney:	33° 50′ S, 151° 12′ E
Cape Town:	33° 59′ S, 18° 30′ E
Rio de Janeiro:	22° 50′ S, 43° 44′ W

2. What is the longitude of an observer who finds that his local noon by the sun occurs at 1.20 p.m. G.M.T.?

3. You are shipwrecked on an uninhabited island. How can you determine your position in order to transmit an S.O.S.?

4. A ship leaves the Australian port of Perth (31° 37′ S, 115° 58′ E) at local noon. If the sun is next overhead 40 min later on the following day, what is the longitude of the ship's position?

5. How is it possible to tell approximately local noon-time using a magnetic compass?

(4) Course Calculations

Perhaps the reader has obtained some feeling for the idea of relative motion and time from the very brief details on the last few pages related to early navigation. We now return to the comfort of modern transport and consider some examples of relative motion of today.

Suppose an aeroplane has to fly due North in a South-West wind which blows at 40 km h⁻¹. (A wind is described by the direction *from* which it blows, thus a South-West wind is one which blows from the South-West to the North-East.) To fly due North means due North relative to the ground, so an observer on the ground will look up and see the plane travelling North. This is

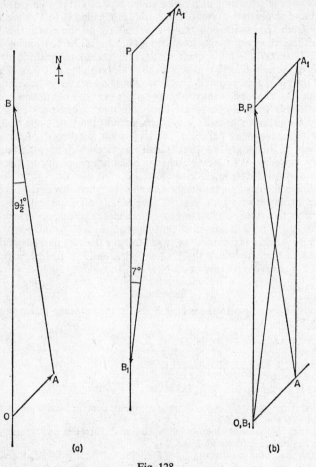

Fig. 128

the **track** of the plane. The **course** which the pilot sets will not be due North since he must allow for the wind; furthermore, he will have to arrange his speed to conform to regulations overland between airports. Suppose he must keep his resultant speed to 200 km h⁻¹. Using the illustration of Fig. 128, let the aeroplane be at point O. Using a scale of 2 mm to 5 km h⁻¹, let OA represent the velocity of the wind and let $\overrightarrow{OB}$ represent the required resultant velocity of 200 km h⁻¹ due North. Join AB.

$\overrightarrow{OB} = \overrightarrow{OA} + \overrightarrow{AB}$, by the triangle of velocities. That is, the resultant velocity $\overrightarrow{OB}$ is the sum of the wind velocity $\overrightarrow{OA}$ and a velocity given by $\overrightarrow{AB}$. Hence $\overrightarrow{AB}$ must represent the speed and course of the aeroplane, i.e. the velocity of the aeroplane relative to the air. Although we could calculate the result, it is more convenient to obtain it graphically as $|\overrightarrow{AB}| = 174$ km h⁻¹ and angle $OBA = 9\frac{1}{2}°$. Hence the pilot sets a course N $9\frac{1}{2}°$ W and a speed of 174 km h⁻¹ in order to travel due North.

We can extend the problem by considering a return flight from P under the same wind and a regulation speed due South of 200 km h⁻¹. Draw $\overrightarrow{PA_1}$ to represent the wind velocity and $\overrightarrow{PB_1}$ to represent the velocity of the aeroplane, 200 km h⁻¹ due South. Join A_1B_1.

$\overrightarrow{PB_1} = \overrightarrow{PA_1} + \overrightarrow{AB_1}$ and we see once again that the resultant velocity $\overrightarrow{PB_1}$ is the sum of the wind velocity $\overrightarrow{PA_1}$ and a velocity given by $\overrightarrow{A_1B_1}$. Hence A_1B_1 must represent the speed and course of the aeroplane. Again by graphical solution, $|\overrightarrow{A_1B_1}| = 230$ km h⁻¹ and angle PBA 7°. So the course the pilot must steer is South 7° West at 230 km h⁻¹.

Note that Fig. 128 (*b*) shows the two triangle of velocities put together in order to form a parallelogram, thereby demonstrating that the two solutions $\overrightarrow{AB}$ and $\overrightarrow{A_1B_1}$ are the diagonals of the parallelogram of velocities.

Example: An aeroplane whose maximum speed in still air is 200 km h⁻¹ wishes to fly due East a distance of 100 km. If a North-East wind blows at 50 km h⁻¹ throughout the flight, calculate the course and resultant velocity of the aeroplane together with the time of flight.

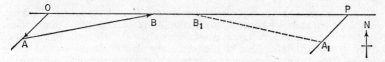

Fig. 129

SOLUTION: Let Fig. 129 represent the problem, with $\overrightarrow{OA}$ representing the velocity of the wind drawn through the starting-point O. We know that the resultant velocity must be due East and represented by $\overrightarrow{OB}$ so that $\overrightarrow{OB} = \overrightarrow{OA} + \overrightarrow{AB}$. We are also told that $|\overrightarrow{AB}|$ is 200 km h⁻¹, consequently with centre A and radius to represent 200 km h⁻¹ we strike an arc to complete the triangle of velocities, $\triangle OAB$. Applying the sine rule to $\triangle OAB$ we get

$$\frac{200}{\sin 135} = \frac{50}{\sin \angle OBA}$$

$$\sin \angle OBA = \frac{50 \sin 135°}{200} = \frac{0\cdot707}{4} = 0\cdot1768$$

$$\therefore \ \angle OBA = 10° \ 11'$$

Since the sum of the interior angles of a triangle is 180° it follows that $\angle BAO$ is 34° 49′.

Applying the sine rule again to $\triangle AOB$, we get

$$\frac{OB}{\sin 34° 49′} = \frac{200}{\sin 135°}$$

$$\therefore OB = \frac{200 \sin 34° 49′}{\sin 135°}$$

$$= 161\cdot4 \text{ km h}^{-1}$$

Time of flight is $\dfrac{100}{161\cdot4} \times 60$ min $= 37$ min (nearest minute).

So the pilot sets a course of E 10° 11′ N giving him a resultant velocity of 161·4 km h^{-1} due East and taking 37 min (nearest minute) to fly the required 100 km (*Answer*).

Exercise 45

1. Using the example of Fig. 128 above, find the velocity relative to the air and the course of the aeroplane if the wind speed decreased to (i) 30 km h^{-1}, (ii) 20 km h^{-1}.

2. Using the example of Fig. 129 above, find the course and resultant velocity together with the time of flight for the return journey of 100 km assuming the wind remains unchanged.

3. It is often said that the straightforward path is 'as the crow flies'. Assuming the crow appreciates the idea of relative velocity, how long will he take to reach his nest 200 m away due West by flying horizontally in a straight line against a South wind of 2 m s^{-1} when his maximum effort in still air is 3 m s^{-1}. What course must the crow set?

4. A fisherman wishes to row his boat across a river 150 m wide with parallel banks and carrying a current of 1 m s^{-1} assumed to be uniform throughout the width of the river. Calculate his course and time of crossing if he wishes to land 150 m downstream and row at a uniform effort which would take him through still water at 1 m s^{-1}.

5. Four buoys set at the vertices of a square $ABCD$ mark the course of a canoe race. A current runs North-east in the direction A to C throughout the race. Assuming the winner managed to canoe at a uniform effort (i.e. his speed in still water would have been constant) and he steered a course East 10° South when travelling due East from A to B, what was his course on each of the other sections of the race?

6. A child in a train travelling at 36 km h^{-1} rolls a marble across the compartment at right-angles to the direction of the train. If it takes 3 seconds to roll the 3-m width of the compartment, calculate the velocity of the marble relative to the track.

When we write a vector such as $\overrightarrow{AB}$ we are expressing a connection or relation between two points A and B. If we say that $AB = 4$ m and direction A to B is North-east, we know immediately that the relation between A and B can be represented by a diagram such as Fig. 130 (*a*) which shows the position of B relative to A. We say 'relative to A' because the measurement and direction of the vector is directed from A to B.

In Fig. 130 (*a*) $\overrightarrow{AB}$ is clearly a displacement of B relative to A. However, the same reasoning applies to any vector according to the units of the vector, as for example $\overrightarrow{AB}$ representing a velocity of 4 m s^{-1} in a direction North-east which is given by Fig. 130 (*b*). Geometrically the two vector diagrams are identical, but now Fig. 130 (*b*) represents a velocity of *B* which is 4 m s⁻¹ in the direction North-east relative to *A*.

Note that all our statements above are independent of whether *A* is a fixed point or not, although it would be more convenient for any intended calculations if *A* was a fixed point. Furthermore, the statement about the magnitude of $\overrightarrow{AB}$ might only be true for a very short period of time. This, however, is unimportant, for we are only concerned with representing the truth of the statement at that particular time.

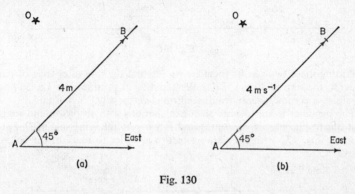

(a)   (b)

Fig. 130

If $\overrightarrow{AB}$ is a vector which expresses something (like displacement, velocity, acceleration) about *B* relative to *A*, then, by similar reasoning, $\overrightarrow{BA}$ must be a vector which expresses something about *A* relative to *B*. Returning to Fig. 130 (*a*), our interpretation of $\overrightarrow{BA}$ at this time is that *A* is displaced 4 m from *B* in a direction South-west. This last statement is just what we would expect from our knowledge of vectors, since $\overrightarrow{BA} = -\overrightarrow{AB}$. Fig. 130 (*b*) shows that $\overrightarrow{BA}$ represents *A* having a velocity of 4 m s⁻¹ in a direction South-west relative to *B* at that time.

If we introduce another point, say *O*, into the diagrams, in each case we can write $\overrightarrow{AB} = \overrightarrow{AO} + \overrightarrow{OB}$ and this enables us to describe the displacements in Fig. 130 (*a*) with reference to the point *O* as follows:

> The displacement of *B* relative to *A* is equal to the vector sum of the displacement of *B* relative to *O* and the displacement of *O* relative to *A*.

Alternatively we could have written $\overrightarrow{AB} = \overrightarrow{OB} - \overrightarrow{OA}$ and rephrased our description to read as follows:

> The displacement of *B* relative to *A* is equal to the displacement of *B* relative to *O* minus the displacement of *A* relative to *O*.

Example: *OX* is a straight line in the direction due East. Two bodies *A* and *B* are moving East along this line, *A* at 20 m s⁻¹ and *B* at 50 m s⁻¹. Find the velocity of *B* relative to *A*.

SOLUTION: The velocity of B relative to A is given by the rate of change of $\overrightarrow{AB}$ with respect to time. Since $\overrightarrow{AB} = \overrightarrow{AO} + \overrightarrow{OB}$, the rate of change of $\overrightarrow{AB}$ is equal to the rate of change of $\overrightarrow{AO}$ plus the rate of change of $\overrightarrow{OB}$. But the rate of change of $\overrightarrow{OB}$ is the velocity of B relative to O which, being a fixed point, gives 50 m s^{-1} due East. Similarly, since the rate of change of $\overrightarrow{OA}$ is 20 m s^{-1} due East it follows that the rate of change of $\overrightarrow{AO}$ is 20 m s^{-1} due West.

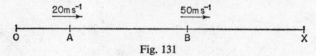

Fig. 131

Putting these two results together we see that the rate of change of $\overrightarrow{AB}$ is equal to the sum of 20 m s^{-1} due West and 50 m s^{-1} due East, i.e. 30 m s^{-1} due East, a common-sense result confirmed vectorially.

The beginner is apt to view such explanations with suspicion and suggest that this result only arises from A and B being on the same side of O. Let us examine Fig. 132 with the data as before in order to reaffirm the solution above.

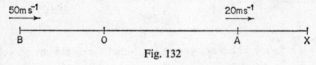

Fig. 132

The vector statement $\overrightarrow{AB} = \overrightarrow{AO} + \overrightarrow{OB}$ remains unchanged, so we see as before that the velocity of B relative to A is the sum of the velocity of B relative to O (50 m s^{-1} East) plus the velocity of O relative to A. Now the velocity of A relative to O is 20 m s^{-1} East, so the velocity of O relative to A is 20 m s^{-1} West and the result follows as before.

Example: AB and OX are two parallel lines running due East. A man walks East along AB at a speed of 5 km h^{-1} and a woman walks East along OX at a speed of 3 km h^{-1}. Find the velocity of the man relative to the woman.

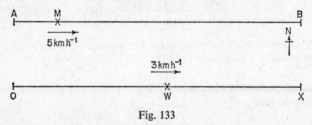

Fig. 133

SOLUTION: Let Fig. 133 represent the problem, with the woman at W and the man at M at the moment of observation. Since A and O are fixed points, the rate of change of the displacement $\overrightarrow{AM}$ is 5 km h^{-1} due East and the rate of change of the displacement $\overrightarrow{OW}$ is 3 km h^{-1} due East.

We want to find the rate of change of $\overrightarrow{WM}$ and we start by making the usual vector statement which relates the given velocities to $\overrightarrow{WM}$. Thus $\overrightarrow{WM} = \overrightarrow{WO} + \overrightarrow{OA} + \overrightarrow{AM}$, so the rate of change of $\overrightarrow{WM}$ is equal to the sum of the rates of change of $\overrightarrow{WO}$, $\overrightarrow{OA}$, and $\overrightarrow{AM}$.

But O and A are fixed points, so $\overrightarrow{OA}$ is a fixed or constant vector and therefore cannot change, i.e. the rate of change of $\overrightarrow{OA}$ is zero.

The velocity of M relative to W = velocity of O relative to W + velocity of M relative to A = 3 km h^{-1} due West + 5 km h^{-1} due East = 2 km h^{-1} due East.

Thus the man appears to the woman to be walking at 2 km h^{-1} due East (*Answer*). Again we have expressed the common-sense result in terms of vectors.

Exercise 46

1. Using the example of Fig. 131 above, find the velocity of B relative to A if A reverses his direction and moves West at 20 m s^{-1}. If they were 40 m apart at the moment of A's change in direction, how long would it be before they were 390 m apart?

2. Using the example of Fig. 133 above, find the velocity of the woman relative to the man assuming their present velocities remain unchanged.

3. $ABCD$ is a square with B due East of A and with D due North of A. A man at M is walking North along AD with a constant speed of 4 km h^{-1} and a woman at W on AB is walking East at 3 km h^{-1}. Obtain an expression for the rate of change of WM in terms of their velocities.

4. A fly crawls along the straight rim of one side of a rectangular tea-tray at a rate of 0·04 m s^{-1} at the same time as the tray is being carried at 0·5 m s^{-1} in a direction at right-angles to the sides of the rectangle. Calculate the velocity of the fly relative to the ground when crawling along each side of the tray.

(5) Moving Reference Point

We have not yet paid very much attention to the influence of a moving reference point even though we have suggested that the vector statement of the velocity remains unchanged. Certainly it will introduce one or two difficulties of actual computation even if the relation is easily expressed by vectors. The nature of the problem is best seen in an actual worked example.

Example: Two cars A and B travel along two straight roads OX and OY which intersect at right-angles. At the time of observation A is at P with $OP = 100$ m, and B is at Q with $OQ = 300$ m. If the velocity of A is 30 km h^{-1} along OX and the velocity of B is 40 km h^{-1} along OY, calculate the velocity of A relative to B assuming the velocities to be uniform.

SOLUTION: Clearly we shall need two diagrams, one to represent displacements and indicate position and the other to represent velocities. It is more convenient and certainly economical to put these diagrams on to one figure, but it does mean that we must use different scales for displacements and for velocities. Let Fig. 134 represent the problem, with a scale of 1 mm to 10 m for displacement, and 1 mm to 1 km h^{-1} for velocity. In all graphical work the scales refer to the magnitude of the vectors.

Using Fig. 134 for descriptive purposes we see that we require the rate of change of $\overrightarrow{BA}$, i.e. velocity of A relative to B. Since

$$\overrightarrow{BA} = \overrightarrow{BO} + \overrightarrow{OA}$$
$$= (-\overrightarrow{OB}) + \overrightarrow{OA}$$

it follows that the velocity of *A* relative to *B* = the reversed velocity of *B* + the velocity of *A*. In other words:

$$\text{rate of change of } \overrightarrow{BO} = \text{minus rate of change of } \overrightarrow{OB}$$
$$= \text{velocity of } B \text{ reversed.}$$

Therefore the velocity of *A* relative to *B* is the resultant of *A*'s velocity and the reversed velocity of *B*. Combining the reversed velocity of *B* with the velocity of *A* is, of course, equivalent to having suggested that we impose a velocity on the whole problem to bring *B* to rest.

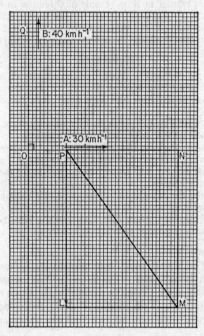

Fig. 134

We now construct the velocity diagram *PLMN*, where $\overrightarrow{PL}$ represents the reverse velocity of *B* and $\overrightarrow{PN}$ represents the velocity of *A* (i.e. the true velocity). The velocity of *A* relative to *B* is given by $\overrightarrow{PM}$ where $\overrightarrow{PM}$ = 50 km h^{-1} and $\angle NPM$ = 53° 8′ (*Answer*).

In simple terms our answer means that, according to *B*, throughout the motion *A* always appears to be moving at 50 km h^{-1} in a direction E 53° 8′ S.

Example: *A* and *B* are two ships steaming with the following velocities: *A*, due North at 20 km h^{-1}; *B*, due West at 10 km h^{-1}. At the moment of observation *A* is at *P* 60 km South of some point *O*, and *B* is at *Q* 30 km East of *O*. Find the velocity of *B* relative to *A* assuming that the velocities are constant.

SOLUTION: Before looking at any diagrams we recall that the easiest way of finding the required velocity of *B* relative to *A* is to bring *A* to rest. We therefore impose a velocity of 20 km h^{-1} South on the whole problem: in other words, we find the resultant of B's true velocity and the reversed velocity of A. Now let us examine Fig. 135. (The displacement scale should read 1 mm to 1 km.)

The rectangle $QCDE$ represents the velocity diagram, with $\overrightarrow{QC}$ the true velocity of B and $\overrightarrow{QE}$ the reverse velocity of A; consequently the resultant $\overrightarrow{QD}$ represents the velocity of B relative to A.

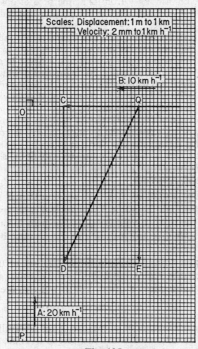

Fig. 135

By measurement: $|\overrightarrow{QD}| = 22.5$ km h^{-1} and $\angle CQD = 63°$
By calculation: $QD^2 = QC^2 + QE^2 = 100 + 400$

$$QD = \sqrt{500} = 22.4 \text{ km h}^{-1} \text{ (one decimal place)}$$

$$\tan \angle CQD = \frac{CD}{QC} = 2$$

$$\angle CQD = 63° 24'$$

In simple terms, to an observer on A, the ship B appears to be steaming at 22.4 km h^{-1} in a direction W 63° 24' S (*Answer*).

Note that if we produce QD it is seen to pass through the point P. Does this mean something? We think again about A being at rest as B 'appears' to move in direction QD. This must indicate therefore that the ships will collide. On further thought, they can only collide at O. Examine the speeds and distances and find where they will be in 3 hours time.

Exercise 47

1. A and B are two cars travelling towards the intersection O of two straight roads OX and OY. A travels with velocity 50 km h^{-1} West on OX, and B travels with

velocity 40 km h^{-1} South on OY. At the time of observation A is 2·5 km from O, and B is 2 km from O. Find the velocity of B relative to A, and comment on its direction.

2. Two aeroplanes flying at the same altitude are heading for the same point O. Aeroplane A is travelling South-east at a speed of 500 km h^{-1} and is 100 km away from O at the time of observation. Aeroplane B is travelling North-east at a speed of 800 km h^{-1} and is 150 km away from O at the time of observation. Find the velocity of B relative to A, and state whether the planes will collide if these velocities are maintained.

3. A man is running with a speed of 2·5 m s^{-1}, and rain is falling vertically with a speed of 8 m s^{-1}. Find the direction in which rain appears to fall to the man.

4. A car is being driven with a velocity of 20 m s^{-1} North in a straight line, and hailstones are being blown from the North at 30° to the vertical with a speed of 8 m s^{-1}. Find the direction in which the hailstones appear to be falling to the driver.

5. Given that the velocity of A is 10 m s^{-1} North, and that the velocity of B relative to A is 15 m s^{-1} due East, find the true velocity of B.

6. Question 5 is a reverse of the usual velocity problem and may have caused the reader some difficulty. If so, examine the solution and try the following question. To an observer on a ship A steaming due North at 10 km h^{-1}, a ship B appears to be travelling due West at 15 km h^{-1}. Find the true velocity of the ship B.

(6) Positions of Closest Approach

In the problem represented by Fig. 135 we saw that if the velocity of B relative to A passes through the point P (at which A is stopped), A and B will collide. But what if the relative velocity does not pass through the reference point such as P? Can we give any meaning to the distance by which it misses P?

A moment's thought should indicate the likelihood of the distance of A from the relative–velocity vector representing the shortest distance apart of the two ships in the subsequent motion. Let us put this idea to the test in a few examples.

Example: Consider the relative motion of two cars A and B travelling in perpendicular directions along straight roads which intersect at O. A passes point P, 100 m from O, with constant speed 30 km h^{-1}. At the same time, B passes point Q, 200 m from O, with a constant speed of 50 km h^{-1}.

SOLUTION: The situation, with both cars heading towards O, is shown in Fig. 136. We shall first find the velocity of A relative to B. This means that we treat the whole problem as if B was at rest, and in order to bring B to rest we impose on the whole problem a velocity of 50 km h^{-1} equal and opposite to the velocity of B. (This is just another way of thinking about the method of combining the reversed velocity of B with the true velocity of A.) Thus, relative to B, A now has a velocity of 30 km h^{-1} along PO and a velocity of 50 km h^{-1} along PN parallel to OQ. We can find the resultant of these two velocities either by calculation or by drawing a parallelogram of vectors (velocities) choosing a suitable scale. In Fig. 136 the parallelogram of velocities is $PNRM$, so the magnitude of the resultant velocity of A is given by $\overrightarrow{PR}$, where

$$|\overrightarrow{PR}|^2 = PN^2 + NR^2$$
$$= 50^2 + 30^2$$
$$= 3400$$
$$|\overrightarrow{PR}| = 58·31 \text{ km h}^{-1}$$

The direction of $\overrightarrow{PR}$ is given by θ, where

$$\tan \theta = \tfrac{50}{30} = 1{\cdot}6667$$
$$\theta = 59^\circ\ 02'$$

So the velocity of *A* relative to *B* given by $\overrightarrow{PR}$ is 58·31 km h⁻¹ in a direction 59° 02′ to *PO* as shown in Fig. 136 (*Answer*). Clearly the velocity of *B* relative to *A* is given by $\overrightarrow{RP} = -\overrightarrow{PR}$.

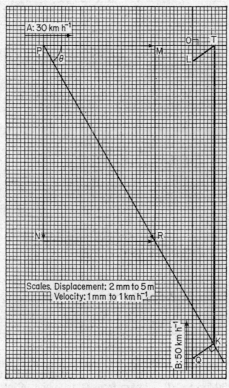

Fig. 136

We do, however, obtain more reward for our method by considering the shortest distance between the two cars in the subsequent motion—for we have reduced *B* to rest at *Q* so that, relatively, *A* now travels in the direction *PR*. Hence, the shortest distance between the cars in the subsequent motion is the perpendicular distance of *Q* from the direction *PR*. Let *QK* be this shortest distance. By measurement from the graph, using the displacement scale of 2 mm to 5 m we obtain *QK* = 15 m to scale. This tells us that the shortest distance apart is 15 m, but it does not tell us where *A* and *B* really are at the moment of closest approach. To obtain the actual position of *A* and *B* at this moment draw *KT* parallel to *QO* and complete the parallelogram *QKTL* to show that when they are nearest *A* is at *T* and *B* is at *L*.

The proof of this last part is as follows. Relative to B, A moves from P to K, and $\overrightarrow{PK} = \overrightarrow{PT} + \overrightarrow{TK}$; but $\overrightarrow{PT}$ is the true displacement of A along the road PO. $\overrightarrow{TK}$ is the displacement due to the reversed velocity of B. Since $\overrightarrow{QK}$ gives the direction and magnitude of the shortest distance apart, all we need do to find the position of B is draw TL parallel to KQ to yield the required position of B at the moment of closest approach. Let us try again with a second example.

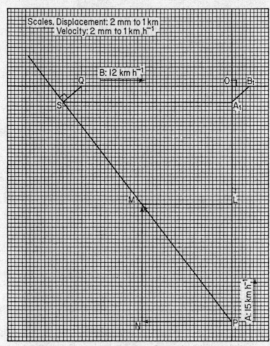

Scales. Displacement: 2 mm to 1 km
Velocity: 2 mm to 1 km h⁻¹

B: 12 km h⁻¹

A: 15 km h⁻¹

Fig. 137

Example: Two ships A and B are sailing towards a point O with the following velocities: A, 15 km h⁻¹ due North; B, 12 km h⁻¹ due East. At the time of observation A is 30 km and B is 20 km from O, the point of intersection of their tracks. Find the velocity of A relative to B and the shortest distance between the ships in the ensuing motion, assuming that the given velocities are maintained.

SOLUTION: Let Fig. 137 represent the problem, with the choice of scales as indicated. At the time of observation let A be at P with $PO = 30$ km, and let B be at Q with $QO = 20$ km. Since we require the velocity of A relative to B we find the resultant of the true velocity of A and the reversed velocity of B, so let $PLMN$ be the velocity diagram. $\overrightarrow{PN}$ represents the reversed velocity of B, and $\overrightarrow{NM}$ represents the true velocity of A. Hence $\overrightarrow{PM}$ represents the velocity of A relative to B. By calculation

$$|\overrightarrow{PM}| = \sqrt{(15^2 + 12^2)} = \sqrt{369} = 19{\cdot}21 \text{ km h}^{-1}$$
$$\tan \angle LPM = \tfrac{12}{15} = 0{\cdot}8$$
$$\angle LPM = 38° \ 39'$$

So the velocity of A relative to B is 19·21 km h⁻¹ in the direction N 38° 39′ W (*Answer*).

If ship B had been anchored at Q we would not have hesitated to say that, as A sailed by along PM produced, the shortest distance would be given by the perpendicular distance from Q to PM produced. But our method of solution has of course reduced B to rest at Q, so we are justified in suggesting that QS is the shortest distance apart. By measurement, the shortest distance between the two ships is $QS = 3$ km. However, we still do not know where the ships are at the time of closest approach, and to find their position we reason as follows. At the moment of closest approach, relative to ship B, the ship A will be at a position given by the displacement $\overrightarrow{QS}$. We therefore require a vector equal to $\overrightarrow{QS}$ joining B on QO to A on PO. To obtain this vector, draw through S a line parallel to QO to intersect PO at A_1. Through A_1 draw $\overrightarrow{A_1B_1} = \overrightarrow{SQ}$ to complete the parallelogram QSA_1B_1. So the ship A is in position A_1 and ship B in position B_1 at the time of closest approach. By measurement, $A_1O = 2$ km, and $OB_1 = 2·5$ km. We observe incidentally that the ship A passes behind ship B.

Each of our worked examples has been simplified by having true velocities which are perpendicular. To experience the difficulties of non-perpendicular directions consider the following example:

Example: Two aeroplanes flying at the same altitude are travelling along straight lines which intersect at O at an angle of 60°. Both aeroplanes are travelling towards O, A at 600 km h⁻¹ and B at 500 km h⁻¹. At the time of observation A is 80 km from O and B is 50 km from O. Find the velocity of A relative to B and the positions of closest approach.

SOLUTION: Let Fig. 138 represent the problem, with scales as indicated. At the moment of observation let A be at P, where $PO = 80$ km, and let B be at Q, where $QO = 50$ km.

To obtain the velocity of A relative to B we find the resultant of the reversed velocity of B and the true velocity of A. Let $PLMN$ be the velocity diagram. $\overrightarrow{PN}$ represents the true velocity of A and $\overrightarrow{PL}$ represents the reversed velocity of B, so $\overrightarrow{PM}$ represents the velocity of A relative to B.

By measurement: $\overrightarrow{PM} = 560$ km h⁻¹, $\angle NPM = 51°$ (*Answer*).

The shortest distance apart is given by $\overrightarrow{QS}$. By measurement $\overrightarrow{QS} = 15·5$ km (*Answer*).

The positions of closest approach are given by A_1 and B_1 where $\overrightarrow{A_1B_1} = \overrightarrow{SQ}$. By measurement: $\overrightarrow{OA_1} = 6$ km, $\overrightarrow{OB_1} = 11·5$ km (*Answer*).

Exercise 48

1. Using the example of Fig. 134, find the shortest distance apart and the position of the cars at that moment if both velocities are reversed.

2. A ship A is steaming due West with a uniform velocity of 20 km h⁻¹. At the same time a ship B is steaming North with a velocity of 25 km h⁻¹. At the time of

observation both ships are steaming towards O, the point of intersection of their tracks, and A is 40 km from O while B is 60 km from O. Find the velocity of A relative to B and the positions of closest approach.

3. Two straight roads intersect at right-angles at O. A car A starts at P, 10 km from O on one road, and travels at 70 km h^{-1} towards O. A car B starts from Q, 5 km from O on the other road, and travels at 50 km h^{-1} towards O. Find the velocity of B relative to A and the positions of closest approach.

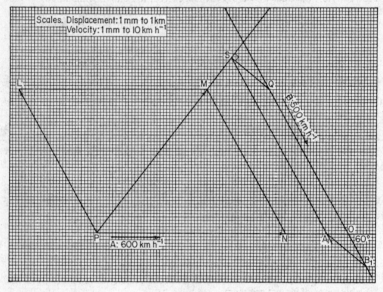

Fig. 138

4. The captain of a fishing boat receives news of a shoal of fish heading South with a velocity of 1 km h^{-1}. His position at the time of the message is 20 km South-west of the shoal. Assuming the fish continue with uniform velocity, in what direction should the boat travel at its maximum speed of 15 km h^{-1} in order to intercept the shoal?

5. In a fairground rifle-range the target moves in a horizontal line at 0·5 m s^{-1}. A customer stands directly opposite the point where the target first appears, i.e. looking at right-angles to the line of motion of the target. In what direction must he point the rifle in order to score a hit if he fires a pellet at 30 m s^{-1}?

6. Using the example of Fig. 138 above, find the positions of closest approach if the speed of the aeroplanes is halved.

7. Two straight roads intersect at O at an angle of 45°. Two cars A and B are each 10 km from O and travelling towards O, A with velocity 80 km h^{-1} directed South-west and B with a velocity of 100 km h^{-1} directed West. Find the velocity of A relative to B and the position of closest approach.

We now use a simple example in order to summarize the ideas we have employed throughout the chapter. Instead of relating the relative velocity to the positions of closest approach only, let us relate the relative velocity to the actual motion by examining the positions at regular intervals.

Example: *OX* and *OY* are two perpendicular straight roads. A woman *W*, 80 m from the crossroads, is walking with a uniform speed of 5 km h⁻¹ towards *O* in direction *OX*. Similarly a man *M*, 360 m from the crossroads, is running with a uniform speed of 10 km h⁻¹ towards *O* in direction *OY*. Show graphically the relation between the velocity of *W* relative to *M* and their actual positions at five regular intervals of 40 m each for *W*.

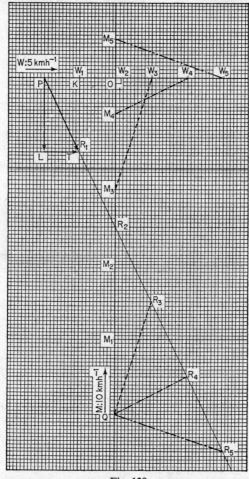

Fig. 139

SOLUTION: Let Fig. 139 represent the problem, with the scales of 1 mm to 4 m for displacement and 2 mm to 1 km h⁻¹ for velocity. At the time of observation let *W* be at *P* and *M* at *Q*, where *PO* = 80 m and *QO* = 360 m. Let *PLTK* be the velocity diagram, with $\overrightarrow{PK}$ representing the velocity of *W*, $\overrightarrow{PL}$ representing the reverse velocity of *M*, and hence $\overrightarrow{PT}$ representing the velocity of *W* relative to *M*. *PT* is produced as shown.

Since the speed of the man *M* is twice the speed of the woman *W* it follows that for each 40 m she walks, he will run 80 m. Therefore when she is at W_1, W_2, W_3, W_4, W_5, he will be at M_1, M_2, M_3, M_4, M_5, respectively.

If the man is kept at rest, then, relatively, the woman appears to travel along PT. During the same distance intervals she therefore appears to be at R_1, R_2, R_3, R_4, and R_5 relative to him.

Consider the positions W_3, M_3, R_3. $\overrightarrow{QR_3}$ represents the displacement vector of the woman relative to the man. But he is really in the position M_3 and she is really in the position W_3, and we see immediately that $\overrightarrow{M_3W_3} = \overrightarrow{QR_3}$. Similarly, $\overrightarrow{M_4W_4} = \overrightarrow{QR_4}$, $\overrightarrow{M_5W_5} = \overrightarrow{QR_5}$.

Exercise 49

1. Using the worked example above, what is their true position when the position of the woman relative to the man is midway between R_1 and R_2?

2. Find the positions of closest approach in Fig. 139.

3. Draw a graph corresponding to Fig. 139 for the positions W_1, W_2, W_3, and W_4, if the man used a cycle and travelled at 15 km h^{-1}. Find the positions of closest approach.

CHAPTER NINE

PROJECTILES

One of the ideas we discussed in Chapter Seven was the replacement of any vector by two perpendicular components, each component being obtained by what was described as *resolving* the vector in the required direction.

To recall the method again, let us examine the results of resolving $\overrightarrow{OP}$ in Fig. 140 in the direction given by the straight line AB.

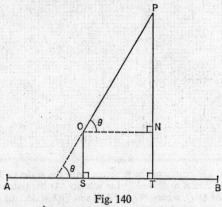

Fig. 140

From each end of $\overrightarrow{OP}$ we draw perpendiculars to AB at S and T. The component of $\overrightarrow{OP}$ in the direction AB is $\overrightarrow{ST}$. To obtain the numerical relation between the vector $\overrightarrow{OP}$ and its component $\overrightarrow{ST}$ we draw ON parallel to AB

and produce *PO* to intersect *AB*. If the angle between *OP* and *AB* is θ, then angle *PON* is also θ. In the right-angled triangle *PON* we have

$$\frac{ON}{OP} = \cos \theta, \ ON = OP \cos \theta$$

Furthermore, the component of $\overrightarrow{OP}$ which is perpendicular to *AB* is given by $\overrightarrow{NP}$ and, once again from triangle *PON*, we have

$$\frac{NP}{OP} = \sin \theta, \ NP = OP \sin \theta$$

(1) Vector Components

Let us suppose that we have a body which is moving through the air with a velocity of 15 m s⁻¹ in a direction 20° above the horizontal at the time of observation. Since velocity is a vector we may resolve the vector into two components: one horizontal and the other vertical. The magnitude of these components can be seen from Fig. 141.

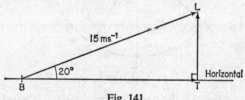

Fig. 141

The horizontal component of velocity is

$$15 \cos 20° = 15 \times 0.9397 = 14.10 \text{ m s}^{-1} \quad \text{(two decimal places)},$$

and the vertical component of velocity is

$$15 \sin 20° = 15 \times 0.342 = 5.13 \text{ m s}^{-1} \quad \text{(two decimal places)}.$$

The components are vectors and we would therefore clearly attach significance to their sign. In Fig. 141 both components were described as 'positive' simply because we have taken, and we continue to take, the directions upward and left to right as positive (in the manner of Cartesian coordinates). The examples of Fig. 142 illustrate other possibilities.

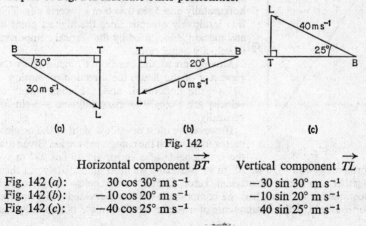

(a)   (b)   (c)

Fig. 142

	Horizontal component $\overrightarrow{BT}$	Vertical component $\overrightarrow{TL}$
Fig. 142 (*a*):	30 cos 30° m s⁻¹	−30 sin 30° m s⁻¹
Fig. 142 (*b*):	−10 cos 20° m s⁻¹	−10 sin 20° m s⁻¹
Fig. 142 (*c*):	−40 cos 25° m s⁻¹	40 sin 25° m s⁻¹

When a body moves along any path, the velocity of the body at any point is in the direction of the tangent at that point (see page 46). A simple illustration is obtained by considering a body swinging by a string in a vertical circle, as shown in Fig. 143. For convenience we shall imagine a horizontal

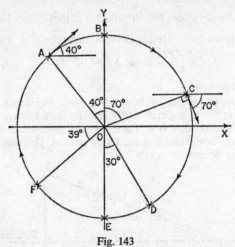

Fig. 143

and a vertical axis through the centre of the circle and labelled OX and OY in the usual manner of Cartesian co-ordinates. We shall suppose that the speed of the body is 5 m s^{-1} and remains constant throughout the clockwise motion. Since the tangent at any point on a circle is perpendicular to the radius to that point, at point A the direction of motion is perpendicular to OA. By the geometry of the figure, the velocity at A is 5 m s^{-1} in the direction 40° to the horizontal. Hence, the components of velocity are 5 cos 40° m s^{-1} horizontally and 5 sin 40° m s^{-1} vertically.

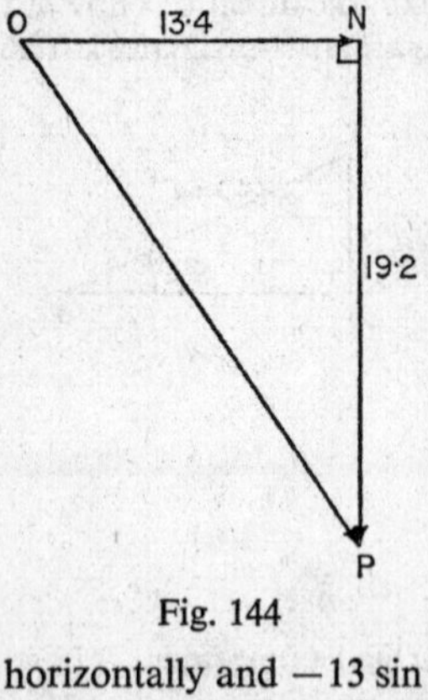

When the body reaches B, the highest point of the circle, the tangent being horizontal means that the components of velocity are 5 cos 0 = 5 m s^{-1} horizontally and 5 sin 0 = 0 m s^{-1} vertically. This was intuitively obvious since the highest point of any motion is indicated by the vertical component of velocity being zero.

Moving on to the position C, again from the geometry of the figure the direction of motion is 70° to the horizontal, and the components of velocity are 5 cos 70° horizontally and -5 sin 70° vertically.

However we must never lose sight of the original vector from which the components arise. Given that the components of a vector are 13 cos 20° m s^{-1} horizontally and -13 sin 20° m s^{-1} vertically we must immediately put them together to produce the original vector, 13 m s^{-1} at 20° below the horizontal. How can we do this when the components are not expressed so obviously? For example, if the components of a vector are 13·4 m s^{-1} horizontally and

Fig. 144

$-19\cdot2$ m s^{-1} vertically, how do we obtain the original vector? As always on these occasions a diagram such as Fig. 144 is very useful.

From the diagram we have $\overrightarrow{OP} = \overrightarrow{ON} + \overrightarrow{NP}$, and since $\triangle ONP$ is a right-angled triangle we apply the theorem of Pythagoras to obtain

$$OP^2 = 13\cdot4^2 + 19\cdot2^2$$
$$= 179\cdot6 + 368\cdot6 = 548\cdot2$$
$$OP = 23\cdot41$$

Since we are finding a vector quantity we must also give a direction, so we find angle *NOP*. Thus

$$\tan \angle NOP = \frac{19\cdot2}{13\cdot4} = 1\cdot433 \text{ (three decimal places)}$$
$$\angle NOP = 55° \ 6'$$

We conclude, therefore, that the original vector was $\overrightarrow{OP}$ of magnitude $23\cdot4$ m s^{-1} and direction $55° \ 6'$ below the horizontal.

Exercise 50

1. Using Fig. 143 find the magnitude of the horizontal and vertical components of the velocity of the body when passing through the points (i) *D*, (ii) *E*, (iii) *F*.

2. If the body in Fig. 143 had been swung round in an anticlockwise direction with a uniform speed of 10 m s^{-1}, what would have been the horizontal and vertical components of the velocity at (i) *C*, (ii) *B*, (iii) *A*?

3. A body travels with a velocity whose horizontal and vertical components are 5 m s^{-1} and 12 m s^{-1} respectively. Find the velocity of the body.

4. A stone is thrown with a velocity of 10 m s^{-1} at an angle of 60° to the horizontal. Find the horizontal and vertical components of the velocity.

5. If the string had snapped at the moment when the body reached the point *A* in Fig. 143, in what direction would the body have moved immediately afterwards?

6. What are the two forces on the body at any time during the circular motion illustrated by Fig. 143? Draw a sketch to show the forces on the body when it is at the point *C*.

(2) Acceleration

Acceleration, being a vector, may also be expressed as the sum of two components, the directions of which will usually be most convenient when at right-angles. But the acceleration a of a body of mass m is related to the resultant force F on the body by the equation $F = ma$, a result which we have discussed at some length in Chapter Five.

Notice that F is the resultant force on the body. Furthermore, since the above equation is also expressed vectorially as $F = ma$, we are reminded that both the vectors a and F have the same direction. Consider the simple case of a body which is projected into the air whose resistance to the motion will be ignored (a most reasonable assumption for the firing of small bullets and the throwing of glass marbles). We see that the only force on the body is its weight. Not only do we know the magnitude of this force but we also know its direction throughout the entire motion, i.e. vertically downwards. To illustrate this point consider Fig. 145 to represent the path described by a steel ball-bearing which is projected into the air with an initial velocity of 100 m s^{-1} at an angle of 60° to the horizontal.

Through O, the point of projection, we draw the now familiar horizontal and vertical axes OX and OY. The initial conditions (i.e. when $t = 0$) are indicated by the tangent line at O being inclined to the horizontal at 60°. At each point of the path such as A, B, C, D, and E, the weight of the body is acting vertically downwards and is the only force on the body. Consequently we have $F = mg = ma$ throughout the entire free motion. In other words, the acceleration of a body projected in a non-resisting atmosphere is always vertically downwards and of magnitude g (see page 59).

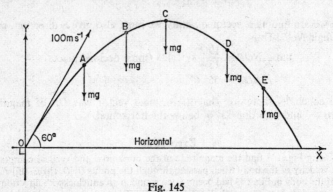

Fig. 145

Returning to the velocity of the body, it must be reasonably obvious that this is changing throughout the motion. Perhaps the fact that the magnitude of the velocity is changing is not too clear, but certainly an inspection of Fig. 145 should convince us that the *direction* of the velocity is always changing. For example, compare the directions of motion (i.e. the direction of the tangent to the curve at the point concerned) at A and D. Moreover, not only is the velocity changing, but so also is the displacement.

Putting together all the relevant vectors in the motion we see that:

(a) Acceleration is constant, being of magnitude g, vertically downwards.
(b) Velocity is not constant, neither in magnitude nor in direction.
(c) Displacement is not constant.

With all this variation, how are we to discuss the motion of a body projected under these conditions? Thinking back to the idea of components of vectors we remember that if we know the components we know the original vectors. With this in mind we decide to consider the displacement, velocity, and acceleration of the body in two parts: one vertical and the other horizontal.

(3) Equations of Motion

Since the acceleration is constant we may use the equations of motion that we deduced in Chapter Four. We recall the following notation and results:

Initial velocity is u
Velocity at time t is v
Acceleration is a
Displacement is s
$v = u + at$; $v^2 = u^2 + 2as$; $s = ut + \frac{1}{2}at^2$

These equations were discussed in terms of motion in a straight line, and we know that the path of a projectile is *not* a straight line. However, there is no difficulty here as we shall be applying the equations to the horizontal and vertical components of the motion separately.

We start by considering a numerical case, that of a body projected freely under gravity only, with an initial velocity of 100 m s⁻¹, at an angle of 60° to the vertical. Its path is represented by Fig. 146. The correct name for such a curve is a **parabola** and the path described by the projectile is called its **trajectory**.

First of all we identify the notation of the question:

	Horizontally	*Vertically*
Displacement:	x	y
Initial velocity (u):	100 cos 60°	100 sin 60°
Velocity at time (v):	v_x	v_y
Acceleration (a):	0	$-g$

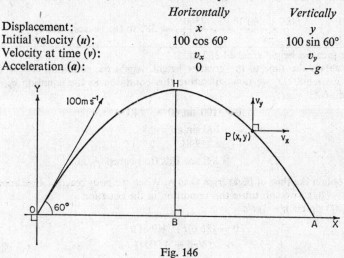

Fig. 146

Thus after time t the body is at P, whose position is given by (x, y) and, the components of the velocity are v_x and v_y. Notice that v_y will be negative since the body is descending at P.

The result of substituting the details above into the equations of motion are as follows:

Motion Horizontally

$v_x = 100 \cos 60°$ (v_x is constant because the horizontal com-
    $= 50$ ponent of the acceleration is zero)
$x = (100 \cos 60°)t$
   $= 50\,t$

Motion Vertically

$v_y = (100 \sin 60°) - gt$ ($g = 9\cdot81$ or $9\cdot80$ m s⁻² as discussed in
$v_y{}^2 = (100 \sin 60°)^2 - 2gy$ Chapter Four)
$y = (100 \sin 60°)t - \tfrac{1}{2}gt^2$

These five equations will tell us all we need to know about the motion. All we have to do is choose the right equations for the information we require. The information we would normally require is 'How high will the projectile rise?' or, to put it mathematically, 'What is the greatest height reached above the point of projection?' To ask 'How far does it go?' is too vague a question because it does not mention the direction concerned. So we ask 'What is the horizontal range of the projectile?' and we would follow this by asking 'When is the horizontal range a maximum?' and so on. We shall obtain the required answers as follows.

To obtain the greatest height. The greatest height is reached when the vertical velocity is zero. Our condition for finding the greatest height is therefore $v_y = 0$.

Using the equation $v_y{}^2 = (100 \sin 60°)^2 - 2gy$ and substituting $v_y = 0$ we find,

$$y = \frac{(100 \sin 60°)^2}{2g} = \frac{7500}{2 \times 9·81} = 382 \text{ m (to nearest metre)}$$

The greatest height reached 382 m.

To obtain the time to the greatest height. Again the condition is $v_y = 0$. Since we require t, we now substitute this condition in the equation $v_y = (100 \sin 60°) - gt$:

$$0 = 100 \sin 60° - 9·81t$$

$$t = \frac{100 \sin 60°}{9·81}$$

$$= 8·8 \text{ seconds (to nearest } \tfrac{1}{10} \text{ s)}$$

To obtain the time of flight from O to A. When the body reaches A we know that $y = 0$, so we substitute this condition in the equation $y = (100 \sin 60°)t - \tfrac{1}{2}gt^2$:

$$0 = (86·6)t - \tfrac{1}{2}(9·81)t^2$$
$$= t(86·6 - 4·905t)$$

This condition is satisfied by two values of t:

$$t = 0 \text{ and } t = \frac{86·6}{4·905} = 17·6 \text{ seconds (to nearest } \tfrac{1}{10} \text{ s)}$$

We recognize this second answer as being twice the time to reach the greatest height—a fairly obvious result in view of the symmetrical look of the path about the vertical line through the point of greatest height. If we go back farther we see that the result comes from

$$t = \frac{100 \sin 60°}{\tfrac{1}{2} \times 9·81} = \frac{200 \sin 60°}{9·81}$$

To obtain the horizontal range. The horizontal range is the distance OA. As in the previous paragraph, the condition that the body has reached A is given by $y = 0$ or, as we have just discovered, by $t = \dfrac{200 \sin 60°}{9·81}$. But we are concerned now with the horizontal displacement x, so we use the equation $x = 50t$ which, on substitution for t, yields

$$x = 50 \times \frac{200 \sin 60°}{9·81}$$

$$= 883 \text{ m (to nearest metre)}$$

Example: A shell is projected with a velocity of 100 m s^{-1} with an elevation of 30° to the horizontal. Obtain the equations of the motion.

SOLUTION: Let Fig. 147 represent the trajectory of the projectile, with horizontal range OA. Take x and y axes as shown.

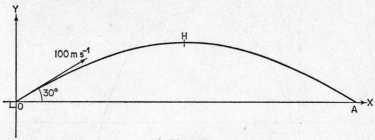

Fig. 147

The equations of motion are as follows:

Horizontal	Vertical
$v_x = 100 \cos 30°$	$v_y = (100 \sin 30°) - gt$
$\quad = 86·6$	$\quad = 50 - 9·81t$
$x = v_x t$	$y = (100 \sin 30°)t - \frac{1}{2}gt^2$
$\quad = 86·6t$	$\quad = 50t - \frac{1}{2} \times 9·81t^2$
	$v_y^2 = (100 \sin 30°)^2 - 2gy$
	$\quad = 2500 - 2 \times 9·81y$

Exercise 51

1. Using the example of Fig. 147 above, calculate the following:

   (i) The greatest height of the projectile.
   (ii) The time of flight from O to A.
   (iii) The horizontal range.
   (iv) The position of the projectile and the components of velocity when $t = 10$ seconds.
   (v) The velocity with which the projectile reaches A.

2. A shell is projected with a velocity of 100 m s^{-1} at an angle of 45° to the horizontal. Calculate the greatest height reached and the horizontal range. Also calculate the velocity of the shell when $t = 1$ second. ($g = 9·80$ m s^{-2})

3. A man and his dog are standing side by side on a horizontal plane when the man throws a ball with velocity 5 m s^{-1} at an angle of 30° to the horizontal. What is the average speed of the dog who sets off immediately and catches the ball just before it hits the ground?

4. Using the example of Fig. 147, find the position of the shell when $t = 11$ seconds.

5. A boy stands on a cliff of height 100 m above sea-level and throws a stone out to sea with an initial velocity of 15 m s^{-1} in a horizontal direction. How long does the stone take to reach the sea? ($g = 9·80$ m s^{-2})

(4) General Equations for Projectile Motion

The numerical examples quoted so far, although giving a more readily accepted feel to the equations of motion, are not the best guide to producing the relations between v, s, and t for *any* trajectory. So we now consider the general equations which apply to all projectiles travelling with unrestricted motion. (This is a fair assumption to make with small heavy bodies at low speeds.)

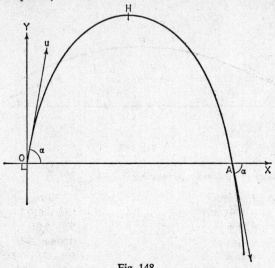

Fig. 148

Suppose a body is projected as shown in Fig. 148 with an initial velocity u at an elevation α. Its equations of motions are as follows.

From $v = u + at$ we have the horizontal component:

$$v_x = u \cos \alpha \tag{i}$$

and the vertical component:

$$v_y = u \sin \alpha - gt \tag{ii}$$

From $s = ut + \frac{1}{2}at^2$ we have the horizontal component:

$$x = ut \cos \alpha \tag{iii}$$

and the vertical component:

$$y = ut \sin \alpha - \frac{1}{2}gt^2 \tag{iv}$$

From $v^2 = u^2 + 2as$ we have the vertical component:

$$v_y^2 = (u \sin \alpha)^2 - 2gy \tag{v}$$

Time of flight for the horizontal range. The horizontal range is OA, and at both O and A we know that $y = 0$. We therefore substitute $y = 0$ in the only two equations which include y, i.e. equations (iv) and (v). This substitution yields

$$0 = ut \sin \alpha - \frac{1}{2}gt^2 \text{ and } v_y^2 = (u \sin \alpha)^2$$

to show that at A, $v_y = \pm u \sin \alpha$ and either $t = 0$ or

$$t = \frac{2u \sin \alpha}{g}$$

Now the condition $y = 0$ which produced these results simply says 'when is the body on the horizontal through O?' Clearly the answer is twofold: at O and A. At O we have, from the initial conditions, that $v_y = u \sin \alpha$ and $t = 0$; so the other solutions we obtained must refer to the body being at A. Thus the time of flight from O to A is

$$t = \frac{2u \sin \alpha}{g}$$

and the vertical component of velocity at A is $v_y = -u \sin \alpha$.

Velocity on impact at A. We just found that $v_y = -u \sin \alpha$ at A and we already know that $v_x = u \cos \alpha$ throughout the motion. It follows, therefore, that the magnitude of the velocity at A is u but its direction is at an angle α below the horizontal (see Fig. 148). By the symmetrical feel of the motion this is just as expected.

Greatest height. At the position of greatest height (indicated by H in Fig. 148) the body has momentarily no vertical velocity. So we apply the condition $v_y = 0$ to equations (ii) and (v) in which v_y appears. The substitution in equation (ii) yields

$$0 = u \sin \alpha - gt$$
$$\therefore t = \frac{u \sin \alpha}{g} \text{ giving the time from } O \text{ to } H$$

Substitution in equation (v) yields

$$0 = (u \sin \alpha)^2 - 2gy$$
$$\therefore y = \frac{(u \sin \alpha)^2}{2g} \text{ giving the greatest height}$$

Comparing the time of flight from O to A with the time from O to H we see a further confirmation of the symmetry of the motion about H, since the body takes the same time to travel from O to H as it does to travel from H to A.

Velocity at any time t. Equations (i) and (ii) give v_x and v_y, the two components of velocity at any time t. The true velocity is v, where

$$v^2 = v_x{}^2 + v_y{}^2$$

Substituting $v_x = u \cos \alpha$ and $v_y = u \sin \alpha - gt$ in this expression we obtain:

$$v^2 = (u \cos \alpha)^2 + (u \sin \alpha - gt)^2$$
$$= u^2 \cos^2 \alpha + u^2 \sin^2 \alpha - 2ugt \sin \alpha + g^2t^2$$
$$= u^2 (\sin^2 \alpha + \cos^2 \alpha) - 2ugt \sin \alpha + g^2t^2$$
$$= u^2 - 2ugt \sin \alpha + g^2t^2$$

The more interesting result arises from equations (i) and (v) which give the velocity at any height y. Thus

$$v_x{}^2 + v_y{}^2 = v^2 = u^2 \cos^2 \alpha + u^2 \sin^2 \alpha - 2gy$$
$$\therefore v^2 = u^2 - 2gy$$

Maximum horizontal range (for a speed of projection u). The time of flight from O to A has already been shown to be $t = \dfrac{2u \sin \alpha}{g}$. We now substitute this time in equation (iii) in order to obtain the horizontal range $x = R = OA$.

$$x = ut \cos \alpha = u \left(\frac{2u \sin \alpha}{g} \right) \cos \alpha = \left(\frac{u^2}{g} \right) (2 \sin \alpha \cos \alpha)$$

Since u is being kept the same it follows that only α, the angle of projection, will influence the horizontal range. Using the trigonometry result $2 \sin \alpha \cos \alpha = \sin 2\alpha$, we may express the range as

$$R = \frac{u^2 \sin 2\alpha}{g}$$

Since the maximum value of $\sin 2\alpha$ is given by $\sin 2\alpha = 1$, i.e. $\alpha = 45°$, we see that the maximum range for a speed of projection u is

$$R = \frac{u^2}{g}$$

and occurs when the angle of projection is 45°.

There are in general **two possible trajectories with the same horizontal range and speed of projection.** The horizontal range is given, as we have just seen, by

$$R = \frac{2u^2 \sin \alpha \cos \alpha}{g}$$

where α is the angle of projection. Suppose we had chosen $(90° - \alpha)$ as the angle of projection. The corresponding range would have been

$$R = \frac{2u^2 \sin (90° - \alpha) \cos (90° - \alpha)}{g} = \frac{2u^2 \cos \alpha \sin \alpha}{g}$$

the same result as before. Therefore we could have projected the body with velocity u at an angle α or $90° - \alpha$ to the horizontal to achieve the same horizontal range. As suggested by Fig. 149, the sum of the two possible angles of

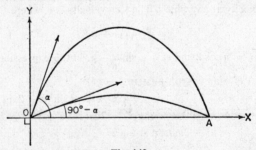

Fig. 149

projection is 90°. Thus angles of projection of 60° and 30° will achieve the same horizontal range from the *same* speed of projection.

The trajectory with $\alpha = 0$. This means that the direction of projection is horizontal, as when throwing stones directly out to sea from the top of a cliff. We can look back to page 170 and amend the equations using the results $\sin 0° = 0$, $\cos 0° = 1$. The amended results are

$$v_x = u \tag{i}$$
$$v_y = -gt \tag{ii}$$
$$x = ut \tag{iii}$$
$$y = -\tfrac{1}{2}gt^2 \tag{iv}$$
$$v_y^2 = -2gy \tag{v}$$

The first thing we notice is that all the vertical vectors are negative. Equation (v) looks strange since it would appear that we shall be trying to find the square root of a negative quantity. This is not so, since the value to be substituted for y will also be negative.

Example: A stone is projected with a horizontal velocity of 10 m s^{-1} from the top of a cliff 50 m above sea-level. Calculate the range at sea-level, and the velocity of impact.

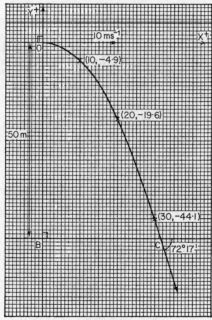

Fig. 150

SOLUTION: Let Fig. 150 represent the trajectory of the stone. Using the equations of motion amended as above, we have:

$$v_x = 10 \tag{i}$$
$$v_y = -9\cdot81t \tag{ii}$$
$$x = 10t \tag{iii}$$
$$y = -4\cdot905t^2 \tag{iv}$$
$$v_y^2 = -19\cdot62y \tag{v}$$

The condition that the stone has reached sea-level is $y = -50$. Observe that since the positive y axis at O is directed vertically upwards, the y co-ordinate of C must be negative.

Substitution in (iv) gives the time of flight from O to C as

$$t^2 = \frac{50}{4 \cdot 905}$$

$$\therefore \ t = \sqrt{10 \cdot 22} = 3 \cdot 197 \text{ seconds}$$

The horizontal distance BC is obtained from eqn. (iii) with $t = 3 \cdot 197$:

$$BC = x = 10t = 31 \cdot 97 \text{ metres} \quad (Answer).$$

The velocity of impact will be obtained from v_x and v_y, either evaluated for $t = 3 \cdot 197$ or simply using equations (i) and (v).

$$v^2 = v_x{}^2 + v_y{}^2 = 10^2 + (19 \cdot 62 \times 50) = 100 + 981$$
$$= 1081$$
$$\therefore \ v = 10 \cdot 4 \text{ m s}^{-1} \quad (Answer).$$

The direction of v is given by

$$\tan \theta = \frac{v_y}{v_x} = \frac{-\sqrt{981}}{10} = -3 \cdot 132.$$

So the stone hits the water surface at an angle of 72° 17′ as indicated on Fig. 150.

Note that we can draw the trajectory shown in Fig. 150 by using equations (iii) and (iv) to give the position (x, y) at any time t. Since we know that the time of flight is only $3 \cdot 197$ seconds we need only find the position of the stone at times $t = 0, 1, 2, 3$ seconds.

$$t = 0, \quad x = 0, \quad y = 0$$
$$t = 1, \quad x = 10, \quad y = -4 \cdot 90 \text{ (three significant figures)}$$
$$t = 2, \quad x = 20, \quad y = -19 \cdot 6 \text{ (three significant figures)}$$
$$t = 3, \quad x = 30, \quad y = -44 \cdot 1 \text{ (three significant figures)}$$

From these results the full trajectory may be drawn.

Example: Find the angle of projection which gives a body a horizontal range of 8 km when projected with a speed of 400 m s^{-1}.

SOLUTION: From the discussion in page 172 we know that there will be two possible angles of projection. The horizontal range is given by

$$R = \frac{u^2 \sin 2\alpha}{g}$$

Substituting $R = 8000$ m, $u = 400$ m s^{-1}, this equation gives

$$8000 = \frac{400 \times 400 \sin 2\alpha}{g}$$

$$\therefore \ \sin 2\alpha = \frac{8000 \times g}{400 \times 400} = \frac{9 \cdot 8}{20} = 0 \cdot 49 \qquad (g = 9 \cdot 8 \text{ m s}^{-2})$$

But $0 \cdot 49 = \sin 29° 20′$ or $\sin 150° 40′$

$$\therefore \ \alpha = 14° 40′ \quad \text{or} \quad 75° 20′ \quad (Answer).$$

Exercise 52

1. Using the example of Fig. 150 above, find the horizontal range at sea-level and the direction of impact if the stone is projected horizontally with velocity 100 m s^{-1}.

2. Find the maximum horizontal range for a body projected with a speed of 300 m s^{-1}.

3. A gun which can fire a shell with a speed of 400 m s⁻¹ in any direction is placed in a bunker on a horizontal plane. Find the maximum area which the gunner can command.

4. A coastguard standing on a cliff 100 m above sea-level is trying to fire a life-line to a boat on the sea below. The firing mechanism fires the line bolt with a velocity of 75 m s⁻¹ horizontally. Calculate how far out the line will reach if 10 m short is allowed for the drag of the line.

5. A jet of water is ejected with a speed of 20 m s⁻¹ from a fire hose inclined at 45° to the horizontal. Calculate how far the water will reach on horizontal ground, assuming no air resistance is offered to the water.

6. At what angle must the hose in Question 5 be inclined in order to direct the same jet of water horizontally through a window at a height of 4·9 m in a vertical wall? Also calculate where the fireman must be standing on the ground. (Hint: 4·9 m is the greatest height of the water.)

7. A body is dropped from rest at a height of $\frac{u^2}{2g}$ above the ground. Calculate the velocity of the body when it is at a height y above the ground.

8. A body is projected with velocity u at an elevation α. Calculate the velocity of the body when it has reached a height y above the point of projection.

9. If you have answered Questions 6 and 7 correctly compare the answers and try to draw a conclusion.

10. A cricketer is standing 50 m from the batsman when he catches him out. He estimates that the ball came to him at an angle of 30° to the horizontal. Calculate the velocity with which the ball left the bat.

11. A diver on a fixed high board runs forward with a velocity of 2 m s⁻¹. How far forward will he strike the water 10 m below and how long will he be in the air? If he now returns to the high springboard 10 m above water-level, how much longer will he have in the air in order to execute a somersault and twist by leaping up with a velocity of 2 m s⁻¹ at 80° to the horizontal?

12. Prove that the trajectory of a projectile which gives the maximum horizontal range R also has a greatest height of $\frac{1}{4}R$.

13. Find the angle of projection which gives a body a horizontal range of 10 km when projected with a speed of 400 m s⁻¹.

(5) Motion Down an Inclined Plane

All the preceding work on projectiles arose from the simple fact that the acceleration of the body was given by g. In order to vary our elementary experience of Newton's equations of motion we shall now consider the motion of a body in which only a component of its weight will be able to contribute to the acceleration of the body. Such a condition is obtained by confining the body to slide down a straight-line groove which offers negligible resistance to the motion. That is, we shall consider the motion to take place over an inclined smooth surface so that the only force on the body, apart from its own weight, is the reaction between the surfaces.

Consider a body sliding down a fixed smooth slope inclined at an angle θ to the horizontal, as shown in Fig. 151. The forces which act on the body are its weight mg vertically downwards, and the reaction N of the surface on the body at the point of contact. Since the contact is smooth the force N must be perpendicular to the surface; in other words, it is a *normal* reaction, hence the use of letter N. (We shall have more to say about such reactions in the next chapter.) By Newton's Third Law the force N is equal and opposite to the force of the body acting on the surface, so we have a force N acting on the plane as shown in Fig. 151 (*a*).

A small gap is shown between the body and the surface, indicating that the two forces N and N act one on the body and one on the plane. Using the geometry of the figure we express the weight mg of the body as two components: $mg \cos \theta$ perpendicular to the plane, and $mg \sin \theta$ along or parallel to the plane. Since we are only interested in the *moving* body, the forces acting on it are shown separately in Fig. 151 (*b*). Finally we suggest com-

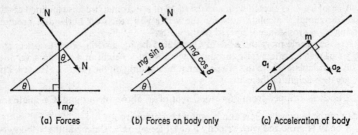

(a) Forces   (b) Forces on body only   (c) Acceleration of body

Fig. 151

ponents of acceleration a_1 and a_2 as shown in Fig. 151 (*c*). We now recall the result of Newton's Second Law in the form $\mathbf{F} = \mathbf{ma}$, a vector equation relating in both direction and magnitude the resultant force $\mathbf{F}$ on the body to its mass-acceleration. But this equation may be applied in any direction, thus

$$F \cos \theta = ma \cos \theta$$

or, to put it another way, the component of $\mathbf{F}$ in any direction is equal to m multiplied by the component of $\mathbf{a}$ in the same direction.

Equating forces to mass acceleration along the plane:

$$mg \sin \theta = ma_1 \qquad \text{(i)}$$

Equating forces to mass acceleration perpendicular to the plane:

$$mg \cos \theta - N = ma_2 \qquad \text{(ii)}$$

Examining eqn. (i) we see that $a_1 = g \sin \theta$.

Giving more thought to eqn. (ii) we realize that, since the inclined plane is fixed, the mass m cannot move perpendicular to the plane: so a_2 must be zero. (Note that if the plane were free to move, a_2 would *not* be zero.) Thus from eqn. (ii) we get $N = mg \cos \theta$.

Let us put this information to use in the following problems.

Example: A particle of mass m is released from a point on a fixed smooth slope inclined at $60°$ to the horizontal. Find the time it takes to travel 2 metres measured along the slope.

SOLUTION: Let Fig. 152 represent the problem, with forces and mass-accelerations shown separately as discussed above.

Equating forces to mass-accelerations we have:

$$mg \sin 60° = ma_1 \qquad \text{(i)}$$
$$mg \cos 60° - N = ma_2 \qquad \text{(ii)}$$

Eqn. (i) gives $a_1 = g \sin 60°$.

Since the inclined plane is fixed, $a_2 = 0$ and eqn. (ii) gives

$$N = mg \cos 60° = \tfrac{1}{2}mg$$

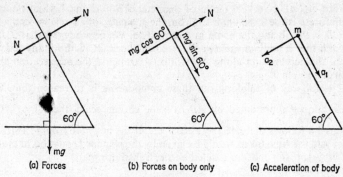

(a) Forces   (b) Forces on body only   (c) Acceleration of body

Fig. 152

The initial conditions of the problem are $t = 0$, $u = 0$, and we are to find t when $s = 2$ m, $a = g \sin 60°$. We therefore substitute in the equation of motion which relates s and t for constant acceleration, i.e.

$$s = ut + \tfrac{1}{2}at^2$$
$$2 = 0 + \tfrac{1}{2}(g \sin 60°)t^2$$
$$t^2 = \frac{4}{9\cdot81 \times 0\cdot866} = 0\cdot47$$
$$\therefore \; t = 0\cdot69 \text{ seconds} \quad (\textit{Answer}).$$

Exercise 53

1. In each of the following cases state the acceleration of a body which slides down a smooth fixed groove which is inclined to the horizontal at an angle of (i) 60°, (ii) 45°, (iii) 30°.

2. A body slides 5 m from rest down a smooth slope of inclination 30° to the horizontal. Calculate the time taken and also the velocity attained.

3. A body is projected along a smooth groove of inclination 30° with an initial velocity of 10 m s^{-1}. Find how far the body will travel up the groove before coming to rest. What is its velocity on returning to the point of projection?

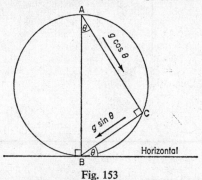

Fig. 153

(6) Motion Down a Chord of a Circle

We close this chapter with one or two observations on the motion suggested by sliding down smooth chords of a vertical circle. We choose chords which either begin at the highest point or end at the lowest point of the circle, the reason being that the results are an unexpected curiosity. In Fig. 153 we have a chord AC of a vertical circle of diameter $AB = d$ and highest point A. (Therefore B is the lowest point.) From the geometry of a circle we know that $\angle ACB = 90°$, being the angle in a semicircle. We now suggest that $\angle BAC = \theta$ and that a body of mass m slides down chord AC without any resisting force. The acceleration along AC will be $g \cos \theta$ and the acceleration along CB will be $g \sin \theta$.

A useful way of thinking out these components is to reason that, with acceleration g represented by $\overrightarrow{AB}$ to some chosen scale, then since $AC = AB \cos \theta$ it follows that $|\overrightarrow{AC}| = g \cos \theta$, etc. If the body fell from rest straight down AB, the time taken would be given by the standard equation of motion $s = ut + \frac{1}{2}at^2$ for uniform acceleration a. Substituting:

$$d = \tfrac{1}{2}gt^2$$

$$t^2 = \frac{2d}{g} \tag{i}$$

Now suppose the body slides down AC, the acceleration is constant and of magnitude $g \cos \theta$. Substitution in the same equation gives

$$s = AC = \tfrac{1}{2}(g \cos \theta)t^2$$

But $AC = d \cos \theta$, therefore $d \cos \theta = \frac{1}{2}gt^2 \cos \theta$ and the time of descent is $t = \sqrt{\dfrac{2d}{g}}$, which is the same as result (i) above. This means that the time of descent from rest at A along *any* chord of the circle is the same.

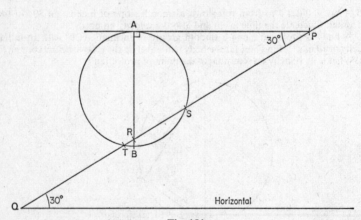

Fig. 154

(7) Lines of Quickest Descent

The above results provide an idea for an interesting exercise called lines of quickest descent. The problem is to find the quickest way of descending in a straight line from a point to a line. For example, imagine that we have to descend in a straight line from the point A in Fig. 154 to the line PQ. In what direction should we travel in order to take the least time?

Following the argument of the previous section, let us draw a circle through *A* as highest point to intersect the line *PQ* in two points *T* and *S*. Just for reference we draw the vertical diameter *AB* of length *d*, say. We proved in the previous section that the time to travel from *A* to *B* is the same as the time to travel from *A* to *S* (or from *A* to *T*), and both of these times must be greater than the time simply to fall from *A* to *R*. Clearly, neither *AS* nor *AT* will give lines of quickest descent. We leave the solution to the reader in the following exercise.

Exercise 54

1. Find the time to slide freely down the chord *CB* in Fig. 153.
2. Using the example of Fig. 154 above, answer the following questions:

   (i) Where is the centre of any circle having *A* as its highest point?
   (ii) Take $AP = 10$ m to a suitable scale, bisect $\angle APR$ and let the bisector intersect *AR* at *O*. Draw the circle centre *O* with radius *OA*. Comment on the result and measure or calculate *OA*.
   (iii) Find the time of quickest descent from *A* to line *PQ* by travelling in a straight line.

3. *ABCDEF* is a regular hexagon of side 2·45 m placed in a vertical plane with *A* as the highest point. A body initially at rest at *A* slides to *B*, where it sets another body in motion from rest. This second body reaches *C*, where it sets in motion a third body. From rest at *C* the third body slides to *D*. How many seconds have passed between the first body starting at *A* and the third body arriving at *D*? (*BC* is vertical.)

CHAPTER TEN

FRICTION

So far we have confined our attention to motion on **smooth** surfaces. A smooth surface can only exert a reaction force which is normal (i.e. perpendicular) to itself, and therefore does not influence the motion assumed to take place across it. No surface is perfectly smooth, even when lubricated, but consideration of such a surface is mathematically convenient for establishing the basic ideas leading to equations of motion. By comparison, therefore, we define a **rough** surface as one which can exert a reaction force which is not normal to itself. To illustrate this point, consider a coin placed on the usual metric wooden ruler resting on a table. Now gradually lift one end of the ruler, keeping the other end steady on the table. The coin remains at rest until the angle of slope of the ruler has become so great that equilibrium is broken and the coin starts to slide. Now what exactly enabled the coin to stay on the ruler? Would any coin remain on the ruler up to this angle of slope? Would a box of drawing-pins stay on for longer? We shall find after experimenting with various objects that everything seems to depend on the type of surfaces in contact, i.e. whether coin on wood, cardboard on wood, wood on wood, and so on. But what of the forces concerned—how do they change during the build up which eventually breaks equilibrium?

(1) Force of Friction

Whatever forces are brought into play at the surfaces in contact will have only one force to balance, and that is the weight of the body on the plane. In Fig. 155 (*a*) we see a body (such as a coin) of mass *m* resting in equilibrium on a plane inclined at an angle θ to the horizontal. The reaction *R* of the inclined plane on the body must be equal and opposite to the weight *mg*. We shall prepare for a study of the motion of the coin down the plane by resolving all the forces concerned into components both along and perpendicular to the plane, as in Fig. 155 (*b*). The weight *mg* is resolved into two components *mg* sin θ and *mg* cos θ. The reaction *R* is resolved into two components *R* sin θ and *R* cos θ.

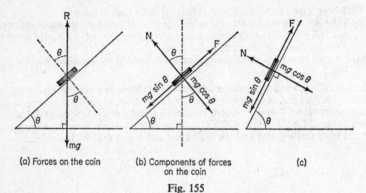

(a) Forces on the coin   (b) Components of forces   (c)
                          on the coin

Fig. 155

We call *R* cos θ the normal reaction *N*, and it is reasonable to assume that such a reaction would exist whether the plane was rough or smooth. A smooth contact, however, would not have supplied the component *R* sin θ, which we now call *F*. Since the body is in equilibrium, we argue that this force *F* must be due to the roughness of the surfaces in contact and we call *F* the **force due to friction,** or simply the **frictional force.** It is this force which stops or resists the body sliding down the slope, i.e. it always opposes any possible motion.

But will *F* always be able to stop the motion? In other words, is there no limit to the magnitude of *F*? We can put this to the test by increasing the angle θ as in Fig. 155 (*c*). By doing so we increase the magnitude of sin θ, and since the body is in equilibrium we now have

$$R \sin \theta = F = mg \sin \theta$$

So we see that *F* must be increasing as θ increases. Eventually the body will begin to move, and when it does it follows that *F* must have reached its limit. When the body is just about to move on the inclined plane we describe it as being in **limiting equilibrium,** and we have now seen that this position is associated with the angle θ.

Exercise 55

1. A body of mass 0·25 kg is placed on a horizontal board, one end of which is then gradually raised in the manner of Fig. 155. If the magnitude of *g* at the place of

experiment is $9 \cdot 80$ m s^{-2}, calculate the forces F and N when the board is inclined to the horizontal at the following angles:

(i) 20°, (ii) 30°, (iii) 45°.

2. If the mass of the body in Question 1 had been 0·5 kg, what would have been the corresponding frictional force F and normal reaction N for the same angles of inclination of the board?

3. A mass of 1 kg is placed on a sheet of glass and begins to slide when the angle of inclination of the glass is 30° to the horizontal. Calculate the frictional force F and normal reaction N at this position. Determine also the direction and magnitude of the resultant of F and N. ($g = 9 \cdot 8$ m s^{-2})

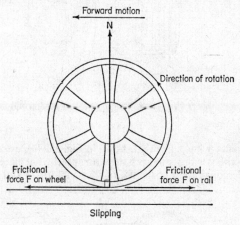

Fig. 156

A practical application of frictional force is commonly seen at airports, where baggage is carried by conveyor belts inclined from one level to another. Needless to say, the angle θ must suit all types of surface-to-surface contact with the variety of suitcases in use.

A typical metal-to-metal contact which generates enormous frictional forces is that between the driving wheels of a train and the rail. Train wheels can often be seen slipping under heavy load when just about to leave a station. Fig. 156 illustrates the forces which apply in this case. As long as the wheel slips on the rail a maximum frictional force will be applied, but as soon as the wheel rolls forward the frictional force required to keep the train moving decreases.

In motor-car brakes a steel disc fixed to the wheel axles is saddled by a pair of special fibre pads. When actuated by the foot-pedal the pads press on the disc in a pincer-like movement (Fig. 157) and exert a considerable frictional force to slow the car down. In the older type of drum-brake (Fig. 158) two curved brake linings are pushed hard against the inside of a drum attached to the wheel, thereby exerting two frictional forces F_1 and F_2. Unfortunately since the drum brakes are enclosed in the drum, they quickly heat up and the change in surface temperature decreases the frictional force which can be applied to the drum, a complaint which is called 'brake fade'. This sometimes happens when the brakes of heavy vehicles are applied for long periods during the descent of a steep hill. Since the disc brake is not housed in a container at all, heat generated by the frictional forces is lost more rapidly, so fading seldom occurs. All high-performance cars are fitted with disc brakes.

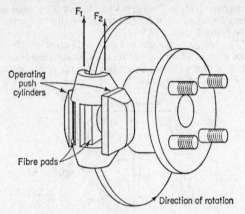

Fig. 157

Even an elementary knot, like that shown in Fig. 159 (*a*), uses the frictional forces at each contact in order to keep the string pinched in place. For an even more striking demonstration of the use of frictional forces and the ease with

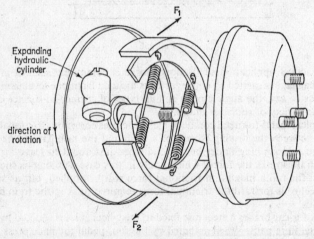

Fig. 158

which they can be called into play, watch the smallest of dockers wrap a rope line from a large liner round a bollard and simply hold on to one end (Fig. 159 (*b*)). The friction increases the tension in the rope as it is wrapped round the bollard, so enabling the docker to hold a load of about 20 000 N by simply wrapping the rope once round the bollard and pulling on the free end with a force of perhaps 200 N.

Friction does not always serve such useful purposes, however, and considerable study is made of the damage it causes by wear. The ordinary car engine experiences a power loss of up to 20 per cent as the result of friction. Since the comparable loss by a turbo-jet is only 2 per cent, one can understand the demise of the piston engine in high-speed aircraft for reasons of friction alone. But with so many important uses for friction we must inquire if there

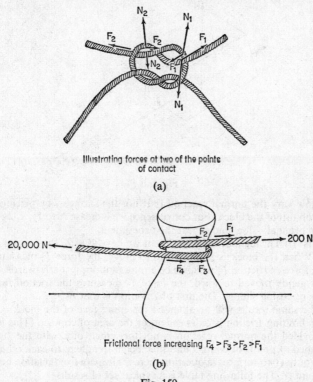

Illustrating forces at two of the points of contact

(a)

Frictional force increasing $F_4 > F_3 > F_2 > F_1$

(b)

Fig. 159

are any laws which we can discover in order to measure and predict its effects. For example, in braking a car the greater the force on the foot-pedal the greater the force of contact of the brake linings, and the sooner we come to a halt: does this mean that doubling the normal force N will also double the frictional force F? Let us try to answer this and similar questions by means of an experiment.

(2) Experimental Investigation of Static Friction

A small block of wood of mass M is placed on a rough horizontal board, and a spring-balance is attached to the block as shown in Fig. 160. The spring is gradually pulled to one side at right-angles to the edge of the block to which it is attached until the block begins to move. Readings are taken from the spring both at the time of just moving and immediately afterwards when the block is being pulled along very slowly. The experiment is repeated about ten times and averages taken of the different readings T_1 for start of motion and T_2 during motion. In order to ensure that the same surfaces are in contact, the block must be returned to its original position each time a reading is taken. The block is now turned over so as to present a different area of the same material in contact with the board, and the experiment is repeated.

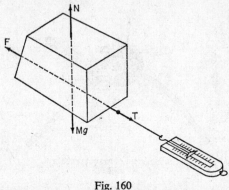

Fig. 160

We now vary the normal reaction (but not the surfaces) by placing extra masses on top of the block. For convenience we increase N to $1\frac{1}{2}$, 2, $2\frac{1}{2}$ and 3 times its original value, and repeat the experiment.

Results: (i) Throughout the experiment we have $F = T$, the tension in the spring. When the block is just starting to move, the force T_1 measures the limiting force of friction F_1 which is opposing motion; just afterwards, as the block is gently moved forward, the force T_2 measures the friction force F_2 which is opposing sliding. The first observation is that $F_2 < F_1$.

(ii) The same results will be obtained for either face of the block, i.e. the force of limiting friction is independent of the area of contact. (This is only true provided the normal reaction is not too great, otherwise the surfaces are crushed at points of contact and they begin to adhere to each other.)

(iii) For the two surfaces in question there is always a fixed relation between F_1, F_2, and N. The following table is a typical set of results:

N newtons	F_1 newtons	F_2 newtons	$\dfrac{F_1}{N}$	$\dfrac{F_2}{N}$
2	0·8	0·7		0·35
3	1·2	1·1		0·37
4	1·7	1·6	0·43	
5	2·1	2·0	0·42	
6	2·5	2·4	0·42	

The results given by $\dfrac{F_1}{N}$ and $\dfrac{F_2}{N}$ are found to be approximately constant for the two surfaces concerned (two significant figures is all that the nature of the experiment will justify). If we now change the horizontal wood board to one of glass, a different result is obtained for $\left(\dfrac{F}{N}\right)$, but nevertheless it will again be approximately constant.

Clearly this relation of limiting frictional force to normal reaction is a very useful means of predicting the frictional force to be overcome in future surface contacts. For example, if we place a block of the same wood of weight 10 N on the same board we shall immediately expect to overcome a maximum frictional force of 4·2 N in order to start the block moving and a frictional force of 3·5 N to keep it sliding slowly.

We call the fraction $\frac{F_1}{N}$ the **coefficient of static friction**, and the fraction $\frac{F_2}{N}$ the **coefficient of sliding** (or kinetic) **friction**. Both these coefficients are referred to by the Greek letter μ (mu). Experiment shows that μ is not a very reliable quantity since it varies according to temperature, humidity, direction of the grain in the wood, whether the surfaces are clean, and so on. Typical values of μ are

Wood on wood	0·2–0·5
Wood on metal	0·2–0·6
Glass on glass	0·7–0·9
Copper on copper	0·8–1·0
Solid sliding on ice	0·02

The results of similar experiments with a variety of materials are summarized in the following **laws of static friction**:

(*a*) The frictional force is in the opposite direction to that in which the body tends to slide.

(*b*) The magnitude of the frictional force is equal to the force tending to cause motion.

(*c*) There is a maximum frictional force which can arise and this is called the limiting frictional force, or limiting friction.

(*d*) The magnitude of the limiting frictional force depends upon the two surfaces in contact and is directly proportional to the normal reaction between the surfaces. The constant of proportion is called the coefficient of friction for the two surfaces concerned and is written as μ. This may be expressed more concisely as $F = \mu N$, where F is the limiting frictional force and N is the normal reaction.

(*e*) The frictional force is independent of the area of surface in contact, if the normal reaction remains unaltered.

(*f*) The frictional force still opposes the sliding if and when it takes place, and remains proportional to the normal reaction between the surfaces and dependent on their nature, but its magnitude is less than the limiting frictional force.

These results and the laws of friction with which we shall be concerned are for the contact of *dry* surfaces only. We shall not be discussing the deeper results of lubrication. The separation of rough surfaces by liquid films reduces the frictional forces considerably. Even water will effectively halve some frictional forces, but unfortunately water evaporates under the heat generated and its lubricating presence is quickly lost. The use of oil, such as castor oil, will reduce ordinary sliding friction dramatically. For example, $\mu = 0·5$ for some surfaces will be reduced to $\mu = 0·05$ by the use of oil, so the frictional force is reduced to one-tenth of its previous magnitude.

Although the values of μ above are claimed for dry contact, the usual conditions which prevail ensure that practically all surfaces are contaminated by a film of dirt-laden moisture, hence the variation in the results obtained for μ.

Example: A glass paperweight has a weight of 10 N and is placed on a horizontal piece of paper. If $\mu = 0.2$ for the two surfaces, calculate the horizontal force necessary to move the paperweight.

SOLUTION: Let Fig. 161 represent the forces on the paperweight.

We have $\dfrac{F}{N} = \mu$, $\therefore F = \mu N$ in limiting equilibrium.

Since the normal reaction is 10 newtons and $\mu = 0.2$, it follows that the limiting frictional force is $F = 0.2 \times 10 = 2$ newtons   (*Answer*).

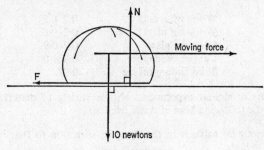

Fig. 161

Exercise 56

1. Complete the table of results on page 184, and calculate the approximate numerical values for F_1 and F_2 when the normal reaction is increased to (i) 8 N, (ii) 12 N.

2. Two blocks of wood A and B are of weight 10 N and 12 N, respectively. If $\mu = 1.4$ for the contact surfaces, calculate the force of limiting friction when (i) A is about to slide on B, (ii) B is about to slide on A.

3. Five copies of the same book, each of weight 4 N, are placed one on top of the other. If $\mu = 0.36$, calculate the horizontal force necessary to remove either (i) the second book up or (ii) the fourth book up, keeping the others in place.

4. Two brake-pads press on a disc with a normal force of 200 N. If the coefficient of friction between pad and disc is 1.2, calculate the resultant frictional force applied to the disc.

During the increase of the tension of the spring in the experiment above, the frictional force increases while the normal reaction remains constant. A diagram such as Fig. 162 which shows this relationship is going to be very useful in future problems, so let us consider it in detail. The resultant of the normal reaction N and the frictional force F is represented by OQ. We represent the increasing tension in the spring by letting T take the values T_1 then T_2, then T_3, and so on, and the corresponding increased values of F are given by F_1, F_2, F_3, and so on. For each value of F we get a different triangle of vectors, but since N (i.e. the weight of the block) is constant, the triangle of vectors for F and N is based on OP to represent N. We have PQ_1 to represent F_1 so that the resultant of F_1 and N is given by OQ_1. Increasing T to T_2 gives OQ_2 as the resultant of N and F_2; and so we go on until eventually, when limiting equilibrium has been reached and motion is about to take place, we have PQ representing the limiting frictional force F and the angle POQ a maximum. This maximum is called the **angle of friction** and is denoted by the Greek

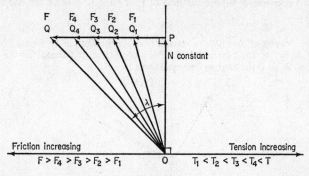

Fig. 162

letter λ (lambda). Using the diagram we see at once that $\tan \lambda = \dfrac{PQ}{OP}$. But PQ represents the limiting frictional force F, and OP represents the normal reaction N, so the resultant reaction of the surface on the block is given by $\overrightarrow{OQ}$. Therefore

$$\mu = \tan \lambda = \frac{F}{N} \text{ or } F = \mu N$$

If we turn back to the situation in Fig. 155 we see a suggestion which enables us to determine the coefficient of static friction without the use of a spring or the need to determine the mass of the body. To find μ between the surfaces of our block of wood and the slide board we only need to raise the board to the greatest angle of inclination possible without motion taking place. As illustrated in Fig. 163 (a), at any time before equilibrium is broken the

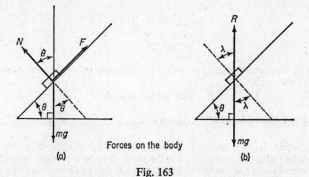

Forces on the body

Fig. 163

three forces N, F, and mg must be in equilibrium; this means that the resultant of N and F must be equal and opposite to the weight mg. In Fig. 163 (b) this resultant is given by R and the geometry of the figure shows that R is inclined at an angle θ to the normal, i.e. the same as the angle of inclination of the slope. Finally, when the block does begin to move, we know that since the body is now in limiting equilibrium the resultant R must be inclined to the normal at the angle of friction λ. Again by geometry $\lambda = \theta$, so the measurement of θ provides a simple method of obtaining $\mu = \tan \lambda$ for the surfaces in question.

Example: A body of mass 0·5 kg rests in limiting equilibrium on a rough plane inclined at 30° to the horizontal. Find the coefficient of friction for the contact surfaces. What is the limiting frictional force?

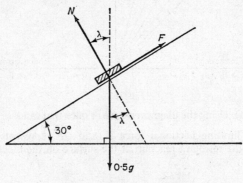

Fig. 164

SOLUTION: Let Fig. 164 represent the problem, with N the normal reaction and F the limiting frictional force (which is directed up the plane because it opposes the possible motion). Since the body is in limiting equilibrium we know that the resultant of N and F must be equal and opposite to the third force (i.e. the weight $0·5g$) and must be inclined at λ (the angle of friction) to the normal reaction N. By the geometry of the figure $\lambda = 30°$ (*Answer*).

This last result now tells us that in the limiting position of equilibrium we have

$$\mu = \tan 30° = \frac{F}{N}$$

hence
$$F = N \tan 30°$$

But we do not need this relation to find F: we merely resolve our forces along the plane to get

$$F = 0·5g \sin 30° = 0·25g$$

where g is 9·80 or 9·81 m s⁻¹   (*Answer*).

Example: A crate is being hauled up a wooden gang-plank of slope 45° by a rope tied to its leading edge and kept parallel to the plank. If the crate weighs 300 N and the coefficient of static friction is 0·4, calculate the force required to start the crate in motion.

SOLUTION: Since the crate is about to move it is in a position of limiting equilibrium. In Fig. 165 (*a*) we see four forces acting on the crate: the normal reaction N, the force of friction F, the tension T in the string, and the weight (300 N) of the crate. There are several alternative methods for finding T which the reader will only appreciate through practice. We shall resolve the forces parallel and perpendicular to the plane as in Fig. 165 (*b*).

Resolving along the plane we have

$$T = F + 300 \sin 45° \qquad \text{(i)}$$

Resolving perpendicular to the plane we have

$$N = 300 \cos 45° \qquad \text{(ii)}$$

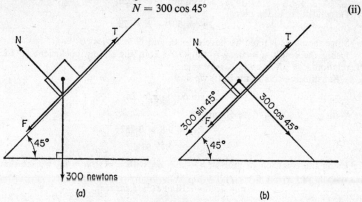

(a)   (b)

Fig. 165

Since the crate is in limiting equilibrium we also have

$$F = \mu N = 0·4N = 120 \cos 45°$$

Substitution of this result in (i) yields

$$T = 120 \cos 45° + 300 \sin 45°$$
$$= 420 \times 0·707$$
$$T = 297 \text{ N} \quad (Answer).$$

Example: A body of mass 2 kg is held at rest on a plane inclined to the horizontal at 60°. If the angle of sliding friction is 45°, calculate the velocity of the body 4 seconds after it is released.

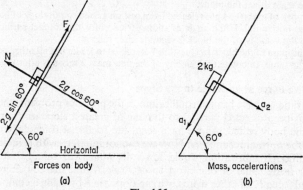

Forces on body   Mass, accelerations

(a)   (b)

Fig. 166

SOLUTION: We know that the body will slide because the angle of inclination of the plane is greater than the angle of friction. Let the forces on the body be represented by Fig. 166. To obtain the velocity we must first find the acceleration. The force of sliding friction F acts along the plane and opposes the motion. The weight $2g$ has been resolved into its two components: $2g \cos 60°$ perpendicular to the plane and $2g \sin 60°$ parallel to the plane. In Fig. 166 (*b*) the accelerations are suggested as a_1 parallel to the plane and a_2 perpendicular to the plane.

We now apply Newton's Second Law and equate forces to mass × acceleration. Along the plane,

$$2g \sin 60° - F = 2a_1 \qquad \text{(i)}$$

Perpendicular to the plane,

$$2g \cos 60° - N = 2a_2 \qquad \text{(ii)}$$

Since the plane is fixed we have $a_2 = 0$, and $N = 2g \cos 60° = g$.
Using the result $\mu = \tan \lambda = \tan 45° = 1$ in the sliding coefficient of friction relation $F = \mu N$, we get $F = N = g$.

Substituting this result in (i) we find

$$2g \sin 60° - g = 2a_1$$

With $g = 9·80$ m s^{-2} we obtain

$$2a_1 = 9·8 \times 0·732$$
$$\therefore \ a_1 = 3·59 \text{ m s}^{-2}$$

Using the equation of motion $v = u + at$, with $u = 0$, $a = a_1$, we see that the velocity at $t = 2$ s is

$$v = 2a_1 = 7·18 \text{ m s}^{-1} \quad (\textit{Answer}).$$

Exercise 57 ($g = 9·80$ m s^{-2})

1. Find the angle of friction when μ is (i) 1, (ii) 0·364, (iii) 1·6.
2. Find the coefficient of friction when λ is (i) 30°, (ii) 60°, (iii) 22°.
3. A block of mass m rests in equilibrium on a rough slope inclined at 30° to the horizontal. Find the normal reaction and the frictional force.
4. If $\mu = 1$ for the contact surfaces in Question 3, to what angle may the slope be slowly increased before motion takes place? Find the limiting frictional force.
5. A crate of weight 300 N rests on a rough plane inclined to the horizontal at an angle of 30°. If $\mu = 2$ for the contact surfaces, calculate the force parallel to the plane which will just begin to move the crate up the plane.
6. In Question 5 calculate the force parallel to the plane which will just begin to move the crate down the plane.
7. A block of mass 0·5 kg is released from rest on a fixed rough slope of inclination 60° to the horizontal. If the angle of sliding friction for the contact surfaces is 45° calculate the velocity of the block 2 seconds after release.
8. If the mass of the block in Question 7 is trebled to 1·5 kg but all other conditions remain the same, calculate the velocity of the new mass 2 seconds after release.

(3) Motive Force at an Angle to the Slope

Each time we have broken equilibrium in the previous problems the motive force has been directed parallel to the line of greatest slope of the plane on which the body rested. As we have seen, the frictional force depends on μ and on the normal reaction N. Now we cannot interfere with μ in our problems (although lubrication would lower its numerical value), but we can consider using some of our effort to decrease N and thereby decrease the frictional force F. In other words, we can decrease F by using a component of the motive force which tends to lift the body off the plane, as the following examples will explain.

Example: A block of mass 2 kg rests on a rough plane of inclination 30°. If the angle of friction is 50°, calculate the least force necessary to move the block up the plane.

SOLUTION: Let Fig. 167 represent the forces in the position of limiting equilibrium. In this position we combine the normal reaction N and the limiting frictional force into the resultant force R acting at 50° (the angle of friction) to the normal. Since we do not know the direction of T we suggest it acts at angle θ to the slope as

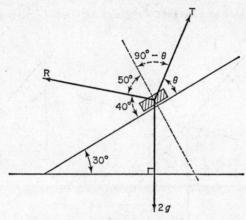

Fig. 167

shown. The object of the problem is to discover the angle θ which makes T the least force possible. Since we have three concurrent forces in equilibrium we apply Lami's Theorem (see page 105) to get

$$\frac{T}{\sin 100°} = \frac{2g}{\sin (140° - θ)}$$

i.e. $(140° - θ)$ is the angle between R and T, hence

$$\therefore \ T = \frac{2g \sin 100°}{\sin (140° - θ)} \qquad \text{(i)}$$

This last result shows that the greater we make $\sin (140° - θ)$ the smaller we make T. But the greatest numerical value for $\sin (140° - θ)$ is 1, and this is obtained when $(140° - θ) = 90°$, i.e. when θ = 50°.

Substituting this result in eqn. (i) we have

$$T = 2g \sin 100° = 2 \times 9·8 \times 0·9848$$
$$= 19·3 \text{ N}$$

So the least possible force to pull the block up the plane is 19·3 N inclined at 50° to the line of slope   (*Answer*).

Example: A block of mass 5 kg is held at rest on a slope inclined at 60° to the horizontal. If the coefficient of friction for the contact surfaces is μ = 0·325, calculate the least force which will *support* the block in equilibrium.

SOLUTION: Here we are only just supporting the block; if we fail to do this, equilibrium will be broken by the block sliding down the plane. So the frictional force is helping us to support the block. The forces on the block are shown in Fig. 168. Since μ = 0·325 we find λ = 18° from tables. As before, the frictional force and the normal reaction have been combined into one resultant force R which acts at 18° to the normal (since the block is in limiting equilibrium). The weight $5g$ acts vertically as usual. We now have the unknown third force P acting through the point of intersection of the other two forces. The direction of P increases the normal reaction which in turn increases the limiting frictional force which helps to support the block on the slope. We shall use the method of the triangle of forces.

Since the three forces are in equilibrium we may represent the forces by the sides of a triangle taken in order. Let $\overrightarrow{OK}$ represent the weight $5g$. SO is the direction of

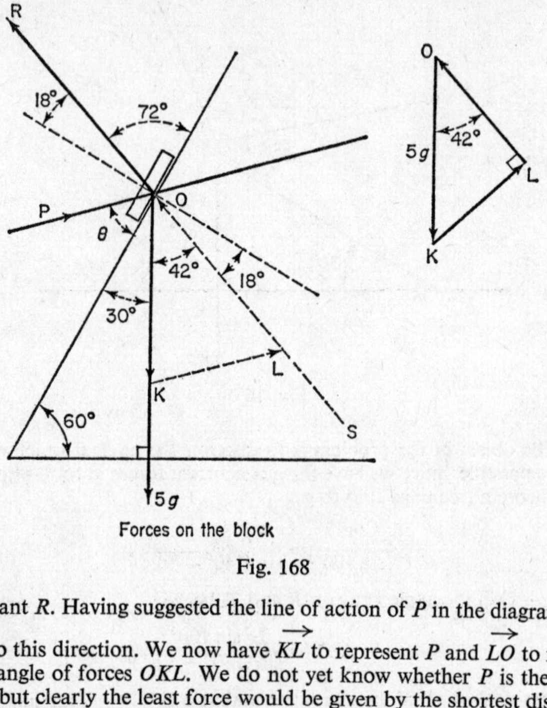

Forces on the block

Fig. 168

the resultant R. Having suggested the line of action of P in the diagram draw KL parallel to this direction. We now have $\overrightarrow{KL}$ to represent P and $\overrightarrow{LO}$ to represent R in the triangle of forces OKL. We do not yet know whether P is the least force possible, but clearly the least force would be given by the shortest distance from K to OS, i.e. the perpendicular distance. The inset triangle of Fig. 168 shows that $KL = OK \sin 42°$ and, on the same scale,

$$P = 5g \sin 42° = 5 \times 9 \cdot 8 \times 0 \cdot 6691$$
$$= 32 \cdot 8 \text{ N} \quad (Answer).$$

Finally, by geometry, with P at right-angles to R we have

$$\theta = \lambda = 18° \quad (Answer).$$

Notice that in both the worked examples above the least force is applied at right-angles to the direction of R the resultant of the normal reaction and limiting frictional force.

Example: Find the least force required to pull a heavy crate of mass 100 kg down a rough gangway inclined at 20° to the horizontal if $\mu = 1 \cdot 6$.

SOLUTION: With $\mu = \tan \lambda = 1 \cdot 6$ we find $\lambda = 58°$. As the crate is in limiting equilibrium we are able once again to combine the normal reaction and the limiting frictional force into the resultant R acting at 58° (the angle of friction) to the normal. Now all we have to do is suggest a direction for the force T with which to pull the crate. Let us suppose this force makes an angle θ with the normal, as shown in Fig. 169.

We can now construct a triangle of forces as in the previous example, with OK representing the weight $100g$ to some chosen scale and with $\overrightarrow{KL}$ representing T. We obtain the least force T when angle $OLK = 90°$. The final solution of this problem is left as an exercise for the reader.

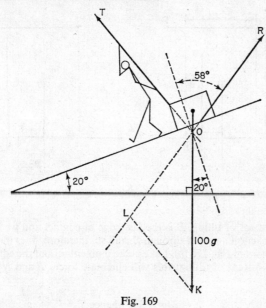

Fig. 169

Exercise 58

1. Complete the example above by finding KL and angle LOK.

2. Use Lami's theorem as an alternative method for solving the example above.

3. In each of the last three worked examples find the magnitude of the least forces concerned when each moving mass is doubled with no change in μ.

4. A crate of mass 200 kg rests on a rough horizontal floor. Calculate the least force necessary to move the crate if $\mu = 1$.

5. Find the least force parallel to the plane which will move a mass of 100 kg down the plane when $\mu = 1.5$ and the angle of slope is 30° to the horizontal.

(4) Toppling before Sliding

We have always assumed that equilibrium will be broken by sliding, and to this end all forces have been drawn concurrent at a point. But, depending on the position of the centre of gravity, there is every possibility that the body might topple over before it slides. For example, consider the block of wood in Fig. 170 being pulled by a horizontal string attached at A. The frictional force opposing the possible motion will be in the opposite direction but will be applied at the point of contact on the surface. The only other forces on the block are its weight Mg acting vertically through the centre of gravity G, which we will assume to be at the geometric centre of the block, and the normal reaction N. With the block in equilibrium we have

$$F = T \text{ by resolving horizontally}$$
$$N = Mg \text{ by resolving vertically}$$

Now the maximum value of F is given by limiting friction where $F = \mu N = \mu Mg$; if we increase T beyond the magnitude μMg the block will slide. On the other hand, the block may topple or tip about the edge through B before it slides.

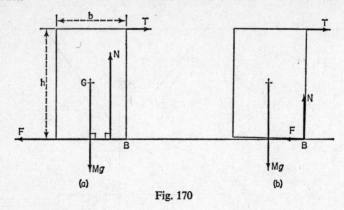

Fig. 170

At the moment of tilting, B is the last edge of contact and we know that the normal reaction must act through B, and so, therefore, does the frictional force F as shown in Fig. 170 (b). Let us take moments about the edge through B just at the instant of tilting; this will eliminate forces F and N which are not required.

$$\therefore\ T \times h = Mg \times \tfrac{1}{2}b, \text{ whence } T = \frac{Mgb}{2h}$$

We now have two results for T: one for sliding and the other for toppling about B.

Suppose we start afresh and, on taking up the string, gradually increase the tension T from zero. If $\mu Mg < \dfrac{Mgb}{2h}$, i.e. if $\mu < \dfrac{b}{2h}$, the tension T will reach the value μMg first—so the block will slide before it topples. But if $\dfrac{b}{2h} < \mu$, the tension T gradually increasing from zero will reach the magnitude $\dfrac{Mgb}{2h}$ first—so the block will topple over about B before it slides.

Example: A block of wood $400 \times 100 \times 50$ mm is placed with one of its smallest faces on a rough inclined plane so that a shortest edge is lowermost and at right-angles to the line of greatest slope. If $\mu = 0.7$, determine whether the block will slide or topple first as the angle of the plane is gradually increased from zero.

SOLUTION: Let Fig. 171 represent the problem, with the plane inclined at the angle θ at the moment of observation. The forces on the block are its weight Mg, the normal reaction N, and the frictional force F. The resultant (R) of F and N is equal and opposite to Mg. The purpose of this remark is to remind us that, since F must intersect the line of action of Mg at P, then N must also act through the same point P. So we are able to fix the point at which the normal reaction must apply, as given in Fig. 171 (a).

As we have already seen, if the body rests freely on the plane it will begin to slide down as soon as the slope of the plane is equal to the angle of friction. But $\tan \lambda = 0.7$, which gives $\lambda = 35°$, so that all we have to establish is whether the block topples over before the inclination of the plane is 35°. As the angle of slope is increased so the point P gets nearer to the lower edge through A, and this lower edge through A is as far as it can go. Immediately the line of action of the weight Mg (the vertical) passes through A the block is about to topple. From the geometry of Fig. 171 (*b*) we see that $\tan \theta = \frac{1}{4}$ at this position. Therefore when $\theta = 14° 02'$ the block will topple—unless it has already slid along the plane, but sliding will only take place when $\theta = 35°$. So the block topples before it slides (*Answer*).

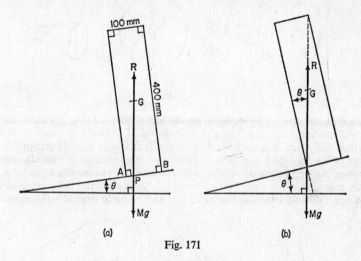

Fig. 171

Exercise 59

1. A uniform solid cube of dimensions $100 \times 100 \times 100$ mm rests on a rough horizontal plane. If $\mu = 0.9$ and the angle of slope of the plane is slowly increased, determine whether the equilibrium of the cube will be broken by sliding first or toppling first.

2. A uniform solid block of dimensions $100 \times 200 \times 50$ mm is placed on a rough plane. If $\lambda = 40°$ determine whether the block slides or topples first as the inclination of the plane is increased from zero with (i) the smallest face of the block in contact, (ii) the largest face of the block in contact.

3. A thick uniform lamina in the shape of an equilateral triangle is placed on a rough plane whose angle of inclination is gradually increased from zero. If the triangle topples before it slides what is the minimum measure for the angle of friction?

(5) Stable and Unstable Equilibrium

Closely associated with the idea of toppling is the question of stability. We all have a vague notion of what stability means, and most people would appreciate that the blunt-nosed cone in Fig. 172 (*a*) is unstable in the sense that a small displacement will cause it to fall over. On the other hand, if we place the cone with its base on a horizontal surface as in Fig. 172 (*b*), a small displacement will not cause it to fall over: if we lift the base slightly at A and then release it, the cone will fall back to its original position.

In Fig. 172 (*c*) we have a bead threaded on a smooth circular wire in a vertical plane. A small displacement either way will set the bead in motion, and the bead will *not* return to its original position of equilibrium. On the other hand, a circular cylinder placed with its curved surface in contact with a horizontal plane will remain wherever it is placed: if the cylinder is at rest and then rolled into a new position, it will stay in this new position. We therefore have three different states of equilibrium.

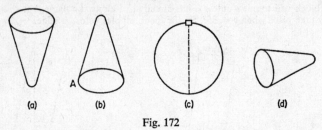

Fig. 172

Bodies are said to be in **stable equilibrium** if they tend to return to their equilibrium position after being slightly displaced. If the body tends to move farther away (i.e. increases the displacement) from the equilibrium position after being disturbed, it is said to have been in a position of **unstable equilibrium**. An egg balanced on one end is an example of unstable equilibrium; such cases are often described as being 'top heavy'. If the body remains in the new position after displacement, it is said to be in **neutral equilibrium**.

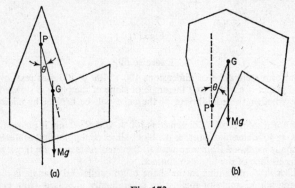

Fig. 173

The blunt-nosed cone illustrates all three states very well. When placed in equilibrium on a horizontal plane, it is unstable on its nose, stable on its base, and neutral on its side. In most cases the type of equilibrium is easy to see and is usually associated with the position of the centre of gravity relative to the point of suspension or support.

The method of finding centre of gravity (see Chapter Two) is illustrated in Fig. 173. The centre of gravity *G* is vertically below the point of support *P*, and if we give the body a small angular displacement θ to one side ('small' means less than 1°) the weight *Mg* has a clockwise moment about *P* which tends to restore the body to its original position. Hence Fig. 173 (*a*) represents a position of stable equilibrium. In Fig. 173 (*b*) the body is in equilibrium with *G* vertically above *P*. A small displacement will produce a moment of *Mg* about *P* which clearly tends to increase the angle θ. So the position in Fig. 173 (*b*) is one of unstable equilibrium.

Exercise 60

State whether the following positions of equilibrium are stable, unstable or neutral:

1. A sphere in a hemispherical bowl.
2. A sphere on a horizontal plane.
3. A heavy body suspended from a fixed point by a string.
4. A lamina with a pivot through its centre of gravity.
5. A uniform cube of wood resting on a rough inclined plane of slope 20° where the angle of friction is 30°.
6. A uniform cube of wood resting on a rough inclined plane of slope 30° where the angle of friction is also 30°.

(6) Freely Hanging Load

In most of the experimental work to determine the coefficient of friction it is convenient to apply forces to the bodies concerned with the aid of pulleys and loaded strings. However, the friction of the pulley and the weight of the string can interfere with the accuracy of the experiment. This is one reason why we used the simpler spring balance to apply force directly to the body in Fig. 160. Nevertheless it is useful to know how pulleys may be used and what assumptions need to be considered in a typical situation.

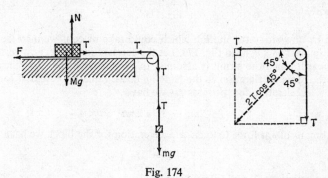

Fig. 174

Suppose, for example, that a block of wood of mass *M* rests on a rough horizontal surface, as shown in Fig. 174. A horizontal force is applied to the block by a light string which passes over a pulley and is then attached to a freely hanging load of mass *m*. (Because the string is light, the tension *T* is the same throughout the string). The hanging mass *m* is at rest under the action of the two forces: *mg* and *T*. It follows that *T* = *mg*.

The block is at rest under the action of four forces: *F*, *N*, *T*, and *Mg*. By resolving perpendicular and parallel to the plane we have

$$N = Mg, \text{ and } F = T$$

If the block just starts to move, i.e. it is in limiting equilibrium, we have

$$F = T = mg$$

But
$$F = \mu N = \mu Mg$$

so we see at once that

$$\mu Mg = mg, \text{ i.e. } \mu = \frac{m}{M}$$

which shows how the apparatus may be used to obtain μ. (We have assumed that no friction applies at the pulley, i.e. the pulley is 'smooth'.)

Finally, the forces on the pulley must be the resultant of T and T, as shown. As usual with two forces of equal magnitude, the resultant bisects the angle between them. Hence the resultant force or thrust on the pulley is $2T \cos 45°$ $= T\sqrt{2} = mg\sqrt{2}$.

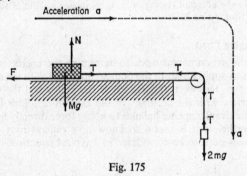

Fig. 175

Let us now study the motion which could take place if we increased the hanging mass to $2m$, as in Fig. 175. Let the acceleration of $2m$ be a vertically downward. Since the string is assumed to be unstretchable, the block will have the same acceleration as the hanging mass. Equating resultant force to mass × acceleration for the mass $2m$ we have

$$2mg - T = 2ma \qquad \text{(i)}$$

Equating resultant force to mass × acceleration for the block we have

$$T - F = Ma \qquad \text{(ii)}$$

Adding equations (i) and (ii) we have

$$2mg - F = a(2m + M) \qquad \text{(iii)}$$

If we now use the fact that $F = \mu N = \mu Mg$, where μ is the coefficient of sliding friction, substitution in (iii) gives

$$2mg - \mu Mg = a(2m + M)$$

which enables us to investigate the ensuing motion.

The following numerical example should help to clarify the above argument, but we shall consider the block to be on an inclined plane just to widen our experience.

Example: A block of mass 6 kg rests on a rough plane of inclination 30° to the horizontal. The block is attached to a 0·5 kg mass by a light unstretchable string which passes over a smooth pulley at the top of the plane and allows the 0·5-kg mass to hang freely. If the block slides down the plane after being released, find its acceleration if the coefficient of sliding friction is $\mu = 0\cdot3$.

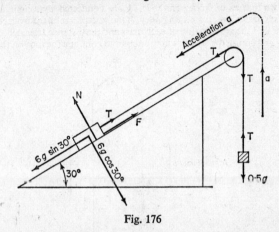

Fig. 176

SOLUTION: Let Fig. 176 represent the problem. As before, the magnitude a of the acceleration is the same for the block and the hanging mass. We note that F acts up the plane directly opposite the direction of motion and that the weight of the block has been resolved into two components: one parallel to the plane and one perpendicular to the plane. Equating resultant force to mass × acceleration we have: for the block of mass 6 kg,

$$6g \sin 30° - T - F = 6a \tag{i}$$

for the mass of 0·5 kg,

$$T - 0\cdot5g = 0\cdot5a \tag{ii}$$

Adding eqn. (i) and eqn. (ii) we have:

$$6g \sin 30° - F - 0\cdot5g = 6\cdot5a \tag{iii}$$

Perpendicular to the plane we have $N = 6g \cos 30°$. Since the block is sliding we know that

$$F = \mu N$$
$$= 0\cdot3 \times 6g \cos 30° = 1\cdot8g \cos 30°$$

which we substitute in eqn. (iii) to get

$$6g \sin 30° - 1\cdot8g \cos 30° - 0\cdot5g = 6\cdot5a$$

or $\qquad 6\cdot5a = g(6 \sin 30° - 1\cdot8 \cos 30° - 0\cdot5)$

With $\sin 30° = 0\cdot5$, $\cos 30° = 0\cdot866$, and $g = 9\cdot80$ m s^{-2}, this last expression gives

$$6\cdot5a = 9\cdot8 (3 - 1\cdot559 - 0\cdot5) = 9\cdot8 (0\cdot941)$$
$$a = \frac{9\cdot8 \times 0\cdot941}{6\cdot5} = 1\cdot4 \text{ (two significant figures)}$$

Therefore the block moves down the plane with an acceleration of 1·4 m s^{-2}
(*Answer*).

What would happen if we increased the inclination of the plane to 90° so as to remove the influence of friction from the problem altogether? In other words, suppose we have *both* masses hanging freely over the pulley as in the following example.

Example: Two masses of 0·5 kg and 0·25 kg are connected by a light inextensible string which passes over a smooth pulley. If the system is released from rest with the string taut, find the acceleration of each mass and the distance travelled in 1 second from rest—assuming the string is long enough to continue the motion for this time.

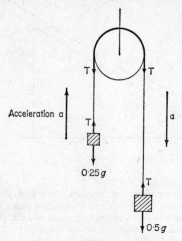

Fig. 177

SOLUTION: Let T be the tension throughout the string, and a the magnitude of the acceleration of each mass. The forces on the separate masses and the pulley are as shown in Fig. 177. Equating resultant force to mass × acceleration (i.e. applying $F = ma$ from Newton's second law) we have:
for mass 0·25 kg

$$T - 0·25g = 0·25a \tag{i}$$

for mass 0·5 kg

$$0·5g - T = 0·5a \tag{ii}$$

Adding eqn. (i) and eqn. (ii):

$$0·5g - 0·25g = 0·75a$$
$$\therefore\ 0·25g = 0·75a$$

i.e. $\qquad\qquad\qquad\qquad a = \tfrac{1}{3}g = 3·27 \text{ m s}^{-1}$ (*Answer*).

Using the equation $s = ut + \tfrac{1}{2}at^2$, we get

$$s = 0 + \tfrac{1}{2} \times 3·27 \times 1^2 = 1·635$$

So the distance travelled in 1 second is 1·635 m (*Answer*).

Exercise 61

(Take $g = 9·80 \text{ m s}^{-2}$ unless told otherwise)

1. Using the example of Fig. 174, find the coefficient of friction for the contact surfaces if a freely hanging mass of 0·1 kg just moves a block of mass 0·5 kg. Find the resultant force on the pulley.

2. Two equal masses of 0·5 kg are connected by a light inextensible string. One mass rests on a rough horizontal plane and the other is held at rest with the string taut over a pulley in the manner of Fig. 175. With the coefficient of sliding friction $\mu = \frac{1}{3}$, find the acceleration of the system when the hanging mass is released. Find also the tension in the string during the motion. ($g = 9·81$ m s^{-2})

3. Given that the body on the horizontal plane in Question 2 is 1 m from the pulley, calculate the time taken for this body to reach the pulley.

4. Two equal masses of 1 kg each are connected by a light inextensible string. The string passes over a smooth pulley fixed to the top of a rough plane inclined at 30° to the horizontal. One mass rests in contact with the plane, the other hangs freely. The system is held at rest with the string taut and then released. If the coefficient of sliding friction is $\mu = 0·5$, find the acceleration of each mass after release.

5. Calculate the acceleration of the masses in Question 4 if the contact surfaces are lubricated so as to reduce the coefficient of sliding friction to $\mu = 0·25$.

6. Two masses of 2 kg and 1 kg are connected by a light inextensible string which passes over a fixed smooth pulley. The masses are released from rest with the string taut. Find the acceleration of each mass and the tension in the string.

7. Two masses of 0·1 kg and 0·3 kg are connected by a light inextensible string which passes over a fixed smooth pulley. Find the acceleration of each mass and the velocity 1 second after release from rest.

8. If the 0·3-kg mass in Question 7 is caught in the hand 1 second after release, for how long will the 0·1-kg mass continue to rise and how much higher will it rise?

9. A woman of mass 50 kg stands in a lift which is moving upwards with an acceleration of 0·1 m s^{-2}. Find the reaction between her feet and the floor.

10. Calculate the reaction in Question 9 if the lift is descending with an acceleration of 0·1 m s^{-2}.

CHAPTER ELEVEN

RIGID BODIES IN EQUILIBRIUM UNDER THE ACTION OF COPLANAR FORCES

Beginners often experience considerable difficulty in applying the principles of moments and resolution to systems of coplanar forces. The difficulty is increased still further when the forces have to be obtained by observing the application of the laws of friction, as well as a general feel for the likely equilibrium position of a rigid body.

(1) Point of Concurrency

Our first concern is with a system of three coplanar forces acting on a body in equilibrium, and we shall describe *either* the system of forces *or* the body as being in equilibrium. Before discussing particular cases let us be quite sure we understand that any three non-parallel forces in equilibrium *must* be concurrent. For example, suppose we have three coplanar forces *P*, *Q*, and *R* in equilibrium. If we find the resultant of *P* and *Q* to be *S* passing through their point of intersection, then we have reduced the number of forces to two, i.e. *S* and *R*, which must be equal and opposite for equilibrium.

This proves that R must pass through the intersection of P and Q and that all three forces are concurrent. We shall now consider applications of this result.

Example: In order to carry a heavy tree-trunk of weight 500 N and length 3 m, two men drive a nail into each end and attach a rope to each nail. They lift the log into a horizontal position with the ropes, which are found to be at 60° to the horizontal at one end and 40° to the horizontal at the other. Calculate the tension in each rope. (We shall also find the position of the centre of gravity, even though the trigonometry is a little involved.)

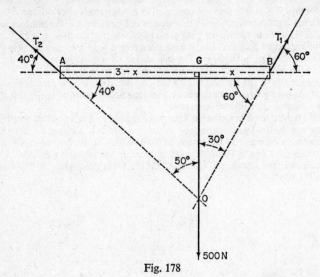

Fig. 178

SOLUTION: The log is in equilibrium under the action of three forces. The forces must be concurrent, so let Fig. 178 represent the problem. We know the direction of the tension T_1 and T_2 but not the line of action of the weight 500 N. However, we do know that it must pass through the intersection O of the forces T_1 and T_2. We draw the vertical line through O to yield the line of action of the weight of the log. The angles in Fig. 178 are found by geometry. We now have three forces acting at the point O, so we can apply Lami's Theorem:

$$\frac{500}{\sin 80°} = \frac{T_2}{\sin 150°} = \frac{T_1}{\sin 130°}$$

Hence,
$$T_1 = \frac{500 \sin 130°}{\sin 80°} = \frac{500 \sin 50°}{\sin 80°} = 389$$

$$T_2 = \frac{500 \sin 150°}{\sin 80°} = \frac{500 \sin 30°}{\sin 80°} = \frac{250}{\sin 80°} = 254$$

So the tensions in the ropes are 389 N and 254 N (*Answer*).

(The reader may wonder how the weight of the log was found to be 500 N when it takes two men to lift it. As we shall see in Question 3 of Exercise 62, we only need a spring balance and a little ingenuity to find the weight of such an object.)

To find the position of the centre of gravity G, we assume it lies somewhere in the log on the vertical line through O. All we know are the angles as indicated and $AB = 3$m.

Let $BG = x$, then $GA = (3 - x)$

$GO = x \tan 60°$ in $\triangle OGB$, and $GO = (3 - x) \tan 40°$ in $\triangle OGA$

Hence $\qquad\qquad x \tan 60° = (3 - x) \tan 40°$

i.e. $\qquad\qquad 1{\cdot}732\, x = (3 - x)\, 0{\cdot}839$

$$2{\cdot}571\, x = 2{\cdot}517$$

$$x = \frac{2{\cdot}517}{2{\cdot}571}$$

$$= 0{\cdot}98$$

So the centre of gravity is $0{\cdot}98$ m from the end B (*Answer*).

Example: A uniform ladder of weight 60 N and length 2 m rests against a vertical shop window. Its base is 1 m from the window and rests in a break in the ground to stop it slipping. If the window it taken as a smooth surface, calculate the force of the ladder on the window. Assume that the centre of gravity of the ladder lies at its midpoint, and that forces lie on a plane at right-angles to the window.

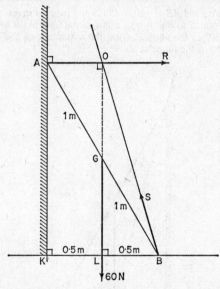

Fig. 179

SOLUTION: Since the window is smooth the forces at the point of contact must be normal to the surfaces. Let the window give a horizontal reaction force R on the ladder as shown in Fig. 179. We also know that the weight acts vertically through the midpoint G of the ladder and thereby intersects the line of action of R at O. Finally, since there are only three forces acting on the ladder and they are in equilibrium, we know that the force S acting at B must also pass through O.

We could use Lami's Theorem for the forces intersecting at O, but we leave this method for the reader in Exercise 62. Since we do not require S we shall take moments about B for the forces acting on the body. Since the forces are in equilibrium, we know that the sum of the moments about any point in the plane will be zero. Alternatively, the sum of the clockwise moments is equal to the sum of the anticlockwise moments about any point in general, including B in particular.

Taking moments about B we have

$$60 \times 0.5 = R \times AK \qquad \text{(i)}$$

We find AK by Pythagoras' Theorem in the right-angled triangle AKB:

$$AK^2 = AB^2 - KB^2 = 4 - 1$$
$$\therefore AK = \sqrt{3} = 1.732$$

Substituting this result in eqn. (i) we get

$$R = \frac{60 \times 0.5}{1.732} = 17.32 \text{ N}$$

Now this is the force of the window on the ladder so it follows from Newton's Third Law that the force of the ladder on the window is a horizontal force of 17·32 N, equal and opposite to R as calculated   (*Answer*).

Note that the ladder would normally be held by a frictional force at the base. In a further example we shall find the increase in R as a man climbs the ladder.

Example: A uniform rod AB of weight 150 N and length 4 m is smoothly hinged to a wall at A and pulled to one side by a horizontal force at the lower end B until AB is inclined at 30° to the vertical. Find the magnitude of the applied force at B and the reaction of the hinge at A.

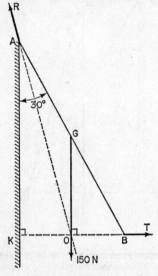

Fig. 180

SOLUTION: The assumption that the hinge is 'smooth' means that the small frictional force which is usually present may be ignored in comparison with the other forces involved. It is reasonable to assume also that the rod hangs in a straight line or that the bending which takes places is small enough be to ignored. Let Fig. 180 represent the problem, with T the horizontal force applied at B, and R the reaction force on the rod at A.

We know that (i) since the rod is uniform, its weight acts through the midpoint G; (ii) since the force T is horizontal, it intersects the line of action of the weight at O so that $\angle GOB = 90°$; (iii) since the rod is in equilibrium under the action of three forces (which are therefore concurrent) the line of action of R must pass through O.

We shall use a triangle of forces to demonstrate a third method of solution. This means constructing a triangle, using an appropriate scale, such that the three sides of the triangle taken in order represent in magnitude and direction the three forces in equilibrium. Before constructing any triangle of forces it is a good idea to see if there is already one available. The triangle AKO in Fig. 180 will serve our purpose, because $\overrightarrow{AK}$ may represent the weight 150 N vertically downwards, $\overrightarrow{KO}$ may represent T, and $\overrightarrow{OA}$ may represent R. All we need to do now is find the lengths of the sides of $\triangle AKO$. Since $\angle KAB = 30°$ and $AB = 4$ m, we have $AK = 4 \cos 30°$.

But,
$$KO = \tfrac{1}{2}KB = \tfrac{1}{2} \times 4 \sin 30°$$
$$AK = 3\cdot464, KO = 1$$

Using Pythagoras' Theorem we have
$$AO^2 = AK^2 + KO^2 = 12 + 1 = 13$$
$$AO = 3\cdot606$$

Since AK represents 150 N, KO represents
$$\frac{150}{3\cdot464} = 43\cdot3 \text{ N}$$

and OA represents
$$\frac{150}{3\cdot464} \times 3\cdot606 \text{ N} = 156\cdot1 \text{ N}$$

The direction of R is given by $\angle KOA$
In the right-angled triangle KAO we have
$$\tan \angle KOA = \frac{AK}{OK} = 3\cdot464$$
$$\therefore \angle KOA = 73° 54'$$

So T is a horizontal force of $43\cdot3$ N, and the force R on the rod due to the hinge is $156\cdot1$ at $73° 54'$ to the horizontal   (*Answer*).

Exercise 62

1. A uniform steel girder of length 3 m and weight 1000 N is held horizontally by two steel ropes attached to the hook of a crane hoist. If the ropes are each 3 m long and are fixed to the two ends of the girder, calculate the tension in each rope.

2. A uniform bar is suspended horizontally by two light strings, one at each end. Say why both strings must be equally inclined to the horizontal.

3. A spring balance which only registers up to 150 N is being used to find the weight (about 250 N) of a non-uniform log of length 2·6 m. A nail is driven into each end of the log. Leaving one end A on the ground, the other end B is lifted vertically about 10 mm by the spring balance hooked on to the nail. The reading is noted and the procedure repeated at the other end. If the reading at A is 140 N and the reading at B is 120 N, find the weight of the log and the position of its centre of gravity.

4. Using Fig. 179 state which triangle, already drawn, may be used as a triangle of forces. State which vectors represent the forces R and S.

5. Complete each of the following equations for the forces in Fig. 179:

(i) Resolving horizontally, $S \cos \angle OBL = ?$
(ii) Resolving vertically, $S \sin \angle OBL = ?$

6. A uniform pipe AB, of length 4 m and weight 100 N, rests with one end on a rough horizontal floor and the other end against a smooth tiled vertical wall. If the angle of inclination of the pipe is 45°, calculate the force on the pipe at the end A on the wall and the direction of the force at the end B.

7. Using the example of Fig. 180 above, complete each of the following equations:

(i) Resolving vertically, $R \cos \angle KAO = ?$
(ii) Resolving horizontally, $T = ?$
(iii) Taking moments about A, $150 \times KO = ?$

8. Calculate the force T necessary to pull the rod in Fig. 180 to an angle of inclination 45° to the vertical.

9. A uniform beam AB has a length of 2 m and a weight of 100 N. It is smoothly hinged to a vertical wall at A and supported in the horizontal position by a cable joining B to a point C on the wall 2 m above A. Assuming that the cable rests in a straight line CB, calculate the tension at B and the magnitude and direction of the force on the beam at A.

(2) Force Normal to a Surface

Whenever two bodies are in smooth contact we know that the action and reaction forces are normal to both surfaces at the point of contact. The elementary nature of our work confines us to one of these surfaces being a plane so that the normal is easily identified. Whether the contact is smooth or not, it is useful to obtain the normal at the point of contact, especially when considering positions of limiting equilibrium, in order to indicate the position of the angle of friction. For the moment we shall concentrate on identifying the direction of the normal forces and we shall leave positions of limiting equilibrium for later. The circle provides the simplest curved surface because we know that the normal at any point passes through the centre.

Example: A rough uniform rod of length 0·8 m and weight 20 N rests with one end on a rough horizontal floor. A point on the rod rests against the smooth curved surface of a fixed hemispherical cylinder. If the radius of the curved surface is 0·2 m and the rod rests inclined at 20° to the horizontal in a plane perpendicular to the axis of the cylinder, calculate the magnitude of the reaction between the cylinder and the rod.

SOLUTION: Let Fig. 181 represent the problem, with the rod in contact with the curved face at D. Since the contact at D is smooth, we know that the force on the rod is normal to the surfaces at the point of contact. The line of action of this normal force N is CD and intersects the line of action of the weight at O. The midpoint of the rod is G.

Since the rod is in equilibrium under the action of three forces, it follows that the third force at A must also act through O. Since we only need to find N we must try if possible to produce some equation or equations which do not involve R. The simple way to do this is to take moments about a point on the line of action of R, and clearly A is the most convenient point to choose.

Taking moments about A we have

$$20 \times KA = N \times DA \qquad \text{(i)}$$

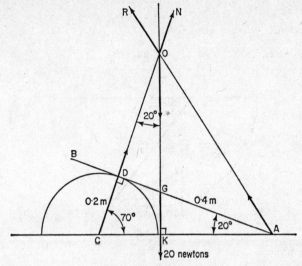

Fig. 181

From the right-angled triangle GKA we have $KA = 0.4 \cos 20°$
From the right-angled triangle CDA we have $DA = 0.2 \tan 70°$
Substitution of these results in eqn. (i) yields

$$20 \times 0.4 \cos 20° = N \times 0.2 \tan 70°$$

$$N = \frac{40 \cos 20°}{\tan 70°}$$

We may now use logarithms to evaluate this result in the usual manner, or we may use the fact that

$$\frac{1}{\tan 70°} = \tan 20° = \frac{\sin 20°}{\cos 20°}$$

so that

$$N = 40 \sin 20° = 40 \times 0.342$$

Hence

$$N = 13.68 \text{ N} \quad (\textit{Answer}).$$

Example: A rough uniform rod of length 200 mm and weight 30 N rests inside a smooth hemispherical bowl of radius 150 mm which is symmetrically fixed with its rim in a horizontal plane. If the rod rests at an inclination of 20° to the horizontal with the upper end A in contact with the smooth surface and the lower end B resting on a piece of rough glass-paper stuck to the inside of the bowl, calculate the reaction at A on the rod assuming that the forces lie in the same vertical plane as the centre of the bowl.

SOLUTION: Let Fig. 182 represent the problem, with N the normal reaction at A acting through the centre C and intersecting the line of action of the weight at O, with G the midpoint of AB. The third force keeping the rod in equilibrium arises from the contact at B and must therefore pass through O.

Again we shall take moments about B because we do not require the force R. We shall, however, require the perpendicular distance from B to the line of action of N. This distance is $AB \sin \angle BAC$, and since we already have $AB = 200$ mm we only need to find $\angle BAC$. We know from geometry that, since C is the centre of the circle and G the midpoint of the chord, $\angle CGA = 90°$ and $\triangle CGA$ is a right-angled triangle.

Therefore $$\cos \angle GAC = \frac{AG}{AC} = \frac{100}{150} = \frac{2}{3}.$$

From tables, $\angle GAC = 48° \ 12' = \angle BAC$.

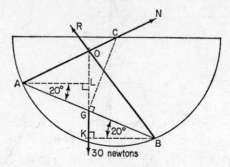

Fig. 182

Taking moments about B we have

$$N \times 200 \sin 48° \ 12' = 30 \times 100 \cos 20°$$

$$\therefore N = \frac{15 \cos 20°}{\sin 48° \ 12'} = 18\cdot91$$

So the reaction on the rod at A is 18·91 N in the direction AC (*Answer*).

Example: A garden roller is a uniform circular cylinder of weight 400 N and cross-section radius 0·25 m. Calculate the minimum force applied at the axle of the cylinder which will pull the roller over a horizontal step of height 0·1 m.

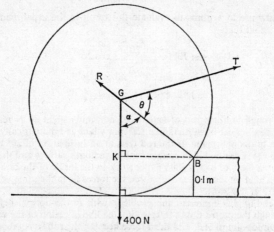

Fig. 183

SOLUTION: Let Fig. 183 represent the problem in cross-section, with the contact line represented by B. At the instant of 'lift off' the roller is no longer in contact with the ground. The only force we know for certain is the weight of the cylinder acting through G. The applied force also acts on the axle through G, but in a direction we have yet to determine in order that it be a minimum.

The reaction R at B is the third and final force on the roller and must pass through G. Since we do not require R we shall take moments about B to get

$$T \times GB \sin \theta = 400 \times GB \sin \alpha$$

$$\therefore T = \frac{400 \sin \alpha}{\sin \theta} \tag{i}$$

Examining this result we see that, since α is fixed by the dimensions of the roller and step, the minimum magnitude of T will depend on making $\sin \theta$ as great as possible. This is done by making $\theta = 90°$ in which case

$$T = 400 \sin \alpha$$

We find $\cos \alpha$ from the right-angled triangle GKB

Thus $$\cos \alpha = \frac{GK}{GB} = \frac{0 \cdot 15}{0 \cdot 25} = 0 \cdot 6$$

$$\therefore \alpha = 53° \ 8', \text{ and } \sin \alpha = 0 \cdot 8$$

(This result could have been obtained by realizing that $\triangle GKB$ is a '3, 4, 5' triangle.) Putting $\sin \alpha = 0 \cdot 8$ in eqn. (i) we have $T = 400 \times 0 \cdot 8 = 320$ N   (*Answer*).

Exercise 63

1. Using the example of Fig. 181, complete the following statements:

   Resolving horizontally, (i) $R \cos \angle OAC = $?
   Resolving vertically, (ii) $R \sin \angle OAC + $? $= 20$

2. Calculate the force N in Fig. 181 if the rod had rested at 15° to the horizontal instead of 20°.

3. A rough uniform rod 1 m long and weight 20 N rests at an angle of 45° to the horizontal with one end on rough ground and a point of its length against a smooth peg 0·4 m above ground level. Calculate the force on the peg.

4. Using the example of Fig. 182, complete the following statements:

   Resolving horizontally, (i) $R \cos \angle OBK = $?
   Resolving vertically, (ii) $R \sin $? $+ N \sin $? $= $?

5. A smooth rod 100 mm long and weight 10 N rests symmetrically in a hemispherical glass bowl of radius 100 mm with its rim in a horizontal plane. Calculate the reaction at each end of the rod.

6. Using the example of Fig. 183, complete the following statement:

$$\frac{400}{\sin (180° - \theta)} = \frac{R}{\sin \ ?} = \frac{T}{?}$$

Hence find R.

7. Using the example of Fig. 183, calculate the least force necessary to lift the same roller over a step which is 0·125 m high. (Note: $GB = 2GK$)

8. A uniform solid right circular cylinder of weight 100 N with a cross-section of radius of 0·2 m rests with its curved surface on a smooth slope which is inclined at 30° to the horizontal. The cylinder is supported by a block with the face of contact with the cylinder perpendicular to the slope and of height greater than 0·2 m. Find all the forces on the cylinder.

(3) Frictional Force and Normal Reaction

In each of the last three examples the equilibrium positions have been indicated without reference to the possibility of limiting friction. What interests us in limiting equilibrium is that we are able to find the extreme positions of equilibrium in answers to such questions as how far we can tilt a rod, what is the greatest load a beam can bear, or how far up a ladder a cleaner can climb. Our examples for the moment will still be three-force problems and, as before, the forces arise from reactions at points of contact, weights, and applied forces via strings and ropes.

Example: A rough uniform ladder AB, 4 m long and weighing 100 N, rests with A against a vertical wall which may be assumed smooth. The end B rests on rough horizontal ground with the contact surfaces yielding $\mu = 0.5$. Calculate the smallest possible angle of inclination of the ladder to the horizontal.

SOLUTION: As usual we know the line of action of two of the forces and their point of intersection. This tells us the direction of the third force. At A the reaction H on the ladder is horizontal since the wall is smooth. The weight of the ladder acts through G, the midpoint of AB, because the ladder is uniform. Hence the resultant reaction R at B passes through O.

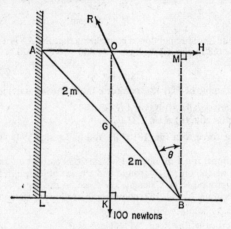

Fig. 184

As yet we do not know the inclination of the ladder given by $\angle ABL$; however, we do know that the farther B is from the wall the greater becomes the angle between the force R and the normal at B. But the greatest possible angle θ is given by λ, the angle of friction for the surfaces in contact at B. Therefore, in the limiting position of equilibrium, $\tan \angle OBM = 0.5$ or $OM = \frac{1}{2}BM$. But O is the midpoint of AM (because G is the midpoint of AB, and GO is parallel to BM) so $AM = BM$.

Hence $ALBM$ is a square, and $\angle ABL = 45°$ (*Answer*).

Example: A heavy uniform floor joist AB, 5 m long and weighing 300 N, is dragged along rough ground by a rope tied to the end A. If the joist is kept just clear of the ground at A, and the coefficient of sliding friction is 0.6 at the end B, calculate the tension and direction of the rope.

SOLUTION: We assume that the joist is horizontal and in contact with the ground at B only, as shown in Fig. 185. We know that the weight acts through the midpoint G and that the resultant reaction at B is inclined at λ (the angle of friction) to the normal at B. These two forces intersect at O and thus indicate that the direction of the rope is along OA. The rest of the solution is left as an exercise for the reader (see Question 3 in Exercise 64).

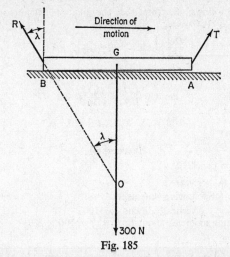

Fig. 185

Example: A uniform rough solid right circular cylinder of cross-sectional radius 0·3 m and weight W is rolled up a rough slope by applying a tangential force to the cylinder parallel to the slope. If $\mu = 0·4$ for the contact surfaces, calculate the greatest possible angle of inclination of the slope along which the cylinder may be rolled by this method.

SOLUTION: Let T be the tangential force parallel to the slope intersecting the line of action of the weight at O. The third force R on the cylinder must act through the contact point C and through the point O, as indicated in Fig. 186.

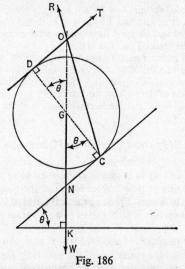

Fig. 186

Let the slope of the plane by θ. We are thus trying to find the greatest possible value of θ. We can see from the figure that if we increase the slope we increase the length DO where CGD is a diameter of the circular cross-section. This in turn increases the angle DCO which we know cannot be increased beyond λ (the angle of friction). It follows, therefore, that the greatest inclination of the plane is achieved when $\angle DCO = \lambda$.

$$\frac{DO}{DC} = \tan \lambda = 0{\cdot}4 \text{ (as given)}$$

$$\therefore \ DO = 0{\cdot}4 \ DC$$

But $$\tan \theta = \frac{DO}{DG} = \frac{DO}{\frac{1}{2}DC} = 0{\cdot}8$$

Hence $$\theta = 38° \ 40' \quad (Answer).$$

Exercise 64

1. In Fig. 184 on page 210

    (a) Find the angle *OBM*
    (b) Complete the following statements:
        (i) Resolving vertically, $R \cos \ ? = 100$
        (ii) Taking moments about B, $H \times \ ? = 100 \times KB$
    (c) Find R and H

2. A rough uniform ladder of length 3 m and weight 150 N rests in limiting equilibrium at 60° to the horizontal with one end against a smooth vertical wall and the other on rough horizontal ground. Find the angle of friction and the forces on the ladder.

3. Using Fig. 185 on page 211

    (a) Find $\angle GOA$
    (b) Resolve horizontally for the forces on the log, and prove that $R = T$
    (c) Resolve vertically to find R

4. Substituting $\theta = 38° \ 40'$ and $\lambda = 21° \ 48'$ in Fig. 186 above, find R by resolving perpendicular to the plane, and find T by taking moments about C.

5. Which triangle in Fig. 186 could have been used as a triangle of forces? Which sides correspond to the forces T, R, and W?

6. A rough uniform beam of length 4 m and weight 200 N rests with one end on a rough horizontal floor and the other against a rough vertical wall. If the beam is in limiting equilibrium at the wall (where $\lambda = 30°$) when inclined at 60° to the horizontal, find the reaction of the wall on the beam.

7. Find the frictional force and the normal reaction at the B in (i) Fig. 184, (ii) Fig. 185.

(4) Equilibrium of a Body under the Action of More than Three Forces

One very obvious way of dealing with such a system has been discussed in Chapter Seven. If the body is in equilibrium, then so is the system of forces: indeed, all we are doing in this chapter is to use the rigid body as an excuse for creating a system of forces. From a practical point of view we like to regard forces as acting on something, rather than as an abstract system as in Chapter Seven.

We could find the resultant of two forces at a time, thus gradually reducing the number of forces in the system until we have only three forces left. These three forces being in equilibrium must be concurrent, a situation with which we are now very familiar. But we have been working towards the much more general method of resolving the forces and taking moments in order to free ourselves from the tedium of reducing all systems to three forces.

The general conditions for equilibrium of coplanar forces are as follows:

(*a*) The sum of the components of the forces in any direction is zero. For example, if we resolve the forces horizontally then the sum of the components in one direction (say, left to right) is equal to the sum of the components in the opposite direction. We must be able to do this in any direction.

(*b*) The sum of the moments of the forces in the system about any point is zero. Alternatively, the sum of the clockwise moments is equal to the sum of the anticlockwise moments.

Either of the above conditions will tell us that the system is in equilibrium. Usually we already know that the body is in equilibrium, so we put the sum of the components in one direction equal to the sum of the components in the opposite direction. Similarly we may put the sum of the clockwise moments equal to the sum of the anticlockwise moments about any point. The equations which these conditions produce enable us to obtain the unknown forces or distances.

Example: A window-cleaner rests his uniform ladder of length 2 m and weight 60 N against a vertical plate-glass window in a plane at right-angles to the window. If his weight is 500 N, calculate the thrust on the window if he stands two-thirds of the way up the ladder without it slipping when it is inclined at 70° to the horizontal.

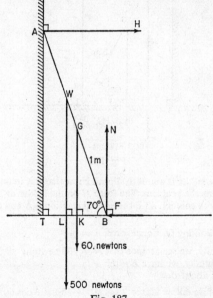

Fig. 187

SOLUTION: Let Fig. 187 represent the problem, with the ladder *AB* resting on the window at *A*, and the window-cleaner standing at *W*. Instead of the one resultant force at *B* it is more convenient to use its two components: the normal reaction *N*, and the frictional force *F* opposing the motion of *B*. This will be easier when it comes to resolving horizontally and vertically for the coplanar system of forces. Now we could have combined *N* and *F* into a resultant force *R*, then found the resultant of the two parallel forces of 500 N and 60 N, and finally applied the usual method for the concurrency of three forces in equilibrium. However, when we examine the question we observe that we do not want either N or F; so let us use a method which avoids finding them.

Taking moments about B we have

$$60 \times 1 \cos 70° + 500 \times \tfrac{4}{3} \cos 70° = H \times 2 \sin 70°$$

$$H = \frac{726\tfrac{2}{3} \cos 70°}{2 \sin 70°} = 363\tfrac{1}{3} \tan 20° = 132\cdot3$$

So the thrust on the window is 132·3 newtons (*Answer*).

Example: A ladder of length 4 m and weight 200 N rests at 60° to the horizontal with the upper end on a smooth wall and the lower end on rough ground. A man of weight 500 N has ascended three-quarters of the length of the ladder, whereupon it is just about to slip. Find the coefficient of friction between the base of the ladder and the ground.

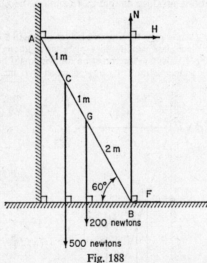

Fig. 188

SOLUTION: Since the ladder is about to slip we know that it is in limiting equilibrium. Fig. 188 represents the problem. The force H on the ladder at A is horizontal because of the smooth contact. At B the normal reaction is N and the frictional force is F.

Since F is the limiting frictional force, $\mu = \dfrac{F}{N}$

Resolving vertically, we see at once that $N = 700$ newtons
Resolving horizontally, we have $F = H$
Taking moments about B,

$$H \times 4 \sin 60° = 200 \times 2 \cos 60° + 500 \times 3 \cos 60°$$

$$4H \sin 60° = 1900 \cos 60°$$

$$H = \frac{475 \cos 60°}{\sin 60°} = \frac{475 \sin 30°}{\cos 30°} = 475 \tan 30°$$

$$F = H = 274\cdot3 \text{ newtons}$$

Hence $$\mu = \frac{274\cdot3}{700} = 0\cdot39 \quad (\textit{Answer}).$$

Example: How far could the man have ascended if $\mu = 0.25$ in the above example?
SOLUTION: Using the same lettering as before we suggest that the man reaches the point C such that $BC = x$.

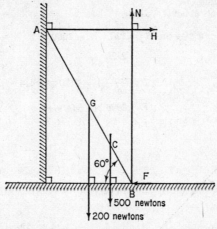

Fig. 189

Since the ladder is in limiting equilibrium,

$$F = \mu N = 0.25N$$

Resolving vertically, $N = 700$
Resolving horizontally, $F = H$
But $F = 0.25N$, hence $F = H = 0.25 \times 700 = 175$.
Taking moments about B,

$$200 \times 2 \cos 60° + 500 \times x \cos 60° = H \times 4 \sin 60°$$

$$200 + 250\,x = 700 \times 0.866$$

$$= 606.2$$

$$\therefore x = \frac{406.2}{250} = 1.625 \text{ m}$$

So the man can only ascend the first 1·625 m of the ladder, or as near to (but always less than) this distance as the rungs of the ladder allow   (*Answer*).

Example: A warehouse swing hoist consists of a uniform horizontal beam AB, 2 m long and weighing 120 newtons. The beam is hinged at A to a vertical wall and supported by a light tie-bar, one end jointed to the wall 1 m above A and the other end jointed to the midpoint of AB. Ignoring any friction at the joints or any bending of the beam, calculate the force in the tie-bar when the hoist is carrying a load of 1000 newtons.
SOLUTION: Let Fig. 190 represent the problem. Again it is more convenient to use two components X and Y for the force on the beam at A. Since the tie-bar is

light we may assume that the force T applied to the beam acts along its direction GC.

By geometry $\angle CGA = 45°$. We could have replaced T by a horizontal component and a vertical component at G but the angle $45°$ will be quite convenient for taking moments.

Taking moments about A we have

$$T \times 1 \sin 45° = 120 \times 1 + 1000 \times 2$$

$$= 2120$$

$$\therefore T = \frac{2120}{\sin 45°} = 2120 \times 1\cdot414$$

$$T = 2998 \text{ N} \quad (Answer).$$

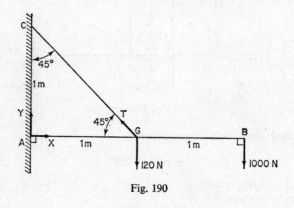

Fig. 190

Exercise 65

1. Find N and F in Fig. 187.
2. If the ladder in Fig. 187 had been in limiting equilibrium, what was μ for the contact surfaces at B?
3. If the man sends his son (who weighs 400 N) up the ladder in Fig. 189, how far can the boy ascend with $\mu = 0\cdot25$ at B?
4. What would be the effect of having a friend to stand on the bottom rung of the ladder in Fig. 189?
5. By taking moments about G in Fig. 190 above, show that Y has been given a wrong direction, and find its magnitude. By resolving horizontally find X.
6. If the breaking tension of the tie-bar GC in Fig. 190 is only 2000 N, what is the maximum load for the hoist?
7. A uniform beam AB of length 2 m and weight 200 N is hinged to a vertical wall at A. The beam is supported in the horizontal position by a tie-rope BC (assumed to be in a straight line) attached to B and a point C on the wall above A such that $\angle ABC = 60°$. If the beam carries a load of 500 N at B, calculate the tension in the tie-rope and the horizontal and vertical components of the reaction at A.
8. A heavy uniform rod AB of length 4 m and weight 200 N is smoothly hinged at its upper end A. A string is attached to the end B and pulls at right-angles to the rod. The rod comes to rest at an angle of $30°$ below the horizontal. Find the tension in the string and the horizontal and vertical components of the reaction at the hinge.

CHAPTER TWELVE

WORK, ENERGY AND POWER

(1) Work

In applied mathematics, work is associated with the movement of bodies from one position to another. Since the movement of a body from rest requires force, and the change in position is represented by displacement, we shall define work by relating it to force and displacement.

If you help to push-start a car with a flat battery on a winter's morning you will readily agree on three things: pushing the car 20 m will be only half the 'work' of pushing it 40 m; directing your push in the line of motion is better than pushing in any other direction; finally, the sooner you can find a downward slope the better. If a young child helps you push she will not do the same amount of work as you, even though she will move just as far. The difference between your work contribution and hers will have something to do with the magnitude of the force applied to the car. We therefore conclude that a definition of work involving force and distance looks promising. In order to keep the early ideas as simple as possible we shall consider first, the work done by a force acting on a particle.

DEFINITION: *If a constant force of magnitude F is applied to a particle which then moves a distance s along the line of action of F, the product Fs is called work and is said to be a measure of the work done by the force on the particle.*

To measure anything we must have a standard unit. The search for a unit need only be brief because we have already obtained units for F and s. Since the standard units are newtons for F and metres for s, it follows that Fs must be measured in newton metres. Unfortunately this is the same measurement as for moment of a force, so it would be helpful to have some means of distinguishing between the two. We therefore refer to the newton metre, when it is a unit of work, as a **joule** (pronounced 'jewel'). (See page 75.)

Thus a force of 5 newtons which moves a particle 6 metres along its line of action does 5 newtons $\times$ 6 metres = 30 joules of work on the particle. Alternatively, a force of 6 newtons which moves a particle 5 metres along its line of action does 6 newtons $\times$ 5 metres = 30 joules of work on the particle. In both these illustrations we have suggested that the particle has moved with the force, so that in each case the work done by the force is positive. Suppose we now consider a particle which moves a distance of 2 metres under the action of two forces of 6 N and 10 N in the same straight line but in opposite directions. Clearly the resultant force on the particle is 4 N in the same direction as the force of 10 N, and the particle will move 2 metres in its direction. So the work done on the particle by this resultant force is 8 joules. The work done by the separate forces is $10 \times 2 = 20$ joules and $(-6) \times 2 = -12$ joules. We see therefore that the work done by a force may be positive or negative.

Exercise 66

1. A force of 200 N is applied to a car and moves it through a distance of 10 m along the line of action of the force. What is the work done by the force on the car?

2. If the same force in Question 1 had pushed a car twice as heavy the same distance of 10 m, how much work would the force have done?

3. A force of 15 N moves a body of mass m through a distance of 4 metres along the line of action of the force. Find the work done when (i) $m = 1$ kg (ii) $m = 3$ kg.

4. A force of 16 N takes t seconds to move a body a distance of 5 metres along the line of action of the force. Find the work done when (i) $t = 1$ second, (ii) $t = 4$ seconds.

5. A car runs down a slope a distance of 5 m in spite of the force of 200 N in the line of motion which I apply to hold it back. What work is done by my force on the car in this distance?

(2) Kinetic Energy

The above definition of work only mentions force and distance and does not exclude the possibility that the particle might already be moving with an initial velocity u at the time of application of the force F. If, therefore, this velocity is changed from u to v after the particle has travelled a distance s in the line of action of the force, then it seems reasonable to expect that the work Fs may be expressed in terms of v and u.

Since F is constant we see from Newton's Second Law that $F = ma$ ensures that the acceleration is also constant, where F is the resultant force acting on the particle. We have already obtained equations of motion in a straight line with constant acceleration, and one equation which involves u, v, and s is $v^2 = u^2 + 2as$ or $v^2 - u^2 = 2as$.

Multiplying this equation by $\frac{1}{2}m$ we obtain

$$\tfrac{1}{2}mv^2 - \tfrac{1}{2}mu^2 = mas \tag{i}$$

But, with F as the resultant force on the particle, we have $F = ma$ to substitute in (i) and give

$$Fs = \tfrac{1}{2}mv^2 - \tfrac{1}{2}mu^2 \tag{ii}$$

We recall that $\frac{1}{2}mv^2$ was defined in Chapter Five as the kinetic energy of the body having mass m and velocity v. So eqn. (ii) shows that the work done on the body by the force F moving it a distance s along its line of action is measured by the **change in kinetic energy.**

Example: A resultant force of 50 N acts on a body of mass 12 kg and changes the magnitude of its velocity from 2 m s^{-1} to 5 m s^{-1}, where all the vectors are directed in the same straight line. Find the work done on the body by the force and the distance the body is moved.

SOLUTION: The work done on the body will be measured by the change in the kinetic energy during the time of application of the force. This change is given by

$$\tfrac{1}{2} \times 12 \times 5^2 - \tfrac{1}{2} \times 12 \times 2^2 = 150 - 24$$
$$= 126 \text{ joules}$$

So the work done by the force is 126 J (*Answer*).

If s is the distance moved by the body in the direction of the force, then $50s = 126$

$$s = 2 \cdot 52 \text{ m} (Answer).$$

Note that if the two given velocities had been interchanged in the same straight line, so that the initial velocity was 5 m s^{-1} and the final velocity 2 m s^{-1}, the change in kinetic energy would have been -126 J. The negative sign indicates that the force must have been applied in the opposite direction to the direction in which the velocity was taken as positive.

Example: A small body of mass 0·25 kg moving in a straight line across a smooth horizontal table with a velocity of 1 m s⁻¹ has its velocity changed to 3 m s⁻¹ in a distance of 1·5 m by a constant force directed along the same straight line. Find the magnitude of the force and its time of application.

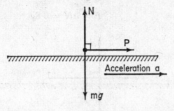

Fig. 191

SOLUTION: Let m be the mass of the body and P the magnitude of the applied force. In Fig. 191 we see that there are three forces on the body: the weight mg, the normal reaction N, and the applied force P. Since there is no motion perpendicular to the table we have, by Newton's Second Law,

$$mg - N = mO$$

Therefore
$$N = mg$$

It follows that P is the resultant force on the body, so the change in the kinetic energy is given by

$$P \times 1·5 = \tfrac{1}{2} \times 0·25 \times 3^2 - \tfrac{1}{2} \times 0·25 \times 1^2$$

$$= \tfrac{1}{2} \times 0·25 \times 8 = 1$$

$$\therefore P = \frac{1}{1·5} = \frac{2}{3} \text{ newtons} \quad (Answer).$$

We know u and v but require t. The simplest equation which comes to mind is $v = u + at$, and since we have just found $P = ma = \tfrac{2}{3}$ we shall be able to substitute for a.

This last result gives

$$a = \frac{2}{3m} = \frac{8}{3}, \text{ where } m = 0·25 \text{ kg}$$

so that on substitution in $v = u + at$ we have

$$3 = 1 + \tfrac{8}{3}t$$

whence
$$t = \tfrac{3}{4} \text{ second} \quad (Answer).$$

To stress the fact that the change in the kinetic energy of a body measures the work done by the constant resultant force on the body, let us now consider an example where the motion is across a *rough* horizontal table.

Example: A small body of mass 2 kg moves in a straight line across a rough horizontal bench under the action of a constant force of 10 N applied over a distance of 3 m in the same line of action. If the initial velocity of the body was 1 m s⁻¹, calculate the change in kinetic energy of the body and its final velocity if the coefficient of sliding friction is $\mu = 0·4$. (Take $g = 9·8$ m s⁻²)

SOLUTION: Let Fig. 192 represent the forces on the body, with N the normal reaction and F the frictional force during sliding given by $F = \mu N$. Since there is no motion perpendicular to the plane, we have

$$N - 2g = 2 \times 0$$
$$N = 2g$$

Hence

$$F = \mu N = 0\cdot 8g$$

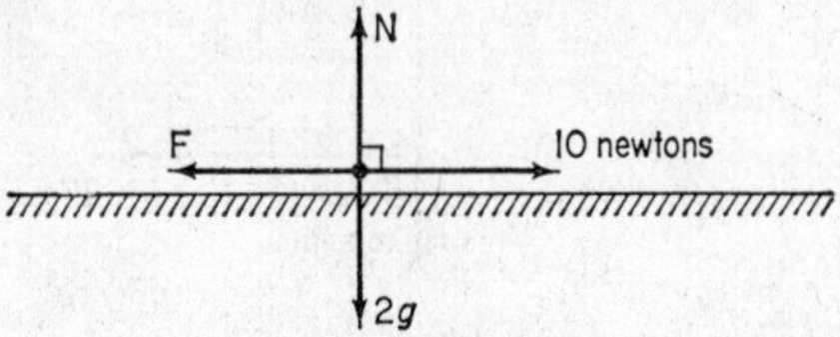

Fig. 192

The resultant force on the body is therefore $(10 - 0\cdot 8g)$ N in the direction of motion. The work done by the resultant force measures the change in kinetic energy, which is therefore

$$(10 - 0\cdot 8g) \times 3 = 6\cdot 48 \text{ J} \quad (Answer).$$

If v is the final velocity, i.e. the velocity when $s = 3$ m, then

$$\tfrac{1}{2} \times 2(v^2 - 1) = 6\cdot 48$$
$$v^2 = 7\cdot 48$$
$$v = 2\cdot 735 \text{ m s}^{-1} \quad (Answer).$$

Exercise 67

1. A particle of mass 3 kg is acted on by a resultant force of 12 N which changes the velocity of the particle from 1 m s^{-1} to 2 m s^{-1} in the line of action of the force. Calculate the work done by the force and the distance moved during this time.

2. If 60 J of work are done by a force moving a particle 3 m along the line of action of the force, calculate the magnitude of the force.

3. A particle of mass 3 kg is pulled in a straight line for 5 s across a smooth horizontal bench by a resultant force of 6 N acting in the same direction. If the initial velocity is 1 m s^{-1}, calculate the gain in kinetic energy of the body.

4. A body of mass 2 kg is projected along a rough horizontal table with an initial velocity of 10 m s^{-1}. If the coefficient of sliding friction is $\mu = 1\cdot 6$, find the frictional force and the work done by this force on the body as it slides for the first 2 m of its motion. Find the velocity when the body has travelled 2 m.

5. A force of 30 N pulls a 2-kg mass for a distance of 3 m across a rough horizontal table with $\mu = 1\cdot 2$. Calculate the change in kinetic energy of the body and its velocity at the end of the 3 m if it had an initial velocity of 4 m s^{-1}.

(3) Motion at an Angle to the Force Producing it

All the problems above have emphasized that the motion has taken place in the line of action of the resultant force producing the motion. But there are many situations in which the motion is not along the line of action of the force which causes the motion. For example, a horse pulling a barge along the canal applies a force at an angle to the direction of motion of the barge; a man pulling a crate along the ground will pull with a force at an angle to the direction of motion along the ground—for reasons we have already discovered (see Question 4 in Exercise 58). How shall we measure the work done by either the horse or the man in these two cases? To find a measure of work done in this manner we shift the emphasis away from the force and distance; it is now more convenient to use the change in kinetic energy of the body as a measure of the work done by the resultant force on the body.

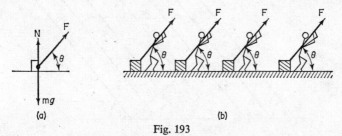

(a)   (b)

Fig. 193

To illustrate this point consider a particle of mass m resting on a smooth horizontal table. The particle is moved in a straight line by an applied force of constant magnitude F with a line of action inclined at an angle θ to the table and always in the same vertical plane. Fig. 193 (*b*) shows what is meant by keeping the force at the same angle θ to the horizontal as the body moves: imagine yourself walking along a straight line on the floor with a rope over your shoulder attached to a box which you are dragging along the same straight line. In Fig. 193 (*a*) we see the forces which act on the particle; N is the normal reaction, and the resultant force on the particle is $F \cos \theta$ horizontally. Perpendicular to the plane we have

$$mg - N - F \sin \theta = m \times 0 = 0$$

since the body remains in contact with the fixed plane.

Equating resultant force to mass × acceleration we get

$$F \cos \theta = ma \qquad \text{(i)}$$

where a is the acceleration of the particle along the table. Furthermore, since F is constant so too is a and we may use the familiar equations of motion for uniform acceleration. The equation we have in mind is of course $v^2 = u^2 + 2as$ But since $u = 0$ in this case, we have $v^2 = 2as$ which, on multiplication by $\frac{1}{2}m$ becomes $mas = \frac{1}{2}mv^2$. Substituting in eqn. (i) we get

$$Fs \cos \theta = \frac{1}{2}mv^2$$

Since $\frac{1}{2}mv^2$ is the change in kinetic energy of the body and measures the work done by the resultant force which caused the change, it follows that our definition of work must now be interpreted as $Fs \cos \theta$.

Examining this situation a little more closely in Fig. 194, we see that if $AB = s$ and the particle is moved from A to B by the force F acting at angle θ to AB as shown, then

$$F \times s \cos \theta = F \times AB \cos \theta = F \times AC$$

Now AC is the displacement of the particle in the direction of the force (or we could describe it as the distance moved by the point of application of the force F in the direction of the force) so we widen our previous definition of work as follows:

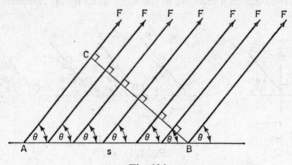

Fig. 194

DEFINITION: *If a constant force of magnitude F is applied to a particle which then moves a distance s along a straight line at an angle θ to the direction of F, we define the work done by the force on the particle by the product $Fs\cos\theta$.*

The product $Fs\cos\theta$ may be thought of in two different ways: either as $F \times (s\cos\theta)$ or as $(F\cos\theta) \times s$. The first of these has just been discussed with the aid of Fig. 194. The second may be thought of as the product of the component of the force F in the direction of motion (i.e. $F\cos\theta$) and the displacement s.

A point to note in this result is that if $90° < \theta \leqslant 180°$ then $\cos\theta$ is negative, in which case the product $Fs\cos\theta$ is negative. (We have already dealt with the example of $\theta = 180°$, $\cos 180° = -1$ on page 90.) If the force F is applied to the body so that $\theta = 120°$ the work done by F is $-\frac{1}{2}Fs$, and we may think of this result as arising in Fig. 194 from a force of $\frac{1}{2}F$ along BA whose point of application is moved through a distance s in the opposite direction. Finally, since $\cos 90° = 0$ we see that no work is done by a force on a body whose point of application moves in a direction at right-angles to the force.

Example: A particle of mass m falls freely from rest until it acquires a velocity v. Calculate the work done by the weight of the particle, and find the distance fallen in acquiring the velocity v.

SOLUTION: The only force on the body is its weight mg, so this must be the force which accounts for the change in kinetic energy measured by $\frac{1}{2}mv^2 - 0$. Therefore work done by the weight is $\frac{1}{2}mv^2$. If h is the distance fallen, we have

$$mgh = \tfrac{1}{2}mv^2$$

$$h = \frac{v^2}{2g} \quad (Answer).$$

Example: A particle of mass m is pulled up a rough plane, inclined at an angle θ to the horizontal, by a force T parallel to the plane. If the coefficient of sliding friction is $\mu = 0.5$ calculate the work done by each of the forces on the particle as it moves a distance s along the line of greatest slope of the plane.

SOLUTION: Our experience in Chapter Eleven should confirm the forces shown in Fig. 195.

Equating resultant force to mass × acceleration we have:

perpendicular to the plane, $mg \cos \theta - N = m0$ since the plane is fixed.

$$N = mg \cos \theta$$

parallel to the plane, $T - F - mg \sin \theta = ma$.

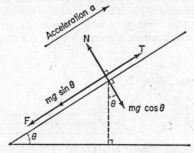

Fig. 195

Multiplying this last equation by s and using the result $mas = \frac{1}{2}mv^2 - \frac{1}{2}mu^2$ we get

$$Ts - Fs - mgs \sin \theta = mas = \tfrac{1}{2}mv^2 - \tfrac{1}{2}mu^2 \qquad (i)$$

Ts is the work done by T, $-Fs = F(-s)$ is the work done by F, and $-mgs \sin \theta = mg(-s \sin \theta)$ is the work done by the weight mg. From eqn. (i) we see that we merely add the work done by each force, using the rules of ordinary algebra, which means that **work is a scalar quantity**. With $F = \mu N = 0.5mg \cos \theta$ we can make a further simplification of the equation if required.

Exercise 68

1. A force of 10 N acts in the direction East to West on a particle of mass m. Find the work done on the particle by this force if the particle moves with the following displacements:

   . (i) 10 m East,   (ii) 6 m West,   (iii) 12 m North,
     (iv) 5 m South,   (v) 8 m North-east.

2. A particle of mass 2 kg moves across a smooth horizontal table under the action of a constant horizontal force of 12 N. If the velocity of the particle changes from 2 m s^{-1} to 3 m s^{-1}, find the work done by (i) the weight of the particle, (ii) the normal reaction, (iii) the force of 12 N.

3. A force of 30 N is applied to a particle of mass 1 kg held at rest on a smooth plane inclined to the horizontal at an angle of 30°. Calculate the work done by (i) the weight of the particle, (ii) the force of 30 N applied parallel to the plane, as the body moves up the plane a distance of 2 m.

4. Two men A and B in argument move a case horizontally through a distance of 3 m. A pulls along the line of motion with a force of 200 N, while B pulls in the opposite direction with a force of 150 N. Find the work done separately by the forces applied by A and B.

5. A particle of mass 4 kg is pulled up a rough plane, inclined at 40° to the horizontal, by a force of 50 N parallel to the plane along the line of greatest slope. If the coefficient of sliding friction is 0·5, calculate the work done by each force on the particle as it moves 2 metres up the plane.

6. A body of mass 3 kg falls freely from rest for 2 seconds. Find the distance the body falls during this time and calculate the work done by the weight of the body.

Although the worked example of Fig. 195 above is not the most general situation from which to draw conclusions, it does show that the change in kinetic energy of a body is equal to the sum of the work done by all of the forces which act on the body. This relation was expressed earlier as the change in kinetic energy of a body being measured by the work done by the resultant force on the body. What may have escaped the reader's notice is the fact that the discussion so far has been restricted to the work done on *particles* and not larger bodies. In the case of particles the distance moved in the direction of the force is easy to determine, but a large body may swing round during the motion so that the displacement received is open to argument.

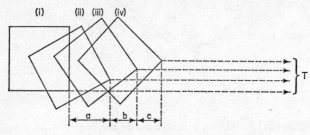

Fig. 196

If we tie a length of string to any point on a crate and pull with a constant force T, it will make no difference to the work done whether we use a long or short string. If we shorten the string right down to the point of application the work done will still be the same and measured by the product of T and the distance moved by the point of application of the force. This situation is represented in Fig. 196 where the crate will clearly turn from positions (i) to (iv) under the action of the constant force T. Suppose we shorten the string as the crate goes from one position to the next by hauling it in. Then

the work done by T from (i) to (ii) is Ta
the work done by T from (ii) to (iii) is Tb
the work done by T from (iii) to (iv) is Tc

The total work done by T is $T(a + b + c)$, where a, b, and c are the distances moved by the point of application in each case.

The work done is independent of the length of the string, but since this may vary we can see that the best way of finding the work done is to use the displacement of the point of application.

Exercise 69

1. A crate is slowly pulled 2 m without rotation across a horizontal ground by a constant force of magnitude 200 N. Calculate the work done on the crate by this force when it is applied by each of the following methods:

  (i) a rope wound on a fixed winch towards which the crate is pulled;
  (ii) a rope of fixed length held by a man who walks forward dragging the crate behind him;
  (iii) a rope held by a man who stands still pulling the crate towards him.

2. Fig. 197 shows a man pulling a crate from A to B, a distance of 4 m, by applying a force of 300 N to the crate with the aid of a smooth rope passed round a smooth bollard and walking in a direction inclined at 40° to AB. Assuming that the tension remains the same throughout the rope, find the work done by the force in the rope.

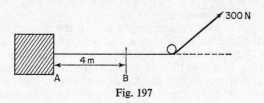

Fig. 197

3. A and B are any two horizontal planes such that B is 2 m vertically above A. A case having a weight of 80 N is slowly raised from A to B. Find the work done by the weight of the case.

4. If the case in Question 3 is now returned to A from B, how much work is done by the weight of the case?

5. A bucket of water of weight 70 N is raised 15 m from a well. Find the work done by the weight of the bucket of water.

(4) Potential Energy

Questions 3 and 4 of the last exercise are of particular interest because we required the work done *by* the weight which was being moved: in other words, we had to find the work done *against gravity*. How is this work done by us in moving the case from its initial position to the final position? If we apply a lifting force T to the case to start it in motion at A, we shall need a different force to bring it to rest at B; so the force we use throughout the lift is not constant. Consequently the definition we have used so far will not enable us to find the exact measure of the work we do to lift the case from rest at A to rest at B. But we do know that we shall have to move the point of application of the weight (i.e. the centre of gravity) a distance s using a force greater than mg, so we (or the force we apply) must do mgs units of work in lifting the case. Similarly for lifting any body of weight Mg to a height h above its original position we must do Mgh units of work.

Examining these two results we see that the product of the weight and the height of a body or particle, has some special significance. But first of all, what do we mean by height? Perhaps in bed your height above the floor is 0·5 m, but relative to the garden your height is probably 3·5 m. If you live in a block of flats, then your height will vary according to where you measure from, and each of your neighbours will regard you as being a different height relative to him. Thus height can be measured from *any* arbitrary level. The surface of the earth is taken as the most convenient level simply because most things will finally come to rest there.

Suppose we drop a body from a height h and let it fall freely under gravity. Since $v^2 = u^2 + 2gs$, we have $v^2 = 2gh$ and $\frac{1}{2}mv^2 = mgh$. After falling to the ground from a height h the body would therefore acquire a kinetic energy of $\frac{1}{2}mv^2 = mgh$. Had we measured heights from some other level, the body would have been at a height H say, and the kinetic energy of the body by the time it fell to this level would then have been mgH.

This is all based on the assumption that we drop from its present position.

Since it is always *possible* for the body to acquire this energy simply by being dropped, we call it the **potential energy** of the body and define it as follows.

DEFINITION: *The potential energy of a body is the work done by the weight of the body if it is moved from its present position to the level from which the height of the body is measured.*

Alternatively, the potential energy of a body is the product of the weight and the height of the body. We use the symbol E_p to represent the potential energy of the body just as we use E_k to represent the kinetic energy.

Thus a particle of weight 10 N which rests on a ledge 2 m above the floor has a potential energy of 20 J. That is the work done by the weight if the body descended to the floor. The floor is clearly taken as the level of zero potential energy.

A particle of weight 2 N at rest on the bottom of a well 7 m below ground-level will have a potential energy measured by the work done by the weight as the particle ascends to ground level. This is $(-2) \times 7 = E_p = -14J$. As expected, any particle below the zero level of potential energy is said to have a negative potential energy.

Example: A block of mass 10 kg slides with a speed of 3 m s⁻¹, without rotation, across a horizontal bench 1·5 m above the floor. Find the sum of the kinetic and potential energy of the block, taking the floor as zero level of potential energy. ($g = 9·80$ m s⁻²)
SOLUTION: The kinetic energy of the block is given by

$$E_k = \tfrac{1}{2}mv^2$$
$$= \tfrac{1}{2} \times 10 \times 3^2$$
$$= 45 \text{ J}$$

The potential energy of the block is given by

$$E_p = mgh$$
$$= 10 \times 9·8 \times 1·5 \text{ J}$$
$$= 147 \text{ J}$$
$$\therefore E_p + E_k = 192 \text{ J} \quad (Answer).$$

Example: A particle of mass 0·1 kg is released from rest at a height of 25 m above ground-level and falls freely under gravity. Taking ground-level as zero level of potential energy, find the sum of the potential energy and kinetic energy of the particle when $t = 2$ seconds. ($g = 9·80$ m s⁻²)
SOLUTION: Using the standard equation $v = u + at$, we have $v = 2g$ as the speed of the particle at $t = 2$.

$$\therefore E_k = \tfrac{1}{2} \times 0·1 \times 2 \times 9·8 \times 2 \times 9·8 = 19·208 \text{ J}$$

Using the standard equation $s = ut + \tfrac{1}{2}at^2$, the distance fallen in 2 seconds is $s = \tfrac{1}{2} \times 9·8 \times 2 \times 2 = 19·6$ m.
Therefore at $t = 2$ s the particle is $25 - 19·6 = 5·4$ m above the ground, so its potential energy relative to ground-level is

$$E_p = 0·1 \times 9·8 \times 5·4 = 5·292 \text{ J}$$
$$\therefore E_p + E_k = 24·5 \text{ J} \quad (Answer).$$

Note that potential energy at the beginning of the motion is given by

$$E_p = 0·1 \times 9·8 \times 25$$
$$= 24·5 \text{ J}$$

A comparison of this result with the sum of the potential and kinetic energies at *any* time is left to the reader in Question 7 below.

Exercise 70

1. A particle of mass 0·2 kg rests on a table 1·5 m above floor level in a room with the ceiling 3 m above the floor. Find the potential energy of the particle relative to the following zero levels of potential energy:

    (i) the floor, (ii) the ceiling, (iii) a shelf 2 m above the floor.

2. Taking ground-level as zero level of potential energy, find the potential energy of each of the following masses:

    (i) 2 kg at height 3 m, (ii) 3 kg at height 2 m, (iii) 6 kg at height 1 m.

3. Calculate the kinetic energy at ground-level if each of the bodies in Question 2 falls from rest.

4. For bodies other than particles the potential energy depends on the height of the centre of gravity of the body above the zero level. A uniform rectangular box with dimensions 1 m × 0·4 m × 0·6 m and weight 50 N can lie with one face on a horizontal table and have three different potential energies. Find the potential energy of the box in each position relative to the table-top as zero level of potential energy.

5. A rectangular trapdoor of size 1 m × 0·5 m and weight 20 N is hinged horizontally along a shorter edge. What is the gain in the potential energy of the door if it is raised from the horizontal to the vertical position?

6. A uniform block of edges 1 m × 0·8 m × 0·6 m and weight 100 N rests with one of the 0·6 m × 1 m faces on a horizontal plane. The block is now rotated without slipping about the edge of length 1 m until the block rests on the plane again. What was the greatest potential energy of the block during the rotation?

7. Find the sum $E_p + E_k$ in the last worked example above when (i) $t = 1$ second, (ii) $t = 1·5$ seconds, (iii) any time t.

8. A particle of mass 0·2 kg is released from rest at a height of 30 m above ground-level and falls freely under gravity to the bottom of a well 10 m below ground-level. Using ground-level as the level of zero potential energy, find $E_p + E_k$ at the following times: (i) $t = 0$, (ii) $t = 1$ second, (iii) $t = 2$ seconds, (iv) when the body strikes the bottom of the well. ($g = 9·80$ m s^{-2})

(5) Conservation of Energy

Questions 7 and 8 in the above exercise suggest that $E_p + E_k = $ constant for any freely falling body. This means that the energy of a body is conserved throughout its motion and depends only on the initial conditions, i.e. the initial velocity and height above the zero level of potential energy. To put it another way, whenever this energy equation is true we can say that what the body loses in potential energy it will gain in kinetic energy. Alternatively we might also conclude that what the body loses in kinetic energy it will gain in potential energy. To put this conclusion to the test we shall examine the motion under gravity again, but this time we choose the example of a general projectile. First let us take the simple example of a particle projected vertically upwards with a velocity v, taking the horizontal through the point of projection as the zero level of potential energy. With the acceleration $-g$ and using the standard equation $v^2 = u^2 + 2as$ we have $v^2 = u^2 - 2gh$ when the particle is at a height h above the point of projection. Multiplying by $\frac{1}{2}m$ and re-arranging this equation we have

$$\tfrac{1}{2}mv^2 + mgh = \tfrac{1}{2}mu^2 \qquad \text{(i)}$$

Now $\frac{1}{2}mv^2 + mgh = E_k + E_p$ when the particle is at height h above the point of projection and $\frac{1}{2}mu^2 + 0 = E_k + E_p$ when $h = 0$. Equation (i) therefore confirms that in any free vertical motion under gravity alone the sum of the kinetic and potential energies is constant.

To find out whether the same result holds good for any projectile moving freely under gravity only (i.e. no resistance by the atmosphere through which it travels) we shall examine the motion of a particle projected with initial velocity u at an angle α to the horizontal, as discussed in Chapter Nine.

The equations of motion giving the components of the velocity at any time t are

$$v_x = u \cos \alpha, \ v_y^2 = u^2 \sin^2 \alpha - 2gy$$

where y is the height of the particle above the point of projection, which is now to be taken as zero level of potential energy.

Putting these two results together we have

$$v^2 = v_x^2 + v_y^2 = u^2 \cos^2 \alpha + u^2 \sin^2 \alpha - 2gy$$
$$\therefore \ v^2 = u^2 - 2gy$$

So $\frac{1}{2}mv^2 + mgy = \frac{1}{2}mu^2$ and once again we see that the sum of the potential and kinetic energy remains a constant.

We can therefore summarize these last two results by what is called the **principle of the conservation of energy**: for any body moving under the force of gravity alone the sum of the potential and kinetic energies is constant. An equation such as $\frac{1}{2}mv^2 + mgy = \frac{1}{2}mu^2$ above is called the energy equation for the body in motion.

Since energy is defined as the capacity for doing work, any form of pent-up or sealed-in energy may be used to perform work; the potential energy due to the height and weight of a body is just one of many ways in which energy may be stored. For example, a watch spring when fully wound has a store of energy or potential energy which is used up in turning the hands to register the time. The old grandfather clocks have large masses suspended by cords which gradually unwind so that the work done by the weight as it descends is used to turn the hands of the clock. When the weight has descended to the bottom of the clock we wind it back up again to replace its potential energy.

A piece of elastic also has potential energy when it is held in a stretched state. Thus a catapult releases the energy of the stretched elastic and in so doing imparts a velocity to a stone, i.e. the potential energy of the elastic is converted into kinetic energy for the stone. There are other forms of energy such as heat, chemical, and electrical energy. One very familiar example is the chemical action in a battery which can be converted to electrical energy to start the engine of a car.

Considering the principle of the conservation of energy in the widest sense, we suggest that although energy can be converted from one form to another it can never be destroyed. This means that the sum of all the different energies in a system remains constant. To illustrate this point consider the energy stored in a gallon of petrol. If this is burned under ideal conditions it will release a certain amount of energy, E say, but if we burn it in a car engine we shall *not* get an output equivalent to E in order to drive the wheels. Suppose, for example, that we are in a car which burns one gallon of petrol and that we travel with a uniform velocity v in a straight line for a distance s. Then, assuming that the frictional force F is constant, the work done by the frictional force is Fs. The burning of petrol has given $(Fs + \frac{1}{2}mv^2)$ joules say, but we find that $Fs + \frac{1}{2}mv^2 \neq E$. So it looks at first as though some energy has been lost or destroyed. However, we have forgotten the work f done against frictional forces in the engine, the energy h lost in heat, and the energy n lost in sound. In fact what we should have written down is

$$Fs + \frac{1}{2}mv^2 + f + h + n + b = E$$

where b accounts for energy lost in overcoming some electrical resistance and so on.

From this we can see that if the efficiency of an engine is measured by the motion of the car, the engine will never be 100 per cent efficient since so much energy will have been lost to forces which do not contribute to the motion of the car.

The efficiency with which one form of energy is converted into another dominates engineering design. The simplest mechanical example is the typical hydroelectric scheme where a lake of water is dammed and the water flow then regulated to feed through tunnels to drive machinery in order to generate electricity. This is a straightforward conversion of the water's potential energy into kinetic energy. But our task here is to summarize these results into a form for applying energy principles to solve our problems. For simplicity we restrict ourselves to two methods of solution.

Method 1: In any motion of a body where the force of gravity is the only force on the body which does any work, the sum of the potential energy and the kinetic energy is constant. We express this in the form

$$E_p + E_k = \text{constant}$$

Method 2: If the body moves under the action of a system of constant forces, let this system have a resultant of magnitude R. Since this acts on the body it follows from Newton's Second Law that $R = ma$, where a is the acceleration of the body. If the body moves a distance s under the action of the resultant R, then the work done by all the forces in the system is the same as the work done by the resultant R of the system, and the work done is Rs.

From the equation $v^2 = u^2 + 2as$ multiplied by $\frac{1}{2}m$ we have

$$mas = \frac{1}{2}mv^2 - \frac{1}{2}mu^2$$
$$Rs = mas = \frac{1}{2}mv^2 - \frac{1}{2}mu^2$$

We see therefore that the **change in kinetic energy of the body is equal to the total work done by all the forces acting on the body.**

We shall now apply these two methods to some problems.

Example: A bullet of mass 15 g is fired into a fixed block of wood with a horizontal velocity of 400 m s^{-1}. If the bullet comes to rest after penetrating a distance of 80 mm, calculate the resistance of the block assuming the resistance to be constant.

SOLUTION: The only two forces on the bullet are its weight and the resistance F of the block. The work done by the weight is zero since the displacement of the bullet is at right-angles to the line of action of mg. The work done by the resistance is $-F \times 0.08$ joules, the negative sign arising from F being in the opposite direction to the displacement of its point of application on the bullet. The change in kinetic energy of the bullet is $0 - \frac{1}{2}mu^2$.

Since the total work done by all the forces acting on the body is equal to the change in kinetic energy we have

$$-F \times 0.08 = -\tfrac{1}{2} \times 0.015 \times 400 \times 400$$
$$F = \frac{0.015 \times 400 \times 400}{2 \times 0.08} = \frac{15 \times 40 \times 400}{2 \times 8}$$
$$= 15\,000 \text{ N} \quad (Answer).$$

In practice the resistance does not remain constant: what we have really found is the *average* resistance of the block throughout the relative motion of the bullet. The change in kinetic energy is always found by subtracting the kinetic energy in the first state from the kinetic energy in the second state.

Example: The shortest braking distance for a car travelling horizontally at 80 km h^{-1} is approximately 38 m. Assuming that the total braking force remains constant, calculate its magnitude if the car has a total mass of 450 kg.

SOLUTION:

$$80 \text{ km h}^{-1} = \frac{80 \times 1000}{60 \times 60} = \frac{200}{9} \text{ m s}^{-1}$$

Change in kinetic energy of the car is

$$0 - \tfrac{1}{2} \times 450 \times \frac{200}{9} \times \frac{200}{9} = \frac{-450 \times 200 \times 200}{2 \times 9 \times 9} \text{ J}$$

There are two forces on the car: the total braking force F and the weight. The total work done by all the forces on the car is $-F \times 38 + 450g \times 0$. The weight does no work since the displacement is horizontal.

Equating the change in kinetic energy of the car to the total work done by the forces on the car we have

$$-F \times 38 = \frac{-450 \times 200 \times 200}{2 \times 9 \times 9}$$
$$F = \frac{450 \times 200 \times 200}{38 \times 2 \times 9 \times 9} = \frac{500 \times 1000 \text{ N}}{171}$$
$$F = 2.9 \text{ kN} \quad (Answer).$$

Example: A pile-driver is being used to build a coffer-dam. A steel section is hammered into the ground by the driver which gives each section plus the driver (a total weight of 1000 N) an initial velocity of 1 m s^{-1} vertically downwards. Driver and section are brought to rest in 0.05 m by the resistance of the ground which is assumed to be uniform. Find the magnitude of this resistance if $g = 9.80$ m s^{-2}.

SOLUTION: The weight of the section plus driver combined is 1000 N, hence their combined mass is

$$\frac{1000}{g} = \frac{1000}{9.8} \text{ kg}$$

Their initial kinetic energy is

$$\frac{\tfrac{1}{2} \times 1000 \times 1^2}{9.8} = \frac{1000}{2 \times 9.8} \text{ joules}$$

Since the final kinetic energy is zero, it follows that the change in kinetic energy is

$$0 - \frac{1000}{2 \times 9.8} \text{ joules}$$

The forces on the moving mass are the resistance R of the ground and the weight of 1000 N acting at the centre of gravity. The total work done by these forces is $1000 \times 0.05 - R \times 0.05$. (Again note that the resistance force R on the body is in the opposite direction to the displacement of its point of application.)

Equating the change in kinetic energy of the body to the total work done by all the forces on the body we get

$$-\frac{1000}{2 \times 9 \cdot 8} = 1000 \times 0 \cdot 05 - R \times 0 \cdot 05$$

$$-51 \cdot 02 - 50 = -R \times 0 \cdot 05$$

$$R = \frac{101 \cdot 02}{0 \cdot 05} = 2020 \cdot 4 \text{ N} \quad (Answer).$$

Example: A uniform piece of string of mass m and length 1 m which offers no resistance to bending is placed on a smooth inclined plane with half on the plane and the other half hanging freely over the top. If the string is released from rest find the velocity of the string at the moment it leaves the incline.

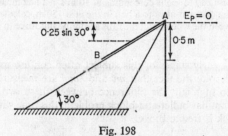

Fig. 198

SOLUTION: Let Fig. 198 represent the problem. Since the reactions between the string and the surface are perpendicular to the direction of motion, the weight of the string is the only force which does any work. Hence $E_p + E_k = $ constant.

Taking the horizontal plane through A as the zero level of potential energy and assuming that the potential energy of a straight piece of string is equal to the potential energy of a particle of equal mass at its midpoint, we have the following results:

at beginning of motion,

$$E_p = -\tfrac{1}{2}mg(0 \cdot 25 \sin 30°) - \tfrac{1}{2}mg(0 \cdot 25), \ E_k = 0$$

when end B reaches A,

$$E_p = -mg(0 \cdot 5), \ E_k = \tfrac{1}{2}mv^2$$

Since $(E_p + E_k)$ is the same in both cases, we have

$$-\tfrac{1}{2}mg(0 \cdot 25 \sin 30°) - \tfrac{1}{2}mg(0 \cdot 25) = -mg(0 \cdot 5) + \tfrac{1}{2}mv^2$$

$$-g(0 \cdot 25 \sin 30°) - g(0 \cdot 25) = -g + v^2$$

$$0 \cdot 625g = v^2$$

$$v^2 = 6 \cdot 125$$

$$v = 2 \cdot 475 \text{ m s}^{-1} \quad (Answer).$$

Exercise 71

1. A rowing boat of mass 100 kg is being rowed in still water at 1 m s^{-1} when the rowing stops. If the boat comes to rest after travelling a further 10 m, calculate the resistance of the water assuming it to be constant throughout the motion.

2. A bullet of mass 0·001 kg is fired with a horizontal velocity of 300 m s^{-1} into a fixed block of wood. If the depth of penetration is 100 mm, calculate the resistance of the block assuming it remains constant throughout the motion.

3. Having calculated the resistance of the block to the motion of the bullet in Question 2, find how far a bullet of the same design and mass would penetrate if it was fired horizontally into the block with velocity of 100 m s⁻¹.

4. A stone of mass 1 kg is dropped from a height of 20 m into a box of sand. If the depth of penetration is 0·2 m, calculate the resistance of the sand assuming this resistance to remain constant throughout the motion. ($g = 9{\cdot}80$ m s⁻²)

5. Find the velocity with which the string in Fig. 198 would slide off the slope if the whole taut string was released from rest on the inclined face with B at its lowest point.

6. Find the velocity of the string in Fig. 198 if it had been placed taut half on a horizontal plane and half hanging freely over the edge. ($g = 9{\cdot}80$ m s⁻²)

7. A car and occupants of total mass 1000 kg is travelling horizontally at a speed of 20 m s⁻¹ when it is braked to a stop in 100 m. Find the braking force between the tyres and the road assuming this force to be uniform.

8. A train of mass 200 tonne is descending at 10 m s⁻¹ a line inclined to the horizontal at 30°. If the driver brakes to a halt in the next 150 m, calculate the frictional braking force between the wheels and the rails assuming this force to be constant.

(6) Momentum

Students in the early stages of this subject often confuse momentum and kinetic energy, despite the fact that mv and $\frac{1}{2}mv^2$ are measured in different units. In order to highlight the difference between these two quantities we now discuss the familiar bullet-and-block problem, this time with the assumption that the block is free to move.

Example: A bullet of mass 0·001 kg is fired with a horizontal velocity of 200 m s⁻¹ into a block of wood of mass 0·01 kg which is free to move on a smooth horizontal table. If the bullet does not perforate the block, calculate the velocity of the block immediately after penetration and the loss in kinetic energy of the bullet and the block assuming that the block does not rotate.

SOLUTION: The bullet and the block move together with the same velocity once the penetration is complete. Let the common velocity of the block and bullet be v.

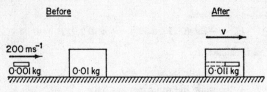

Fig. 199

By the conservation of linear momentum we have

$$0{\cdot}001 \times 200 + 0{\cdot}01 \times 0 = 0{\cdot}011v$$

$$v = \frac{200}{11} \text{ m s}^{-1} \quad (Answer).$$

The kinetic energy before impact is $\frac{1}{2} \times 0{\cdot}001 \times 200 \times 200 = 20$ joules
The kinetic energy after impact and penetration is $\frac{1}{2} \times 0{\cdot}011 \times v^2$
The change in kinetic energy is

$$\frac{0{\cdot}011 \times 200 \times 200}{2 \times 11 \times 11} - 20 = \frac{20}{11} - 20 = -\frac{200}{11} \text{ joules}$$

There has been a loss of $\dfrac{200}{11}$ joules in the kinetic energy of the system   (*Answer*).

So, although we do not lose any momentum, we have lost some kinetic energy. By the principle of the conservation of energy this energy must have been used for something. Normally some would have been lost in generating heat due to the penetration and the rest will be due to the work done by the resistance. If we assume that *all* the energy lost is due to the work done by the resistance, then knowing the depth of penetration will enable us to calculate the resistance.

Exercise 72

1. A lump of mud of mass 0·5 kg is thrown with a horizontal velocity of 10 m s^{-1} to strike a block of mass 3 kg which is free to move on a horizontal plane. If the block and mud move off together without rotation, calculate the common velocity and the loss in kinetic energy.

2. Find the resistance to the motion of the bullet in the worked example above if the depth of penetration is 0·05 m.

3. A bullet of mass 0·001 kg is fired horizontally with a velocity 300 m s^{-1} into a block of wood which is thick enough to prevent perforation. If the mass of the block is 0·009 kg find the depth of penetration when the block is (i) fixed, (ii) free to move without rotation on a smooth horizontal plane, assuming the resistance to be a constant force of 450 N.

4. A hammerhead of mass 2 kg strikes a nail of mass 0·002 kg horizontally with a velocity of 5 m s ^{1}. Calculate the resistance of the wood if the nail penetrates 0·02 m with one blow.

5. A post of mass 6 kg is hammered vertically into the ground by a sledge-hammer of mass 6 kg. If the post descends 0·01 m under the blow of the hammer being dropped from rest at a height of 1 m above the post, calculate the average resistance of the ground.

(7) Power

Whenever a machine is hired to do work one of the main concerns is the time the machine is likely to take on the job. To plan an efficient exercise it is always necessary to know the rate at which both men and machines are able to work. In the case of a man the work rate is variable since he is subject to increasing fatigue as the work day proceeds. Machines, while less versatile than human beings, are more predictable and their work rate usually remains constant whether they are digging a trench or filling bottles. To describe this aspect of machines we use the word **power,** and its precise meaning is given by the following.

DEFINITION: *Power is the rate of doing work.*

Since the basic unit of work is the joule it follows that the basic unit of power will be the joule per second. We give this unit the special name of **watt.** Thus 1 watt = 1 joule per second = 1 J s^{-1}. ($1 \text{ J} = 1 \text{ N m}$, so we could write 1 watt = 1 N m per second or 1 N m s^{-1}, but this is not really necessary or convenient.)

We can see how the unit is used in practice by supposing that we load boxes on to a platform lift. Suppose that each box has a mass of 10 kg and has to be lifted vertically a distance of 12 m. Let us assume that 100 boxes are loaded in 1 h. Since each box goes from rest to rest there is no gain in kinetic energy, so the moving force exerted is certainly not constant throughout the lift. Therefore we cannot measure with our elementary knowledge the work done in lifting the box by examination of the force applied to it. We can only measure accurately the work done by the weight of the box and use this as a measure of the minimum amount of work which we will do for each lift (as discussed in Section 4 of this chapter).

Once the box is in motion the lifting force T is given by $T - mg = ma$, the equation of motion. By putting $a = 0$ we have $T = mg$ as the minimum force we must apply to keep the box moving upwards.

The work done by the force T when the box is lifted 12 m is

$$T \times 12 = 10g \times 12 \text{ J}$$

In 1 hour the minimum work required is $100 \times 10g \times 12$ J. So the average rate of working is $\dfrac{100 \times 10g \times 12}{60 \times 60}$ J s^{-1}.

Taking $g = 9\cdot81$ m s^{-2}, this becomes $\dfrac{120 \times 9\cdot81}{36} = 32\cdot7$ watts

This last result has been found by ignoring any possible gains in kinetic energy by the boxes. This is why we refer to the *minimum* work done in the lift.

Suppose now that the boxes had been raised through the same height of 12 m by a conveyor belt moving at a constant speed of $0\cdot5$ m s^{-1}. We see that the boxes are gaining the same potential energy as before but in this case they are also gaining a kinetic energy which we can calculate as follows.

The kinetic energy of 1 box is $\dfrac{10 \times 0\cdot5 \times 0\cdot5}{2} = 1\cdot25$ J

The kinetic energy given to the 100 boxes is therefore 125 J in 1 hour. Therefore the work done on the boxes by the machine of the conveyor belt in 1 hour is

$$100 \times 10g \times 12 + 125 \text{ J} = 117\ 720 + 125 \text{ J}$$
$$= 117\ 845 \text{ J}$$

We say that the machine does work on the boxes with a power of

$$\frac{117\ 845}{60 \times 60} \text{ J s}^{-1} = 32\cdot73 \text{ watts}$$

Of course we know that the power of the machine is greater than this because it does work by the displacement of its moving parts and overcoming general internal frictional forces. What we have calculated is the rate at which work is being done by the machine in order to increase the energy of the boxes.

We now consider similar problems on finding the power or rate of working of machines.

Example: A pump forces 100 kg of water through a hose every minute. If the water is being raised vertically through 20 m and ejected at the nozzle with a speed of 10 m s^{-1}, calculate the rate at which the pump is working on the water. ($g = 9\cdot80$ m s^{-2})

SOLUTION: The pump not only increases the potential energy of the water but also gives it a kinetic energy for ejection. The question simply asks us to find the rate of change of the energy of the water and identify this as being the work done on the water by the pumping forces supplied by the engine.

The change in energy every minute is $100g \times 20 + \frac{1}{2} \times 100 \times 10^2$ J

$$= 19\ 600 + 5000$$
$$= 24\ 600 \text{ J}$$

The rate of change of energy of the water is $\dfrac{24\ 600}{60}$ J s^{-1}

$$= 410 \text{ watts}$$

So the machine is working with a power of 410 W on the water alone   *(Answer)*.

Example: A locomotive draws a train of total mass 300 t horizontally with a uniform velocity of 15 m s^{-1} against a resistance of 20 000 N. Calculate the rate at which the locomotive forces are working.

SOLUTION: Let T be the tractive or moving force exerted by the engine at the wheel-to-rail contacts. Bearing in mind that the mass of the train is 300×1000 kg, the equation of motion is

$$T - 20\,000 = 300 \times 10\,000a$$

But since the velocity is uniform, the acceleration must be zero, which means that the tractive force is $T = 20\,000$ N.

The point of application of this force moves 15 m every second in the direction of T. Hence the work done is:

$$20\,000 \times 15 \text{ J s}^{-1}$$
$$= 300\,000 \text{ watts or } 300 \text{ kilowatts} \quad (Answer).$$

Normally a uniform velocity is easier to achieve when travelling downhill because under such conditions a component of the weight assists the engine in 'overcoming the resistances'. We consider such a problem in the next example.

Example: A train of total mass 250 tonne brakes with a uniform velocity of 10 m s^{-1} down a slope inclined at 30° to the horizontal. If the engine is working at the rate of 100 kW find the total resistance to the motion. ($g = 9 \cdot 80$ m s^{-2})

SOLUTION: Let Fig. 200 represent the forces on the train. T is the tractive force and R is the total resistance to the motion.

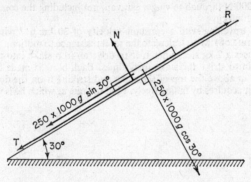

Fig. 200

Equating resultant force to mass × acceleration along the plane, we get

$$T + 250 \times 1000g \sin 30° - R = 250 \times 1000a = 0 \qquad \text{(i)}$$

since the train is travelling with a uniform velocity.

We have T and R both unknown and R to be found. The work done by T every second is given by the power equation:

$$T \times 10 = 100 \times 1000 \text{ J s}^{-1} = 100 \times 1000 \text{ W}$$
$$\therefore T = 10 \times 1000 \text{ newtons}$$

Substituting this result in eqn. (i) we get

$$10\,000 + 250 \times 1000 \times 9 \cdot 8 \times 0 \cdot 5 - R = 0$$
$$R = 10\,000 + 1\,225\,000 = 1\,235\,000 \text{ newtons}$$

The total resistance to motion was 1235 kN   (*Answer*).

Exercise 73 (Take $g = 9 \cdot 80$ m s^{-2})

1. A man of mass 70 kg walks upstairs through a vertical distance of 3 m in 5 s. Find the rate at which he is increasing his potential energy.

2. A factory conveyor belt carries 100 crates of bottles per minute at a speed of $0 \cdot 25$ m s^{-1} through a vertical height of 1 m. If each crate has a mass of 16 kg, calculate the power used by the conveyor on the crates alone.

3. A lorry is travelling horizontally along a road at a constant speed of 18 m s^{-1}. Find the power of the engine at this moment if the total resistance to the motion is 400 N.

4. A horse draws a plough through a field at a constant speed of $0 \cdot 5$ m s^{-1}. If the horse is working at the rate of 750 W, find the resistance to the motion.

5. A water pump ejects 600 kg of water per minute with a velocity of 10 m s^{-1} after raising it through a vertical height of 10 m. Calculate the power of the engine if 25 per cent of its output is wasted by internal frictional forces, etc.

6. A bullock is harnessed at the end of a 5-m beam with the other end fixed into a capstan turning a waterwheel. If the bullock exerts a force of 300 N on the beam at a point 5 m from the centre of rotation and walks round the circle once every minute, at what rate is the bullock working?

7. A car of mass 1000 kg runs downhill with a constant speed of 15 m s^{-1}. If the slope of the hill is 30° and the engine is working at the rate of 1500 W, calculate the resistance to the motion.

8. If the maximum power of a car is 10 000 W, calculate the maximum speed at which it can ascend a slope inclined at 30° to the horizontal against a total resistance to motion of 1000 N through towing a caravan, not including the component of the car's weight of 1000 kg.

9. A ship is travelling with a constant velocity of 30 km h^{-1} with the engines working at the rate of 1 MW. Calculate the total resistance to motion.

10. A man uses a 7-kg sledge-hammer to drive an iron stake into the ground. It takes him 5 min to strike the stake forty times. Each blow is made by raising the hammerhead 3 m above the top of the stake and striking it on the downward path with the speed it acquires by falling freely. Find the rate at which he is working.

CHAPTER THIRTEEN

MACHINES

In applied mathematics a machine is any device or piece of apparatus which enables a resisting force or load applied at one point of the machine to be overcome by a force called the effort applied at some other point of the machine. Normally we use a machine to gain some kind of advantage, which may or may not be measured in terms of increasing the force we can apply. Other machines may be used solely to speed up a process, e.g. the gears which increase the speed of a cyclist riding downhill.

(1) Mechanical Advantage

As always in mathematics we wish to obtain a means of measuring the advantage of using a machine in precise terms. We find that a direct comparison of the magnitudes of the resistance and the effort forces will give a useful measure of this advantage which we define as follows.

DEFINITION: *Mechanical advantage* $= \dfrac{\textit{Magnitude of the resistance or load}}{\textit{Magnitude of the effort}}$

We sometimes abbreviate this statement to

$$\text{M.A.} = \frac{L}{E}$$

but only where the meaning is obvious from the context, since the letters may have other meanings as for example E for Energy as well as Effort.

Thus any machine which enables a resistance of 600 N to be overcome by the application of an effort of 100 N is said to give a mechanical advantage of $600 \div 100 = 6$. If the machine has adjustable parts, it is possible that the mechanical advantage will lie between certain bounds; this is stated as

$$1 < \text{M.A.} < 6$$

(2) Levers

The first and simplest machine is the lever. With only three mechanical factors concerned in its practical use we classify the lever into three different types according to the relative positions of the factors which are load, effort, and fulcrum.

A **lever of the first type** (*LFE*) has the fulcrum between the points of application of the load and effort, and we have already discussed a lever in equilibrium under such an arrangement in Chapter Two.

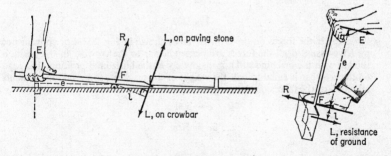

Fig. 201

Using a crowbar or a shovel as a lever of the first type to lift either paving stones or earth, we apply forces in the manner shown in Fig. 201. The suggested directions of the forces are the most likely. In each case, just before movement takes place, the three forces on the lever are in equilibrium and therefore concurrent. Having suggested the directions for both the resistance force L and the effort E, it follows therefore that the reaction force R on the lever at the fulcrum F will pass through the intersection of L and E.

Equally important, the moment of the forces about F are equal and oppo-site, i.e. equal in magnitude but opposite in direction. Thus $Ee = Ll$ in both cases. We can modify this relation to $\dfrac{e}{l} = \dfrac{L}{E}$ so that the mechanical advantage of the machine being used in this manner may be expressed as follows:

$$\text{M.A.} = \frac{\text{Perpendicular distance of fulcrum from line of action of effort}}{\text{Perpendicular distance of fulcrum from line of action of load}}$$

In practice this result is not worth remembering since students usually con-fuse the two expressions for M.A. when relying on memory alone. However, this form does have the merit that we may obtain a rough guide to the maxi-mum possible mechanical advantage without bothering to measure the forces. Thus by placing the fulcrum 0·1 m from the lifting end of a crowbar of length 1 m we know immediately that we shall obtain a maximum mechanical advantage of 9 if we get all the forces perpendicular to the lever.

Example: A machine fixed to a square base 2 × 2 m has a total weight of 1000 N. The machine is to be moved on rollers, and a crowbar of length 1·5 m is being used to lift the leading edge of the base. Taking the centre of gravity of the machine to be symmetrically above the centre of the base, find the position of the fulcrum which enables a man to apply half his weight of 600 N to lifting the machine.

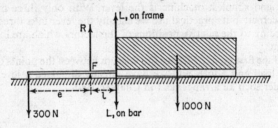

Fig. 202

SOLUTION: Let the forces on the crowbar be as shown in Fig. 202. For convenience we have assumed all the forces are perpendicular to the crowbar in the position shown, with the machine still approximately in the horizontal position. The load on the crowbar is equal to half the weight of the machine, so on taking moments about F for the forces on the crowbar we have

$$300 \times e = 500 \times l$$

$$\therefore 3e = 5l, \; e = \frac{5l}{3}$$

Since $e + l = 1\cdot5$ m, substitution for e yields $\dfrac{5l}{3} + l = 1\cdot5$

$$\therefore l = \frac{3 \times 1\cdot5}{8} = 0\cdot5625 \text{ m} \quad (Answer).$$

Note that the crudeness of this operation in general workshop practice does not justify offering a result to three decimal places. The result $l = 0\cdot56$ m would be satisfactory, but the position is usually found by trial and error.

Using a lever in the above manner we obtain a mechanical advantage which is greater than 1. At first sight it might appear that we are getting more work done by the lever than we are putting in, i.e. more output than input. However, on recalling that work is a product of force and distance, we shall need to find the distances moved by the points of application of the forces before speaking of input and output. Let us therefore examine the general situation in Fig. 203.

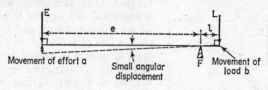

Movement of effort *a* — Small angular displacement — Movement of load *b*

Fig. 203

Not wishing to indulge in tedious proof, we suggest that if $e = 9$ and $l = 1$ and if the angular rotation of the lever about F is very small, then the distance which the effort moves is equal to the distance moved by the load multiplied by 9. To put this in general terms we have

$$\frac{a}{b} = \frac{e}{l} = \frac{L}{E} \text{ (see page 238)}$$

Hence $$Ea = Lb$$

which is simply another way of saying that the work done by E is equal to the work done by the force L on the body moved.

The **principle of work,** for machines in which the weight of the moving parts is ignored, states that *the work done on the machine by the effort (input) is equal to the work done by the force applied by the machine (output).* For a lever used in the manner of the examples above the principle of work is true. But in the majority of machines a large percentage of the input is wasted on internal frictional forces, in which case the principle is restated to suggest that the work done on the machine by the effort (input) is equal to the work done by the force applied by the machine (output) plus the work done by the internal resistance forces of the machine. Unless we are dealing with the ideal frictionless machine, this means that the input is always greater than the output.

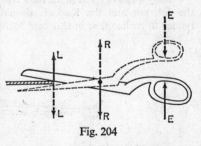

Fig. 204

A pair of scissors is an example of a double lever, and Fig. 204 illustrates the manner in which the forces are usually applied when cutting a piece of paper. The forces are in pairs and we have chosen a simple illustration in which all the forces are parallel, in which case the magnitude of the reaction at the pivot of the scissors is given by $R = E + L$.

With the same notation as before we have $Ee = Ll$ in the equilibrium position. By opening the scissors wider we decrease l so that L is increased considerably and, as experience shows, this is one of the reasons for getting a better cut by starting with the scissors wide open.

We obtain a **lever of the second type** (FLE) by placing the load in between the effort and the fulcrum. This arrangement ensures that the mechanical advantage will be greater than 1. The reader will be able to make a useful comparison with the above scissors if we take as our example of a lever of the second type a double lever in the form of the traditional nutcracker. In Fig. 205 we illustrate the forces which arise, and once again we have drawn them parallel. (The alternative is to have R symmetrical with respect to each lever as shown and then have L and E intersecting on the line of action of R; but we leave these diagrams to the reader in Exercise 74.) Again by taking moments about the fulcrum in the equilibrium position, we have

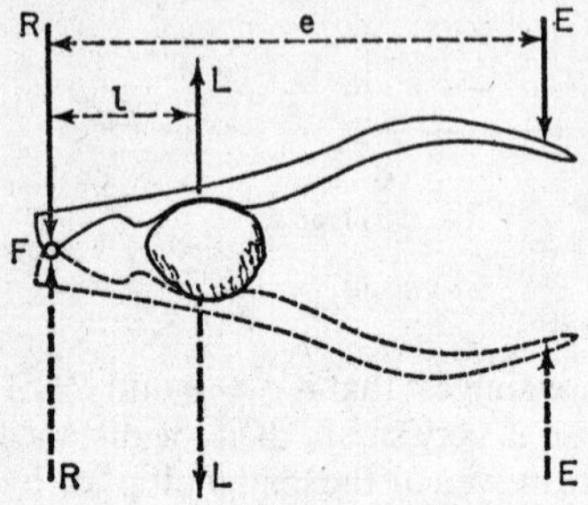

$$Ee = Ll \text{ whence } E = L \times \frac{l}{e}$$

and since $e > l$ we see that $E < L$ which confirms the opening remarks that the M.A. > 1.

Fig. 205

Example: The average man can crack walnuts by forcing one nut against another in the palms of his hands with a force of approximately 300 N. If the nutcracker of Fig. 205 has $l = 30$ mm and $e = 120$ mm, find the effort necessary to ensure that the walnuts crack.

SOLUTION: We have $L = 300$ N. Taking moments about the fulcrum we have

$$E \times 120 = 300 \times 30$$
$$E = \frac{300 \times 30}{120} = 75 \text{ N } (Answer).$$

—a considerable saving on both hands and effort

Finally we have **levers of the third type** (*FEL*), in which the effort lies between the fulcrum and the load. It should only take a little thought for the reader to realize that in this case M.A. < 1, so the reason for using such a

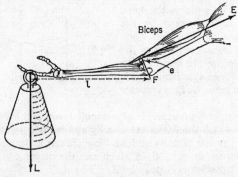

Fig. 206

lever is probably one of convenience, like using tongs to take baked potatoes out of the fire or cubed sugar from a bowl. The human arm is the most successful lever of this type, and the simplified illustration of Fig. 206 should give some idea of its operation. The biceps, to the front of the upper arm, is the principal muscle that bends the elbow and raises the muscle at the shoulder joint. (The triceps muscle at the back of the upper arm is used to straighten the elbow.) From the geometry of Fig. 206, with E the effort in the muscle and taking moments about the elbow pivot F, we see that $Ee = Ll$. Modified to $\dfrac{L}{E} = \dfrac{e}{l}$, this shows the mechanical advantage to be less than 1. Under these conditions we say that there is a 'mechanical disadvantage'.

The average arm operates at M.A. $= \frac{1}{10}$ to $\frac{1}{9}$. Just to get some idea of the strength of the biceps, consider holding the usual airline baggage allowance of 200 N at arm's length for a few seconds. With $l = 320$ mm and $e = 35$ mm, approximately, we see that the effort made by the biceps is given by

$$E \times 35 = 200 \times 320$$
$$\therefore E = \frac{200 \times 320}{35} = \frac{12\,800}{7} = 1830 \text{ N approximately,}$$

which is equal to the weight of about three or four average men. Yet the mechanical advantage is only $\dfrac{L}{E} = \dfrac{35}{320} = \dfrac{7}{64}$ (approximately $\frac{1}{9}$).

Another well-known lever of the third type is the spring clip shown in Fig. 207, which is used in offices to hold a large number of sheets of paper. In this case the spring is in permanent tension and is used to provide both the

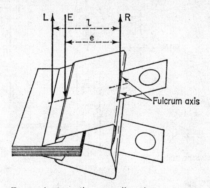

Forces shown on the upper lip only

Fig. 207

effort and the fulcrum at one and the same time. The method of relating the magnitude of the load and effort is exactly the same as the above. Notice once again that because the effort is nearer to the fulcrum than the load, its magnitude will be greater than the load, so the mechanical advantage will be less than 1.

Exercise 74

1. Identify the type of lever in each of the diagrams of Fig. 208.
2. Using the information given in Fig. 208 calculate the mechanical advantage of each machine.
3. The force diagrams for both double levers in Fig. 204 and Fig. 205 were drawn with all parallel forces. Draw force diagrams for a case in which the load and effort are not parallel to the reaction at the fulcrum.

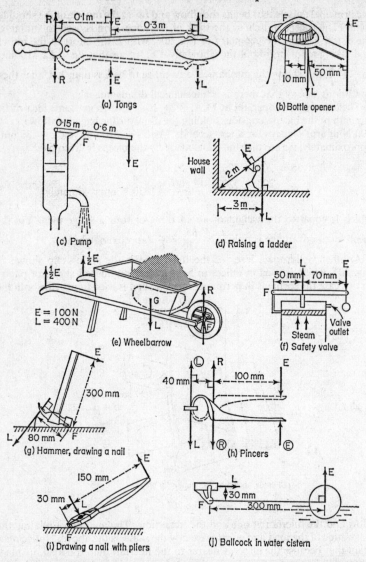

Fig. 208

4. The diagram Fig. 208 (*d*) illustrates one type of lever. Is the ladder always of this type during the motion?

5. Using the example of the human arm in Fig. 206 calculate the effort of the biceps muscle when a load of 350 N is held.

6. A crowbar of length 1·5 m has the lower 0·1 m jammed vertically between two heavy railway timbers each of weight 500 N and resting side by side on rough ground. If the coefficient of friction between the timbers and the ground is 1·2, what is the

least horizontal force applied to the free end of the crowbar which will widen the gap between the timbers?

(3) Wheel and Axle

Although the lever has many applications it is limited by the small distances through which it can move the load. Where a body has to be raised through large distances we can use a device called the wheel and axle. This consists of two right-circular cylindrical drums of common axis but different radii, about which are wound two separate ropes in opposite directions. The arrangement is illustrated in Fig. 209, and it can be seen that an anticlockwise

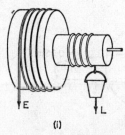
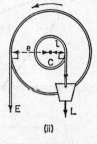

(i)   (ii)

Fig. 209

rotation obtained by pulling the effort rope will cause the load to ascend. Taking moments about C when the system is in equilibrium and ignoring any friction, once again we have $Ee = Ll$ but this time e and l are the radii of the drums. An examination of this result shows that the smaller we make l relative to e, the greater the load which can be lifted for the same effort E.

With the circumference of a circle given by $2\pi r$ where r is the radius, we see that one revolution of the apparatus will give a movement of $2\pi e$ for the effort and $2\pi l$ for the load. Since $e > l$ we see that, as might be expected, raising a heavy load through a small distance is achieved by exerting a light effort through a long distance. To illustrate this numerically suppose $e = 3l$; then $E = \frac{1}{3}L$, but the point of application of E will be moved three times the corresponding distance for L. Since these displacements are made in the same time, it follows that the speed with which the effort rope is moving is three times as great as the speed of the load rope.

The wheel and axle is not very common these days, but a similar arrangement of two pulleys of different radii keyed to the same shaft is very common in belt drives where the main purpose is to increase or decrease the angular speed of a final drive unit.

Example: A main drive shaft rotates at 150 rev/min. This power is transmitted to a machine via two belts as illustrated in Fig. 210. Find the angular velocity of the final drive wheel D assuming that no slipping takes place. Diameters of the four wheels are as indicated in the diagram.

SOLUTION: First, note the wheel-and-axle arrangement of the two pulleys B and C. Since the diameter of B is three times the diameter of A, 150 turns of A will give only 50 turns to B, and thereby 50 turns to C (since B and C are keyed to the same shaft).

Since the diameter of D is half the diameter of C, 50 turns of C will give 100 turns to D.

This arrangement of belts and pulleys therefore converts the original angular velocity of A of 150 rev/min into 100 rev/min of D (*Answer*).

Belt drives like the above rely upon friction to transmit the drive force. Slipping always takes place, so the answer obtained in the example above must be regarded as approximate. That is, since slipping takes place the true angular velocity of D will be less than 100 rev/min—how much less depends on the tightness of the belts.

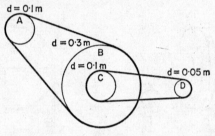

Fig. 210

The domestic sewing machine contains a belt drive similar to A, B, and C in Fig. 210, but C is usually the final drive shaft for operating the up-and-down motion of the needle.

Where slipping of the belts might interfere with the function of a machine, we use cogged wheels which may be connected by chain drives (as in the bicycle or caterpillar tractor) or permanently enmeshed (as in the gearbox of a car).

Example: A lawn mower has a constant-mesh drive which consists of a 74-cog wheel fixed to the roller shaft and a 12-cog wheel fixed to the cutter shaft linked by a dummy wheel with 56 cogs. If the roller has a diameter of 0·14 m, calculate the angular velocity of the cutters when the mower is pushed with a speed of 1 m s⁻¹.

SOLUTION: Let Fig. 211 represent the three cogged wheels of the lawn mower.

For one revolution of the roller, 74 teeth will pass the point A. Therefore 74 teeth on the dummy wheel will be moved past both A and B, thus giving the cutter shaft $74 \div 12 = 6\frac{1}{6}$ revolutions.

If the roller is pushed forward for 1 m in 1 s then, since the circumference of the roller is $\pi d = \frac{22}{7} \times 0.14 = 0.44$ m, the roller will make $\frac{1}{0.44}$ revolutions in 1 s.

But for each revolution of the roller the cutters make $6\frac{1}{6}$ revolutions. Therefore the cutter shaft will make $\frac{74}{12} \times \frac{1}{0.44} = 14$ rev/s (approx.) (*Answer*).

The reader has probably noticed that no calculation has involved the information that the dummy wheel had 56 teeth. Indeed, a dummy wheel of any size would have produced the same result since it is only used to transmit the motion. Its function is to make the blades rotate in the direction necessary for cutting grass.

(4) Velocity Ratio

In each of the two previous examples we have seen that the machine has been used mainly to effect a change in speed from one part to another. So we assess its usefulness not in terms of the ratio of load to effort, i.e. mechanical advantage, but in terms of a **velocity ratio** instead.

DEFINITION:

$$\textit{The velocity ratio of a machine} = \frac{\textit{Angular velocity of the input drive}}{\textit{Angular velocity of the output drive}}$$

We use the abbreviation V.R. for velocity ratio.

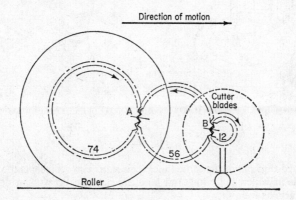

Fig. 211

In the worked example of Fig. 210 the angular velocity of the input drive was 150 rev/min and the angular velocity of the output drive was 100 rev/min. Therefore the velocity ratio of the machine was

$$\frac{150}{100} = 1\cdot5$$

Since we only require the ratio of the angular velocities it is not necessary to find the separate angular velocities explicitly. Thus in the worked example of Fig. 211 we know that the cutter shaft (output) makes $6\frac{1}{6}$ revolutions in the same time that the roller (input) makes 1 revolution. Therefore the velocity ratio of the lawn mower is given by

$$\text{V.R.} = \frac{1}{6\frac{1}{6}} = \frac{6}{37}$$

Expressing this ratio in the form $\frac{12}{74}$, the reader will realize how to obtain the result immediately from an inspection of Fig. 211.

The definition of velocity ratio given above is particularly suitable for the rotary motion of gear wheels and belt drives. We shall require some modification where the motion of either or both the load and effort is linear. To consider this point we now discuss the windlass and the capstan. Fig. 212 shows the windlass in its most favoured position astride a well. We suppose a constant effort E to be applied to the handle of radius e and thereby travelling a distance $2\pi e$ for one revolution of the handle.

At the same time the load L will descend or ascend a distance of $2\pi l$, where l is the radius of the winding drum. Now since the distance $2\pi e$ and $2\pi l$ have

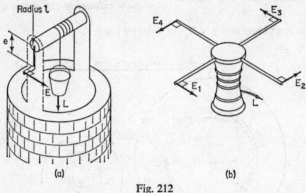

Fig. 212

been moved through in the same time, their ratio also represents the ratio of the speeds at which the effort and load are moving. So we can define the velocity ratio as follows:

$$\text{V.R.} = \frac{\text{Speed of the point of application of the effort}}{\text{Speed of the point of application of the load}}$$

$$= \frac{\text{Distance moved by point of application of the effort}}{\text{Distance moved by point of application of the load}}$$

where the distances are travelled in the same time.

Therefore the windlass has a velocity ratio given by

$$\text{V.R.} = \frac{2\pi e}{2\pi l} = \frac{e}{l}$$

Note that the distances moved by the point of application are measured in the direction of the load and effort forces. This observation will be especially important when discussing the inclined plane.

The capstan Fig. 212 (*b*) is exactly the same mechanical device as the windlass except that its axis of rotation is vertical instead of horizontal. It is also given the added refinement of three or four handles or levers to enable more effort to be applied. The anchor of the old wooden ships was raised and lowered by the use of a capstan.

Example: A windlass with a winding drum of radius 0·1 m rotated by a handle of radius 0·3 m is mounted astride a well. It is used to draw water in two buckets of equal weight and wound separately on the drum so that as one bucket descends empty the other rises full. If the water in a full bucket has a weight of approximately 900 N, calculate the mechanical advantage and velocity ratio of the windlass.

SOLUTION: Let W be the weight of each bucket. The weight of the full bucket is $(W + 900)$ N. A simple representation of the forces on the windlass is given in Fig. 213.

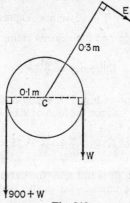

Fig. 213

In the equilibrium position with E perpendicular to the handle, taking moments about C we get

$$E \times 0\cdot3 + W \times 0\cdot1 = (900 + W) \times 0\cdot1$$
$$\therefore E \times 0\cdot3 = 900 \times 0\cdot1$$
$$E = 300 \text{ N}$$
$$\text{M.A} = \frac{L}{E} = \frac{900}{300} = 3 \quad (Answer).$$

The distance moved by the load and the effort in the time it takes for one revolution of the drum is $2\pi \times 0\cdot1$ and $2\pi \times 0\cdot3$ metres respectively.

$$\text{V.R.} = \frac{2\pi \times 0\cdot3}{2\pi \times 0\cdot1} = 3$$

The importance of the last example lies in the result M.A. = V.R. since it is natural to wonder if this relation is always true. We have already seen that the result of assuming that the output is equal to the input (i.e. that we have an ideal machine) is represented by $Ee = Ll$, where e is the distance moved by the point of application of the effort and l is the distance moved by the point of application of the load. This result may also be written as $\dfrac{e}{l} = \dfrac{L}{E}$

But we have

$$\text{V.R.} = \frac{e}{l} \text{ and M.A.} = \frac{L}{E}$$

which means that in the ideal machine M.A. = V.R. (Note that e and l here have different meanings from those given in previous pages of this chapter.)

In the practical situation the input Ee is greater than the output Ll and so we define the **efficiency** of the machine as follows:

DEFINITION: *Efficiency of a machine* $= \dfrac{Output}{Input}$

By 'output' we mean useful work performed by the machine in moving the load or resistance. Thus in the case of the ideal machine, where output = input, we have

$$\text{Efficiency} = 1 \text{ (i.e. 100 per cent efficiency)}$$

Furthermore, since $\dfrac{\text{output}}{\text{input}} = \dfrac{L}{E} \div \dfrac{e}{l}$, we may express the efficiency in terms of the mechanical advantage and the velocity ratio, thus

$$\text{Efficiency} = \frac{\text{M.A.}}{\text{V.R.}}$$

Returning to the windlass of Fig. 213, suppose that because of friction it takes an effort of 350 N to balance the load of 900 N. This leads to the result

$$\text{M.A.} = \frac{900}{350} = \frac{18}{7} = 2\tfrac{4}{7}$$

On the other hand friction does not alter the geometry of the machine: so the effort still travels three times as far as the load, and the velocity ratio remains unchanged at V.R. = 3. Hence the efficiency $= \dfrac{\text{M.A.}}{\text{V.R.}}$ of the windlass is $\dfrac{18}{7 \times 3} = \dfrac{6}{7}$ or approximately 86 per cent.

Exercise 75

1. Using the example of Fig. 210, calculate the angular velocity of the final drive wheel D if the wheel A is reduced to a diameter of 0·05 m and rotates at 90 rev/min. Find the velocity ratio of this arrangement.

2. Using the example of Fig. 211, find the velocity ratio obtained by interchanging the roller and dummy gear wheels.

3. A bicycle has a pedal wheel of 48 teeth with chain drive to a gear of 16 teeth on the back wheel. Find the velocity ratio of the machine and the speed of the bicycle if the pedal wheel is rotated at 1 rev/s and the back wheel has a diameter of 0·63m. $\left(\pi = \dfrac{22}{7}\right)$

4. If a windlass has a drum of radius 0·1 m and a handle turning in a circle of radius 0·25 m, calculate its efficiency if an effort of 300 N is needed to just lift a load of 700 N.

5. The drum of a capstan has a radius of 0·25 m and the 4 spoke levers are each of length 1·5 m measured from the axis of the capstan. If an anchor of weight 1500 N is to be raised, find the minimum equal effort to be exerted by each of four men on the capstan assuming

(i) the machine is ideal,
(ii) the machine is only 80 per cent efficient.

6. If in Fig. 212 (*b*) the spoke levers are each 1·5 m and the drum has a radius of 0·25 m at the point of contact of the rope, find the load which can be lifted when the efforts applied at right-angles to the levers are 100 N, 150 N, 120 N, and 160 N, when the efficiency is 60 per cent.

7. A carpenter's screwdriver has a cylindrical handle of radius 15 mm, and the end of its blade is 6 mm wide. Considering this tool as an ideal machine, find its mechanical advantage and velocity ratio.

(5) Inclined Plane

The inclined plane may be regarded as a machine since it enables us to raise loads by using forces of smaller magnitudes than the load. The arrangement of Fig. 214 shows an effort E being applied to a load of weight W.

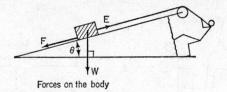

Forces on the body

Fig. 214

If we move the load a distance s along the plane, then the work done by E (input) is Es, while the output is measured by the change in potential energy of the load $Ws \sin \theta$ which we may consider as either $W(s \sin \theta)$ or $(W \sin \theta)s$. We know that the rough plane will not be an ideal machine because we shall have to do work Fs in overcoming the frictional resistance force F.

$$\text{M.A.} = \frac{W}{E}$$

$$\text{V.R.} = \frac{s}{s \sin \theta} = \frac{1}{\sin \theta}$$

Therefore, efficiency is

$$\frac{\text{M.A.}}{\text{V.R.}} = \frac{W \sin \theta}{E} \qquad \text{(i)}$$

By equating the resultant force to mass × acceleration parallel to the plane we have

$$E - F - W \sin \theta = Ma \text{ (i.e. } W = Mg)$$
$$\therefore E = W \sin \theta + F + Ma$$

so that even with $a = 0$ we must have $E > W \sin \theta$ and consequently we see from eqn. (i) that the efficiency must always be less than 1. However, the frictional force can be very useful since it may prevent the load sliding back down the plane as a fresh hold is taken on the hauling rope.

The motion of the **screw** is closely related to the inclined plane, as we shall see in Fig. 216. There are several points of interest about the screw. It enables us to convert a circular motion into a linear motion, for example tightening a bolt with the aid of a spanner. In this case we are using the screw to increase the thrust force in the direction of the bolt. The screw is also used to transmit a motion, as in the car jack shown in Fig. 215 (*a*), where the circular motion of the handle lifts one side of the car vertically.

Both practical experience and Fig. 215 should illustrate how one complete turn of the screw drives the collar on the screw up or down a distance equal to what is called the pitch of the screw thread. In Fig. 215 (*a*) the screw (bolt) is fixed and the collar (nut) moves up or down; in Fig. 215 (*b*) the collar is fixed and the screw moves up or down. In the car jack we use the screw to lift the load, while in the press we use the screw to increase the load or thrust.

To relate the inclined plane to the screw consider Fig. 216. On the left we have the profile of the inclined plane of slope θ. If you draw a similar profile and wrap it round an ordinary pencil you will obtain a profile of a screw. Keeping the paper round the pencil draw a line *DEFGB* parallel to

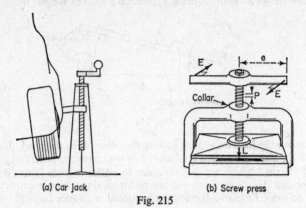

(a) Car jack   (b) Screw press

Fig. 215

the axis of the pencil, as in Fig. 216 (*b*). Now unwrap the paper and you will have four separate parallel lines *PD, QE, RF, BG* of equal length *p*. The length *p* is called the **pitch** of the screw (sometimes called the 'step' of the screw). We can readily see that if we hold the collar and give the screw four clockwise

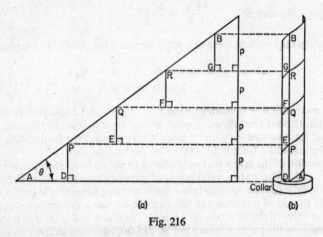

(a)   (b)

Fig. 216

turns (in our diagram) the collar will rise through a vertical distance $4p$. That is, a movement from *A* to *B* along the thread or inclined plane is equivalent to a rise of $4p$. Furthermore, the distances $AD = PE = QF = RG$ measure the circumference of the screw such that if the screw has a diameter of d, then $\pi d = AD$. Therefore the relation between the inclined plane and the screw may be expressed as $AB \sin \theta = 4p$, $AB \cos \theta = 4\pi d$.

Consider lifting a car of weight 4000 N from the side with the jack of

Fig. 215 (*a*). With two wheels off the ground the force exerted by the jack is approximately 2000 N. (Bear in mind that the whole of the force of 2000 N is borne by the stem of the jack, so there is a considerable risk in working under a car which is supported by the jack alone.) One turn of the handle raises the car a distance equal to the pitch of the screw. If the handle has a radius of 150 mm and the pitch is 3 mm and we assume that the machine is ideal, then the work done by the effort E is equal to the work done by the lifting force of 2000 N.

$$\therefore E \times 2\pi \times 150 = 2000 \times 3$$
$$E = \frac{20}{\pi} = 6\cdot4 \text{ N (nearest } 0\cdot1 \text{ N)}$$

The mechanical advantage of the jack is

$$\text{M.A.} = 2000 \div \frac{20}{\pi} = 100\pi$$

The velocity ratio is

$$\text{V.R.} = 100\pi$$

since the machine has been considered as ideal.

Even if the efficiency of the jack was only $0\cdot5$ because of friction, it would still enable the car to be lifted from one side by an effort of 13 N, i.e. no more than the weight of 13 average apples.

Example: The ideal screw press of Fig. 215 (*b*) is operated by applying two forces each of magnitude 50 N to the handle of radius $0\cdot3$ m. Calculate the thrust of the end-plate if the pitch of the thread is 10 mm. Find also the mechanical advantage of the press.

SOLUTION: The work done by the effort in one rotation of the handle is equal to the work done by the force L applied by the end-plate in moving through a distance of 10 mm.

$$\therefore 2 \times 50 \times 0\cdot3 = L \times 0\cdot01$$
$$30 = L \times 0\cdot01$$

i.e. $L = 3000$ N (which is a force equal to three-quarters the weight of the car in the above discussion).

Hence we have the

$$\text{M.A.} = \frac{3000}{2 \times 50} = 300$$

Exercise 76

1. An effort of 400 N is required to move a weight of 700 N along a rough plane inclined at 30° to the horizontal. Calculate the mechanical advantage and the velocity ratio of the plane.

2. If the efficiency of an inclined plane of slope 30° is only $0\cdot45$, calculate the effort needed to move a load of 300 N up the plane.

3. Using Fig. 216, if $AB = 100$ mm and $\theta = 30°$, calculate the pitch p of the thread and the diameter of the screw.

4. A car jack has a turning handle of radius 160 mm and a screw thread with a pitch of 4 mm, calculate its velocity ratio.

5. If the jack in Question 4 has an efficiency of $0\cdot6$, calculate the effort which needs to be applied to the jack in order to lift a car of weight 8000 N from the side.

6. Find the velocity ratio of the screw press in the worked example above. If the press had only been 40 per cent efficient, calculate the effort necessary to exert a thrust of 3000 N.

(6) Pulleys

The pulley system is probably the most widely used machine for the movement of loads through large distances whether by hand or by mechanical means such as cranes and hoists. The pulleys are usually housed in blocks and system operated by applying the effort to one end of a rope. In Fig. 217

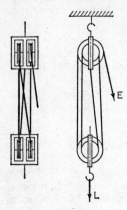

a system of four pulleys in two blocks is used to lift a load L. The usual convention is to draw the system from a side view making the pulleys a *different size for ease of illustration.*

Taking the ideal situation where there is no friction between the rope and the pulleys, the tension will be the same throughout the length of the rope. If the effort is a force E, then this will be the tension throughout the rope. There are four sections of rope supporting the load and the lower pulley block, each section exerting an upward force of E. We may ignore the weight of the moving block since it will be small compared with the load. Therefore, if we consider the machine to be ideal, the equilibrium of the load is given by $4E = L$.

The mechanical advantage of the system is

Fig. 217　　　　　$$\text{M.A.} = \frac{L}{E} = 4.$$

We can find the velocity ratio by considering the load to rise 1 m. This will mean that each of the four sections of the rope will drop 1 m of slack, so the effort will need to move 4 m in order to keep the rope taut.

The velocity ratio if the system is therefore V.R. = 4. (We could of course have deduced this result from the relation M.A. = V.R. for the ideal machine.)

Garage workshops frequently use a system called a **Weston differential pulley** which has the advantage of being controllable by one man without the need to anchor the free end to stop the load from falling back, a point of special importance when removing and replacing car engines. The upper pulley block is fixed and contains two cogged wheels keyed to the same axis but having different radii in the manner of a wheel and axle. An endless chain is threaded over the pulleys and supports the movable cog wheel which carries the load L, as shown in Fig. 218.

To understand how the system works, consider the effort E to be applied as shown and giving the upper wheels one complete rotation. This means that the effort has moved through a distance $2\pi a$ and has done work given by $E \times 2\pi a$. While the larger pulley has wound a length $2\pi a$ out of the loop supporting L, the smaller pulley of radius b has wound a length $2\pi b$ back into the loop supporting L. The combined result of this one revolution is to decrease the loop by a length $2\pi a - 2\pi b = 2\pi(a - b)$, thus raising the load through a vertical distance of $\pi(a - b)$.

The velocity ratio of the system is therefore

$$\text{V.R.} = \frac{2\pi a}{\pi(a - b)} = \frac{2a}{a - b}$$

This result suggests that by making $a - b$ very small we can greatly increase the velocity ratio and thereby obtain a finer adjustment for lowering heavy machinery into small spaces with little margin for error. Notice that the size of the movable pulley is not concerned in the calculations.

Naturally we do not want the load to run back once the effort is removed. In a frictionless or ideal machine this will always happen. It follows that we would be prepared to use a system with considerable internal frictional

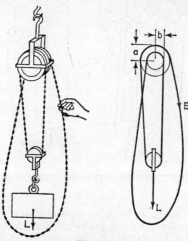

Fig. 218

forces to prevent this happening even at the expense of a low efficiency. We can investigate this point with a comparatively simple equation which follows from the principle of work for any machine:

Input = Output + Work done by internal resistance forces

If the system is left to itself the load will fall back if the work it can do in any descent is greater than the work done by the internal resistance forces. Therefore the output must always be less than the work done by the internal resistance if the load is to remain suspended at rest when the effort E is removed. If

$$\text{Input} = \text{Output} + \text{Work greater than output}$$

then

$$\text{Input} > \text{Output} \times 2$$

so the efficiency must be less than 0·5 in order to prevent fall-back.

Example: A Weston differential pulley block has fixed upper pulleys of diameter 150 mm and 120 mm. A force of 200 N is necessary to raise a load of 800 N. Find the velocity ratio and mechanical advantage of the machine, and determine whether the load can be left unattended without falling back.

SOLUTION: From the discussion above we know that the velocity ratio is given by

$$\text{V.R.} = \frac{2 \times 150}{150-120} = 10$$

The mechanical advantage is

$$\text{M.A.} = \frac{800}{200} = 4$$

Therefore the efficiency is

$$\frac{\text{M.A.}}{\text{V.R.}} = \frac{4}{10} = 0·4$$

Since the efficiency is less than 0·5 the load will not fall back   (*Answer*).

Exercise 77

1. Find the mechanical advantage, velocity ratio, and efficiency of each of the pulley systems in Fig. 219. (Assume that all pulleys are the same size.)

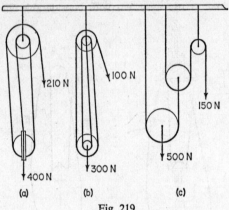

Fig. 219

2. A Weston differential pulley block has the equivalent of 16 teeth on the larger and 12 teeth on the smaller of the two fixed upper pulleys. If an effort of 200 N is needed to raise a load of 350 N, find the efficiency of the machine.

3. A man passes a rope over a smooth fixed pulley. He ties a loop at one end in which he sits and then proceeds to pull on the other end of the rope in order to raise himself. If his weight is W, what force must he exert on the rope in order to pull himself up?

(7) The Law of the Machine

We have discussed the relation between the load and the effort in the ideal machine, and we have speculated on the efficiency of some actual machines, but the individual characteristic of any machine can only be discovered through practical experiment. The relation discovered between the load and corresponding effort is called the **law** of that particular machine. Naturally the law is sometimes difficult to formulate in simple algebraic terms, and we have to draw a graph of the experimental results in order to relate fully the load with the effort.

Very often the relation is approximately a straight-line graph given by an equation of the form

$$E = mL + c$$

(Those readers familiar with co-ordinate geometry will recognize the form $y = mx + c$, in which the slope of the line is given by m and the intercept on the y axis is given by c.) If we find m and c we shall have obtained the law of the machine.

The relation between E and L will give us the mechanical advantage since

$$\text{M.A.} = \frac{L}{E} = \frac{L}{mL + c}$$

The geometry of the machine will give us the velocity ratio. Combining these two results we shall obtain the efficiency of the machine.

Example: The following table gives a set of results relating the effort applied and the load moved during the test of a machine with a velocity ratio of 5.

Load: 100, 200, 300, 400, 500 N

Effort: 34, 62, 89, 115, 140 N

(i) Plot a graph of effort against load and obtain the law of the machine.

(ii) On the same axes plot the graph of efficiency against load.

SOLUTION: Choosing scales of 1 mm to 10 N for the load, and 1 mm to 2 N for the effort, we obtain the straight-line graph of Fig. 220 which best fits the results. Now we know the relation is of the form $E = mL + c$ because the graph is a straight line, so we need two points on this line to give us the numerical value of m and c.

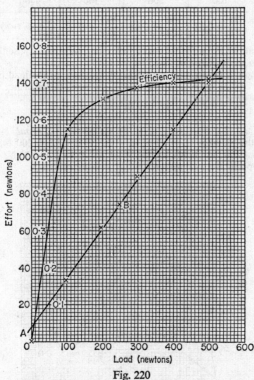

Fig. 220

First point A: when $L = 0$, $E = 8$; substituting in the equation $E = mL + c$ we have $c = 8$. We observe that this means that an effort of 8 N is required to set the machine in motion, i.e. to overcome internal resistance before any load can be moved.

Second point B: when $L = 250$, $E = 74$; substituting in the equation $E = mL + 8$ we have

$$74 = 250m + 8$$

$$m = \frac{66}{250} = 0 \cdot 264$$

The law of the machine is therefore $E = 0 \cdot 264 L + 8$ (*Answer*).

The mechanical advantage for any load is given by

$$\text{M.A.} = \frac{L}{E} = \frac{L}{0\cdot264L + 8}$$

From this result we see at once that the mechanical advantage varies with the load.

The efficiency for any load L is given by

$$\frac{\text{M.A.}}{\text{V.R.}} = \frac{L}{5(0\cdot264L + 8)} = \frac{L}{1\cdot32L + 40} \tag{i}$$

To plot the graph of efficiency against the load we now compute the table of results as follows by substituting the values of L into the efficiency equation (i)

Load:	0	100	200	300	400	500 N
Efficiency:	0	$\frac{100}{172} = 0\cdot58$	$\frac{200}{304} = 0\cdot66$	$\frac{300}{436} = 0\cdot69$	$\frac{400}{568} = 0\cdot70$	$\frac{500}{700} = 0\cdot71$

Using a scale of 1 mm to 0·01 for the efficiency, we superimpose the efficiency–load curve on to the graph of Fig. 220.

From the graph we see that the efficiency increases as the load increases, but only up to a certain value. In this case as L increases the efficiency will never be greater than $\frac{1}{1\cdot32} = 0\cdot76$ (two decimal places).

Example: The law of a machine is given by the equation $E = mL + c$. Obtain m and c from the readings ($L = 50$ N, $E = 30$ N) and ($L = 200$ N, $E = 90$ N). Assuming that this law holds for the machine up to $L = 500$ N, plot a load–efficiency graph on the assumption that the velocity ratio is 5.

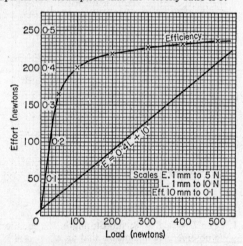

Fig. 221

SOLUTION: Substituting the given readings into the suggested equation we have

$$30 = 50m + c \tag{i}$$
$$90 = 200m + c \tag{ii}$$

Subtracting eqn. (i) from eqn. (ii) we obtain

$$60 = 150m \text{ hence } m = 0\cdot4$$

Substituting this result for m in eqn. (i) we have

$$30 = 20 + c$$
$$c = 10$$

The law of this machine is given by $E = 0.4 L + 10$ (*Answer*).

The mechanical advantage is

$$\frac{L}{E} = \frac{L}{0.4L + 10}$$

Since the velocity ratio is 5, the efficiency is

$$\frac{L}{5(0.4L + 10)} = \frac{L}{2L + 50}$$

As in the previous example we now make a table of results relating the efficiency to the load.

Load:	0	50	100	200	300	400	500 N
Efficiency:	0	$\frac{1}{3}$	0.4	0.44	0.46	0.47	0.48

The graph of these results is given in Fig. 221, from which we see that efficiency improves as the load increases but the maximum possible efficiency of the machine is 0.5 or 50 per cent.

Exercise 78

1. Using the example of Fig. 220, find the mechanical advantage and the efficiency of the machine when the load is 150 N.

2. If the pulley system in Question 1 (i) of Exercise 77 requires an effort of 10 N to set it in motion without any load being supported, find the law of the machine and plot a load–efficiency graph for loads of 0, 200, 400, 600, and 800 N. Find the greatest efficiency with which the system can work.

3. A machine test on a Weston differential pulley block with a velocity ratio of 10 gave the following results:

Load:	100	200	300	400	500	600 N
Effort:	40	60	80	100	120	140 N

Find the law of the machine in the form $E = mL + c$ and plot the load–efficiency graph. Does the machine have sufficient friction to prevent a fall-back of the load?

4. Two readings taken from a machine-test graph of load against effort are $(L = 100$ N, $E = 50$ N) and $(L = 300$ N, $E = 110$ N). Find

(i) the law of the machine in the form $E = mL + c$;
(ii) the efficiency of the machine for a load of 500 N;
(iii) the maximum possible efficiency of the machine if the velocity ratio is 20.

CHAPTER FOURTEEN

HYDROSTATICS I: DENSITY, RELATIVE DENSITY, AND BUOYANCY

Hydrostatics is the study of fluids at rest, or alternatively the study of the system of forces which keeps a fluid in equilibrium. Most people regard a fluid and a solid as two separate substances, but we consider them to be two different states of mass. As suggested by the example of water, there are in fact three different states for a mass: ice (solid), water (liquid), and steam (gas) is the same mass at different temperatures. Most substances exist in these three states. We shall therefore identify a fluid as being either a liquid or a gas. The difference between a fluid and a solid is not only a question of temperature, but also one of ability to retain shape in spite of the applied forces which attempt to change that shape. We can generalize these ideas in the following definitions:

(*a*) A **rigid body** or **solid** is a state of mass which maintains the same shape under the application of any system of forces such that the distance between any two points of the body remains unchanged.

(*b*) A **fluid** is either a liquid or a gas which yields to any applied force. We emphasize *any force* because a fluid is such that it will yield to any force no matter how small the force may be, in other words we are not concerned with the time it takes. Thus cold treacle and honey are fluids just like water and there is no difference between them as far as our definition is concerned, for, given the time, they will all yield to the smallest of forces.

(*c*) A **liquid** is a fluid whose volume remains practically constant under the action of any force however great its magnitude.

(*d*) A **gas** is a fluid which can be made to occupy a volume of any shape or size.

(*e*) Since we know from everyday experience that some fluids offer more resistance than others to the applied forces, e.g. stirring tea is easier than mixing paint, we choose to grade fluids in terms of this resistance by saying that one fluid is more **viscous** than another.

(*f*) A fluid which offers no resistance to any movement through it, is called a **perfect fluid**. This of course is an ideal mathematical creation since there is no such fluid in practice.

(*g*) A fluid which offers resistance to change of shape or any movement through the fluid is called a **viscous fluid**. To say that one fluid is more viscous than another means that it offers more resistance to a movement of some of its parts relative to others.

To summarize: solids are permanent in shape; liquids are incompressible; gases are compressible and able to occupy any container space available.

(1) Density

In statics we only compared the properties of two bodies when we discussed the frictional forces which arose from their contact. Now we find that with fluids we are already speaking of differences of viscosity and whether the fluid is a liquid or a gas. So before studying the effects of forces on fluids we evidently need to study the fluids themselves.

Since we are concerned with mass and volume we shall catalogue both solids and fluids in terms of the relation between their mass and the volume which that mass occupies—a relation that is called the **density**.

DEFINITION: *The density of a body is the mass of a unit volume of the body.*

The basic SI units of mass and volume are the kilogramme and cubic metre. It follows therefore that the basic SI unit of density is the kilogramme per cubic metre, abbreviated to kg m^{-3}. To put this in the form of an equation we have

$$\text{Density of a substance} = \frac{\text{Mass of the substance}}{\text{Volume of the substance}}$$

Thus, if 3 kg of a substance has a volume of 0·004 m^3, the density of the substance is

$$\frac{3}{0·004} = 750 \text{ kg m}^{-3}$$

If the volume is given in cubic millimetres or cubic centimetres the expression of density in SI units occasions no difficulty since the required arithmetic is elementary. For example, suppose 20 g of a substance has a volume expressed in the older but more convenient units of 5 cm^3: then the density is simply

$$\frac{0·020 \text{ kg}}{5 \times (0·01)^3 \text{ m}^3} = \frac{0·02}{5 \times 0·01 \times 0·01 \times 0·01} \text{ kg m}^{-3} = 4000 \text{ kg m}^{-3}$$

Thus 4 g cm^{-3} becomes 4000 kg m^{-3}, and similarly 7·1 g cm^{-3} would become 7100 kg m^{-3}, so the conversion is a simple one to perform. The measurement of volume in cubic metres is very inconvenient in actual laboratory work, so we shall express volumes in cubic centimetres whenever the practical nature of an example makes this necessary.

Densities of Common Substances
kg m^{-3}

Water (at 4 °C)	=	1 000
Sea-water	=	1 024
Aluminium	=	2 710
Copper	=	8 930
Gold	=	19 320
Iron	=	7 860
Lead	=	11 370
Mercury	=	13 600
Silver	=	10 500
Tin	=	7 290
Cork	=	220–260
Methylated spirit	=	830
Turpentine	=	870
Zinc	=	7 100

The density of a solid body varies very little. But with gases the influence of pressure and temperature is very great, and we shall discuss the effects of this later.

Since most liquids and bodies expand when the temperature is raised, it follows that temperature will influence volume and thereby density. This influence is only slight, but we agree to quote the density of water at a temperature of 4 °C.

Examining the table above we see that as the density of water is 1000 kg m^{-3}, or 1 g for every cubic centimetre, this enables a straightforward comparison to be made between water and every other body or liquid. Thus we can see straight away that, taking equal volumes, sea-water is heavier than fresh water, and mercury is 13·6 times as heavy as water, and so on. (We shall only use the term water to mean fresh water.) In this respect we are comparing all other substances to water in the sense that the density of water is used to measure the densities of all the other substances. For example, if we know that tin is 7·29 times as heavy as water, or that zinc is 7·12 times as heavy as water, then we know immediately that the density of tin is 7290 kg m^{-3} and the density of zinc is 7120 kg m^{-3}. That is, by taking the density of water as a basic unit we can express the density of any other substance relative to water simply by giving a number. This number is called the **relative density**.

DEFINITION: *The relative density of a substance is the ratio of the mass of a given volume of the substance to the mass of an equal volume of water at 4 °C.*

Alternatively, the relative density of a substance is the ratio of the weight of a given volume of the substance to the weight of an equal volume of water at 4 °C, since the ratio of the weights is the same as the ratio of the masses.

Note that the relative density (or specific gravity as it used to be called) is a number, so that if we say that the relative density of magnesium is 1·74 this means a density *relative to water*. The density of magnesium is therefore 1·74 × 1000 kg m^{-3} or 1740 kg m^{-3}. Arsenic has a relative density of 5·72, so its density is 5720 kg m^{-3}. Clearly 1 cm^3 of magnesium has a mass of 1·74 g, and 1 cm^3 of arsenic has a mass of 5·72 g. Any substance which has a relative density greater than 1 is 'heavier' than water, and any substance with a relative density less than 1 is 'lighter' than water.

In conclusion therefore, the relative density of a substance with volume V is:

$$\frac{\text{Mass of substance}}{\text{Mass of equal volume of water}}$$

$$= \frac{\text{Mass of substance} \div V}{\text{Mass of equal volume of water} \div V}$$

$$= \frac{\text{Density of substance}}{\text{Density of water}}$$

We observe that as the name implies we are expressing the density of the substance relative to the density of water. Because relative density is purely a number it is independent of any units of measurement.

Example: Given that the relative density of bismuth is 9·81, find the mass of 0·002 m^3 of the metal.

SOLUTION: We have already been informed that the density of water is 1000 kg m^{-3}. Substitution in the definition above gives

$$9 \cdot 81 = \frac{\text{Density of the substance}}{1000}$$

∴ Density of bismuth is 9810 kg m^{-3}

The mass of 0·002 m^3 is therefore 19·62 kg   (*Answer*).

Example: If the relative density of nickel is 8·9, find the volume of 50 g of nickel.
SOLUTION: If we knew the density of nickel we could easily find the volume.

As in the previous example we have

$$8.9 = \frac{\text{Density of substance}}{\text{Density of water}}$$

$$\therefore \text{Density of nickel} = 8.9 \times 1000 = 8900 \text{ kg m}^{-3}$$

This means that 8900 kg of nickel has a volume of 1 m^3.

$$\therefore \text{1 kg of nickel has a volume of } \frac{1}{8900} \text{ m}^3$$

$$\text{1 g of nickel has a volume of } \frac{1}{8900 \times 1000} \text{ m}^3$$

$$\text{50 g of nickel has a volume of } \frac{50}{8900 \times 1000} \text{ m}^3$$

$$= 0.000\,005\,62 \text{ m}^3 \quad (\textit{Answer}).$$

We must here accept that the basic SI unit of kg m^{-3} has become inconvenient, so we shall use the smaller unit of the cubic centimetre to make the arithmetic of ordinary laboratory practice more reasonable.

As already noted a density of 1000 kg m^{-3} is the same as 1 g cm^{-3}. Returning to our expression for relative density we have

$$8.9 = \frac{\text{Density of nickel}}{\text{Density of water}} = \frac{\text{Density of nickel}}{1 \text{ g cm}^{-3}}$$

Density of nickel is 8.9 g cm^{-3}.

It now follows that the volume of 50 g of nickel is

$$\frac{50}{8.9} = 5.62 \text{ cm}^3 \quad (\textit{Answer}).$$

Example: Find the density and relative density of zinc if 15 cm^3 of zinc has a mass of 106.5 g.

SOLUTION: Since density is the mass per unit volume we have

$$\text{Density of zinc} = \frac{106.5}{15} \text{ g cm}^3$$

$$= 7.1 \text{ g cm}^3 \quad (\textit{Answer}).$$

As already noted this is 7100 kg m^{-3}

The relative density of zinc $= \dfrac{\text{density of zinc}}{\text{density of water}}$

$$= \frac{7100}{1000} = 7.1 \quad (\textit{Answer}).$$

Note that density when expressed in g cm^{-3} is numerically the same as relative density.

Exercise 79

1. From the density table on page 259 find the relative density of the following:

   (i) sea-water, (ii) copper, (iii) turpentine, (iv) mercury.

2. From the density table on page 259 find the mass of 0.002 m^3 of the following:

   (i) iron, (ii) lead, (iii) silver, (iv) gold.

3. Express the following densities in kg m^{-3}:

   (i) 1 g cm^{-3}, (ii) 1.024 g cm^{-3}, (iii) 10 g cm^{-3}.

4. Arrange the following in order of heaviness with the heaviest substance last:

(i) lead, (ii) tin, (iii) silver, (iv) gold, (v) iron, (vi) copper.

5. Find the density of a substance whose relative density is 2·71.
6. State whether the following are fluids:

(i) putty, (ii) coal-tar, (iii) a dessert jelly, (iv) sand, (v) gear-box oil.

7. A body has a mass of 10 kg and a volume of 0·075 m³. Find its density and relative density.
8. Find the volume of a mass of substance with density ρ.
9. Find the volume of 10 kg of a substance with a density of (i) 2 g cm⁻³, (ii) 2 g m⁻³, (iii) 2 kg m⁻³
10. Find the volume of 1000 kg of (i) sea-water, (ii) methylated spirit, using the density table on page 259.
11. Find the volume of 10 kg of copper and 10 kg of zinc.

(2) Mixtures

We need no new theoretical knowledge to find the density of a mixture of a number of substances since we only require the total volume and the total mass of the mixture concerned. However, some substances contract by chemical action when they are mixed and this has to be borne in mind when computing the results. We can mix substances by two methods: by mass or by volume, as the following examples will demonstrate.

Example: Find the density of a mixture of fresh water and sea water by (i) equal volumes, (ii) equal weights, assuming that no loss of weight or volume takes place as a result of the mixing.

SOLUTION: (i) Let us take a cubic metre of each type of water. The density of the mixture will be given by

$$\rho = \frac{\text{Total mass}}{\text{Total volume}} = \frac{(1000 + 1024) \text{ kg}}{2 \text{ m}^3} = 1012 \text{ kg m}^{-3} \quad (\textit{Answer}).$$

(ii) Now let us take the mixture by equal mass. It makes no difference whether we take 1, 10, or 1000 kg of each substance since we are involved only in an arithmetical process which is not limited by the mass available. The volume of 1000 kg of fresh water is 1 m³ and the volume of 1000 kg of sea-water is

$$\frac{1000 \text{ kg}}{1024 \text{ kg m}^{-3}} = 0.9766 \text{ m}^3$$

$$\therefore \rho = \frac{\text{Total mass}}{\text{Total volume}} = \frac{1000 + 1000}{1 + 0.9766} = \frac{2000}{1.9766} = 1011.8 \text{ kg m}^{-3}$$

So the density is 1011·8 kg m⁻³ (*Answer*).

It is clear from the above example that the finding of the density of mixtures by mass is arithmetically more tedious than finding density of mixtures by volume. However, a closer inspection of the general principle reveals a slightly easier approach.

Consider mixing equal masses of three substances and finding the density of the resulting mixture. We need to know the separate volumes of each substance in the mixture, so let the density of the substances by ρ_1, ρ_2, ρ_3, and m the mass of each substance.

The volume of each substance is $V_1 = \dfrac{m}{\rho_1}$, $V_2 = \dfrac{m}{\rho_2}$, and $V_3 = \dfrac{m}{\rho_3}$.

If ρ = density of the mixture, the total volume of the mixture is $V = \dfrac{3m}{\rho}$

But assuming that there is no chemical contraction

$$V = V_1 + V_2 + V_3 \text{ so } \frac{3m}{\rho} = \frac{m}{\rho_1} + \frac{m}{\rho_2} + \frac{m}{\rho_3}$$

which simplifies to $\qquad \dfrac{3}{\rho} = \dfrac{1}{\rho_1} + \dfrac{1}{\rho_2} + \dfrac{1}{\rho_3}$

We see therefore that a use of reciprocal tables will render a quicker solution to such a problem.

If we have another look at part (ii) of the example above we see that the required result is given by

$$\frac{2}{\rho} = \frac{1}{\rho_1} + \frac{1}{\rho_2} = \frac{1}{1000} + \frac{1}{1024}$$
$$= 0 \cdot 001 + 0 \cdot 000\ 976\ 6$$
$$= 0 \cdot 001\ 976\ 6$$
$$\therefore \frac{1}{\rho} = 0 \cdot 000\ 988\ 3$$

Whence $\qquad \rho = \dfrac{1}{0 \cdot 000\ 988\ 3} = 1011 \cdot 8 \text{ kg m}^{-3}$ as before.

Example: Brass is an alloy of copper and zinc which varies in content from 95% copper + 5% zinc to 59% copper + 41% zinc. Other metals are sometimes added according to the intended use of the brass. Being flexible and quick-melting it is particularly useful in casting and general decorative work. Find the mass of copper and zinc in 10 kg of brass which has a relative density of 8·5.

SOLUTION: The density of copper is 8930 kg m^{-3} and the density of zinc is 7100 kg m^{-3}. Since we do not know the volume or the mass of either substance in the brass we suggest that there is m kg of copper and thus $(10 - m)$ kg of zinc. We do know, however, that the volume of brass is equal to the sum of the volumes of the copper and zinc.

i.e. $\qquad \dfrac{10}{8500} = \dfrac{m}{8930} + \dfrac{10 - m}{7100}$

Multiplying by 1000 we get

$$\frac{10}{8 \cdot 5} = \frac{m}{8 \cdot 93} + \frac{(10 - m)}{7 \cdot 1}$$

Using reciprocal tables we get

$$1 \cdot 176 = 0 \cdot 112m + (10 - m)\ 0 \cdot 1408$$
$$1 \cdot 176 = 0 \cdot 112m + 1 \cdot 408 - 0 \cdot 1408m$$
$$0 \cdot 0288m = 0 \cdot 232$$
$$m = \frac{232}{28 \cdot 8} = 8 \cdot 056$$

There is 8·056 kg of copper (80·56 per cent) and 1·944 kg (19·44 per cent) of zinc in this particular brass (*Answer*).

Example: Two buckets A and B each contain 10 litres (1 dm³) of liquid. The liquid in A has a relative density of 0·8 and the liquid in B has a relative density of 0·5. Two 1-litre jugs are now filled one from each bucket and then transferred into the other bucket. Find the relative density of the resulting mixture in the buckets.

SOLUTION: We observe that these mixtures are obtained from mixing by volume. Since 1 litre l = dm³ = 1000 cm³, the mass of a litre of water is 1000 g = 1 kg.

After the transfer we have

in A: 1 litre with $d = 0·5$, plus 9 litres with $d = 0·8$

in B: 1 litre with $d = 0·8$, plus 9 litres with $d = 0·5$

If the mixture in A has a relative density d, then (because the total volumes remain unchanged) we have

$$d = \frac{\text{Total mass}}{\text{Mass of equal volume of water}} = \frac{1 \times 500 + 9 \times 800}{10\,000} = 0·77$$

Therefore the relative density of the mixture in A is 0·77 (*Answer*).
Similarly in bucket B,

$$d = \frac{1 \times 800 + 9 \times 500}{10\,000} = 0·53$$

Therefore the relative density of the mixture in bucket B is 0·53 (*Answer*).

Example: Suppose the process of interchanging 1 litre of the contents of buckets A and B is repeated once again. What is the relative density of the mixture in the buckets?

SOLUTION:

$$\text{In } A\colon d = \frac{1 \times 530 + 9 \times 770}{10\,000} = 0·746$$

$$\text{In } B\colon d = \frac{1 \times 770 + 9 \times 530}{10\,000} = 0·554$$

Hence the relative densities of the mixtures in A and B are 0·746 and 0·554 respectively (*Answer*).

Exercise 80

1. A piece of gold of relative density 19·0 has a mass of 10 g and a volume of 1·6 cm³. Obtain the volume it should have and account for the difference.

2. A 10-gramme mixture of two liquids A and B has a relative density of 0·5. Find the mass of each liquid in the mixture if the relative densities of A and B are 0·4 and 0·8 respectively.

3. Find the mass of metal A ($d = 6·0$) and metal B ($d = 10·0$) which make 100 kg of an alloy ($d = 8·0$).

4. Modern bronze is an alloy of 85 per cent copper, 14 per cent tin, and 1 per cent phosphorus. It has very good casting properties and can also be worked and machined cold. An older piece of bronze has a mass of 16·2 kg and contains only copper and tin. Find the mass of copper ($d = 8·9$) and tin ($d = 7·3$) if the bronze has a relative density of 8·1.

5. Using the last of the worked examples above, find the relative density of the two mixtures after the transfer has been repeated for a second time. If this process was to continue indefinitely, what would be the final relative density in each case?

6. Five cubic centimetres of distilled water ($d = 1$) are added to 16 cm³ of sulphuric acid ($d = 1·83$) for use in a battery. Find the loss in volume if the relative density of the mixture is 1·714.

7. A piece of solid jewellery is made from an alloy of gold ($d = 19·3$) and silver ($d = 10·5$). If the relative density of the alloy is 12·7, find the ratio of the volume of silver to the volume of gold in the jewellery.

(3) The Relative-density Bottle

The calculation of results often obscures the interesting experiments which produce them. The main problem in the determination of density is that of finding the volume of the substance concerned. Liquids present no special problem because they can be poured into a measuring jar, but a solid body needs a different method which we shall consider later. For the present we shall consider a standard method of finding the relative density of a liquid using a relative-density bottle (this used to be called a specific-gravity bottle).

Basically the idea is very simple, because all we need is the mass of a volume of the liquid and then the mass of the same volume of water: but we must take care to ensure that we have the same volume in both cases. The bottle shown in Fig. 222 is designed to hold a set volume of liquid, and such bottles are manufactured in various sizes to hold 25, 50, and 100 cm³ of liquid. It is closed by a well-fitted glass stopper which has a hole drilled through it of about 1-mm diameter. The bottle is filled with liquid, then the stopper is pushed in forcing the surplus liquid to overflow through the hole in the stopper. The surplus is wiped off the outside of the bottle by using blotting paper.

Fig. 222

We first find the combined mass of the bottle and stopper. Next, we fill the bottle with water and find the mass of the bottle + stopper + water. By subtracting these two results we are able to find the mass of water the bottle will hold. The volume of liquid the bottle will hold is etched on the glass by the manufacturer, e.g. 25, 50, or 100 cm³. Finally we replace the water by the liquid concerned and, by weighing, find the mass of the same volume of liquid. Since we now have the mass of equal volumes of water and liquid we can find the relative density of the liquid.

Example: A relative-density bottle complete with stopper has a mass of 8 grammes when empty, 33 grammes when full of water, and 35·4 grammes when filled with a mixture of glycerine and water. Find the relative density of the glycerine mixture.
SOLUTION:
Mass of a bottleful of glycerine mixture is $35·4 - 8 = 27·4$ g
Mass of same volume of water is $33 - 8 = 25$ g

Hence relative density of the mixture is $\dfrac{27·4}{25} = 1·096$ *(Answer)*.

(The *density* of the glycerine mixture is therefore 1096 kg m⁻³.)

To obtain a result to two significant figures it is not necessary to have a special bottle. An ordinary narrow-necked olive-oil bottle is good enough. Find the mass of the bottle empty, mass of bottle + water, and mass of bottle plus liquid to the same level, then proceed as above.

To find the relative density of a solid. We can use the relative-density bottle in finding the relative density of a solid such as sand. We know the mass of the bottle + stopper and the mass of water it will contain. Pour sand into the bottle and then find the mass of the bottle + stopper + sand. Subtraction of results gives the mass of the sand alone. Leaving the sand in the bottle, top up with water (making certain that all air bubbles are released by shaking the sand) and then find the mass of the sand + water in the bottle. Let us tabulate these results for clarity.

Mass of bottle + stopper: B
Mass of sand + bottle + stopper: $S + B$
Mass of sand + bottle + stopper + water to fill bottle: $S + B + m$
Mass of bottle full of water + stopper: $B + M$

Now $M - m$ must be the mass of water displaced by the sand, i.e. the mass of an equal volume of water. Hence

$$\text{Relative density of sand} = \frac{\text{Mass of sand}}{\text{Mass of equal volume of water}} = \frac{S}{M - m}$$

An alternative way of thinking about this method is discussed in the following examples.

Example: Two grammes of powder which does not dissolve in water are poured into a relative-density bottle. When topped up with water the total mass of the bottle + stopper + water + powder is 34 g. When the bottle is filled with water only, the mass of water + bottle + stopper is 33 g. Find the relative density of the powder.
SOLUTION: Consider the two total masses on the balance in Fig. 223. When the powder is outside, the total mass on the balance is 33 + 2 grammes.

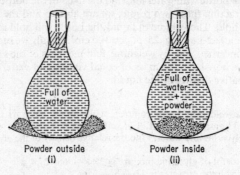

Powder outside   Powder inside
(i)   (ii)

Fig. 223

When the powder is placed inside, the total mass is 34 g, which means that the process of putting the powder inside the bottle has displaced a mass of 1 g of water. Therefore 1 cm³ of the bottle must now be occupied by the powder. In other words 2 g of powder occupy a volume of 1 cm³.

$$\text{Relative density of powder} = \frac{\text{Mass of powder}}{\text{Mass of equal volume of water}} = \frac{2}{1} = 2$$

                                                              (*Answer*).
(Since the relative density is 2 it follows that the density of the powder is 2000 kg m⁻³.)

Example: A marble of mass 2·1 g is placed in an R.D. bottle which is topped up with water. The mass of the full bottle with the marble inside is 63·26 g, and the mass of the bottle full of water is 62 g. Find the relative density of the marble.

SOLUTION: The total mass of the full bottle of water and the marble outside is 64·1 g. The total mass of the bottle and water with the marble inside is 63·26 g. Therefore putting the marble inside the bottle has displaced 64·1 − 63·26 = 0·84 g of water.

$$\text{Relative density of the marble} = \frac{\text{Mass of the marble}}{\text{Mass of an equal volume of water}} = \frac{2{\cdot}1}{0{\cdot}84}$$

$$= 2{\cdot}5 \quad (\textit{Answer}).$$

(Since the relative density is 2·5 it follows that the density of the marble is 2500 kg m^{-3}.)

Exercise 81

Throughout this exercise we shall assume a relative-density bottle having a mass of 8 g complete with stopper and holding 25 g of water when full and stopped.

1. What volume and mass of liquid with a relative density of 0·8 will the bottle hold?

2. When filled with turpentine our bottle has a total mass of 29·75 g. Find the relative density and density of the turpentine.

3. When 1·6 g of powder are poured into the relative-density bottle and the bottle filled with water and stopped, the total mass is 34·1 g. Find the relative density of the powder.

4. Three grammes of ball-bearings are placed inside the bottle and the total mass is 35·6 g. Find the relative density and density of the ball-bearings.

5. (i) How many grammes of methylated spirit ($d = 0{\cdot}82$) are required to fill our bottle?

(ii) When a 1 g chip of wax is dropped into the bottle filled with methylated spirit the total mass is 27·6 g. Find the relative density of the wax.

(4) Archimedes' Principle

The reader will have noticed how the volume of a body is obtained by finding the mass of water displaced and then using the fact that 1 g of water occupies a volume of 1 cm^3 to give the volume of the body. A density bottle is not essential for finding such a volume since we could fill a measuring jar with a known quantity of water, put the body in the jar and note the rise in the level of the water; this would tell us the volume of the body immersed. This method is acceptable for bodies which sink, but what of those which float? Using a pin, the body could easily be held under the water surface and the volume noted as before. However, on doing this we see that a force is necessary to push the body beneath the surface, and we should ask ourselves immediately whether we always need the same force to hold any body under. For example, does it depend on the size, shape, and weight of the body itself or even perhaps the liquid in which it floats?

Archimedes' principle which answers this question is probably the best-known principle in the whole of applied mathematics. To obtain some preparation for an understanding of the statement of the principle, consider the following simple experiment.

Take a heavy body with a shape which can be held in a cotton loop and suspend it from a spring balance to find its weight. Now raise a bowl of water underneath the body until it is completely immersed as shown in Fig. 224 (*a*). As the volume immersed in the water is increased so the readings of the spring will decrease and will continue to decrease until the body is completely immersed.

Let us look at a diagram of the forces on the body when it is in equilibrium completely immersed in the liquid as shown in Fig. 224 (*b*). The forces on the body are the tension T in the spring as registered by the scale on the spring and the weight mg of the body. Since we know that $T \neq mg$ there must be another force on the body due to its presence in the water. Furthermore, since $T < mg$

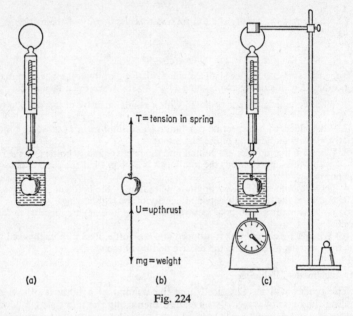

Fig. 224

this other force must be vertically upwards. We call this force the upthrust U of the water on the body. Since the body is at rest the resultant force in the body is zero, or

$$T + U = mg$$

By immersing the body in other liquids we find that U varies according to the liquid and the volume of the body which is immersed, e.g. doubling the volume immersed doubles the upthrust. Knowing the volume of the body and experimenting with other liquids we find that the upthurst U is equal in magnitude to the weight of the liquid displaced.

For example, when the body is a fully immersed cube of volume 1 cm³, the upthrust is found to be equal to the weight of 1 cm³ of water, i.e. $0.001g$ newtons. When the body is a heavy cube of $2 \times 2 \times 2$ cm³, the upthrust is found to be $0.008g$ newtons, i.e. the weight of 8 cm³ of water. Since $T < mg$ the body 'appears' to lose weight, according to the spring balance, so that we sometimes speak of either the upthrust or 'the apparent loss in weight' when referring to U.

The Principle of Archimedes states that when a body is immersed in a liquid the upthrust on the body is equal and opposite to the weight of liquid displaced. (The principle was discovered by Archimedes, a mathematician who lived in Syracuse in Sicily about 250 B.C.)

Looking again at the upthrust on the body in Fig. 224 (*b*), it follows from

Newton's Third Law that there must be an equal and opposite force on the water in the bowl. Fig. 224 (*c*) shows how this can be verified by placing the bowl of water on a pair of ordinary kitchen scales. The decrease in the reading of the spring balance will be 'balanced' by an increase of the same magnitude in the reading on the kitchen scales.

Archimedes' principle also provides us with an alternative method for finding the relative density of a body or a liquid. The relative density of a substance is

$$d = \frac{\text{Mass } m \text{ of the substance}}{\text{Mass } M \text{ of an equal volume of water}}$$

If we multiply this fraction by 1 in the form of g/g we get

$$d = \frac{mg}{Mg} = \frac{\text{Weight of the substance}}{\text{Weight of an equal volume of water}}$$

But we have seen that the weight of an equal volume of water is given by the magnitude of the upthrust when the substance is completely immersed. Therefore

$$d = \frac{\text{Weight of the substance}}{\text{Magnitude of the upthrust when the substance is completely immersed in water}}$$

Example: A body is suspended from a spring balance and registers a weight of 6 N. When the body is completely immersed in water the reading of the balance shows 4 N. Find the relative density of the body.
SOLUTION: The upthrust is 2 N. Therefore the relative density of the body is $\frac{6 \text{ N}}{2 \text{ N}} = 3$ (*Answer*).

Example: If the same body is completely immersed in a liquid of relative density 0·7, what will be the reading on the spring balance?
SOLUTION: We are required to find the upthrust on the body and we know that this is given by the weight of liquid displaced. Since the upthrust in water was 2 N, the weight of the same volume of liquid is $2 \times 0·7 = 1·4$ N.
Therefore the reading on the spring balance is $6 - 1·4 = 4·6$ N (*Answer*).
(Notice that the smaller the density the smaller the upthrust.)

Example: A body of weight 12 N suspended from a spring balance. The reading of the balance is 10 N when the body is totally immersed in water and 10·5 N when the body is immersed in a liquid. Find the relative density of the liquid and the body.
SOLUTION: The upthrust on the body in water is $12 - 10 = 2$ N, and this gives the weight of an equal volume of water. Therefore

$$\text{Relative density of the body} = \frac{12 \text{ N}}{2 \text{ N}} = 6 \quad (\textit{Answer}).$$

The upthrust when the body is immersed in the liquid is 1·5 N. This means that the volume occupied by the body has a weight of 2 N for water and 1·5 N for the liquid. Therefore the relative density of the liquid is $\frac{1·5 \text{ N}}{2 \text{ N}} = 0·75$ (*Answer*).

(The density of the body is 6000 kg m^{-3} and the density of the liquid is 750 750 kg m^{-3}.)

If a body floats in a liquid, the upthrust must be equal and opposite to the weight of the body, and a spring balance on which the body is suspended will give a zero reading. So if we wish to find the relative density of a body which floats we are faced with the problem of measuring its volume. We could apply a known force to the floating body in order to immerse it completely. This known force together with the weight of the body would tell us the volume of liquid displaced. In fact, all we do is to tie a sinker to the floating body and reason in the following way.

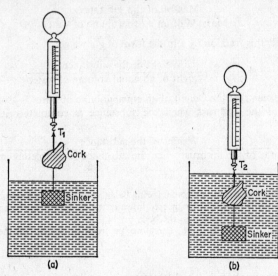

Fig. 225

Suppose we are trying to find the relative density of cork, and our specimen of cork has a weight of 0·5 N. A heavy body is tied to the cork and remains completely immersed in the water throughout the experiment, as shown in Fig. 225. Consequently the resultant force on the sinker remains unchanged. The difference between the readings T_1 and T_2 on the spring balance must be the result of completely immersing the cork, and therefore measures the upthrust on the cork when it is completely immersed. Since the cork is immersed in water this difference $(T_1 - T_2)$ is the weight of its own volume of water. Therefore the relative density of the cork is $\dfrac{0·5}{T_1 - T_2}$.

Exercise 82

1. A body with a weight of 10 N registers 8 N and 7·5 N when suspended from a spring balance and totally immersed in water and a liquid respectively. Calculate the relative density of (i) the body and (ii) the liquid.

2. A body of weight 6 N is suspended from a spring balance which registers 4 N when the body is completely immersed in water. What weight will be registered on the balance when the body is completely immersed in a liquid of relative density 0·9?

3. A piece of cork has a weight of 0·25 N and a relative density of 0·2. The cork is

completely immersed in water and anchored by cotton thread to the base of the containing vessel. What is the tension in the cotton?

4. A tennis ball has a volume of approximately 134 cm³ and a mass of approximately 56 g. Find the force necessary to hold the ball completely immersed in water.

5. A wooden cube floats in water with two-thirds of its volume immersed. Find the relative density of the wood and the vertical force necessary to hold the cube completely immersed if it has a weight of 4 N.

(5) The Hydrometer

Question 5 in the above exercise suggests that we could use a floating body for finding the relative density of a liquid. We know that in any liquid in which the body floats the mass of liquid displaced is not only equal to the mass of the body but is also related to the volume immersed and the density of the liquid. Thus a measurement of the volume immersed should provide us with a means of measuring the relative density of the liquid.

For example, 10 g of a liquid of relative density 0·8 has a volume of 12·5 cm³, therefore if a body of mass 10 g is placed in the same liquid it needs to be able to displace 12·5 cm³ of the liquid in order to float. If the volume of the body is less than this it will sink. Similarly if the body floats in a liquid of relative density 0·9 then $10 \div 0.9 = 11\frac{1}{9}$ cm³ will be immersed. Perhaps these results will prove to be more helpful in the form of a table.

	Volume of body immersed:	
	Body of mass 10 g	body of mass 2 g
R.D. of liquid	cm³	cm³
0·2	50	10
0·4	25	5
0·5	20	4
0·6	$16\frac{2}{3}$	$3\frac{1}{3}$
0·8	$12\frac{1}{2}$	$2\frac{1}{2}$
0·9	$11\frac{1}{9}$	$2\frac{2}{9}$
1·0	10	2
1·2	$8\frac{1}{3}$	$1\frac{2}{3}$

Examining the table we can see immediately that if the volume immersed of the 10-g body lies between 50 cm³ and 25 cm³, then it must be floating in a liquid whose relative density lies between 0·2 and 0·4. This suggests that if we had a scale on the body to indicate the volume immersed we could read off the relative density of the liquid at once. This is the principle behind an instrument called the **hydrometer**. One important point to notice is that equal differences in the relative density do *not* correspond to equal differences in volume immersed.

The common hydrometer consists of a straight glass tube closed at both ends but having bulbs at the lower end, one of which is weighted in order to ensure that the hydrometer floats in a vertical position as shown in Fig. 226. As the table above suggests, a small change in the relative density gives rise to a considerable change in the volume immersed. Hence, to test the whole range of liquids an enormous hydrometer would be necessary and likewise a considerable quantity of liquid to enable it to float. In practice we use different hydrometers for different ranges of relative density. An example of a commercial short-range hydrometer is a lactometer which is adapted for testing whether milk has been mixed with water. The upper and lower points of the scale correspond to pure water and pure milk and the intermediate marks indicate the proportions of milk and water in the mixture. Similarly the hydrometer for testing the strength of the acid solution in a car battery also has a limited range.

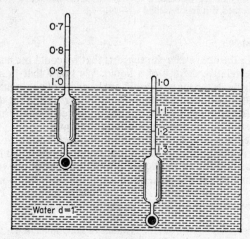

Fig. 226

Example: The cross-sectional area of the straight stem of a hydrometer is 0·2 cm². If the mass of the instrument is 15 g, and the base of its straight column corresponds to a mark of 1·0 for water, find the relative density of the liquid registered by a mark 6 cm farther up the stem.

SOLUTION: Since the instrument has a mass of 15 g it follows that when the mark 1·0 is reached the volume immersed must be 15 cm³. The mark which is 6 cm farther up the stem corresponds to an immersed volume of $15 + 1·2$ cm³. That is, we need to displace 16·2 cm³ of the liquid to make the hydrometer float. So 16·2 cm³ of the liquid has a mass of 15 g.

$$\text{The relative density of the liquid} = \frac{\text{mass of liquid}}{\text{mass of equal volume of water}}$$

$$= \frac{15}{16·2} = 0·926 \quad (Answer).$$

Example: The straight stem of a hydrometer has a cross-sectional area of 0·4 cm² and a mass of 20 g. Find the length of the stem necessary to register relative density readings from 1·00 to 1·25.

SOLUTION: In this case we need the 1·00 reading for water near the top of the stem. At this mark the volume immersed in water is 20 cm³ since the mass of the instrument is 20 g. For the reading of 1·25 we float the instrument in a liquid of relative density of 1·25 so that the mass of the immersed volume is 20 g. Now because 1 cm³ of the liquid has a mass of 1·25 g the immersed volume is

$$\frac{20}{1·25} = 16 \text{ cm}^3$$

Thus a volume of 4 cm³ of the stem lies between the readings. Since the stem has a cross-sectional area of 0·4 cm² the distance on the stem between the readings is 10 cm   (*Answer*).

Example: A common hydrometer of mass 20 g is calibrated to read relative densities from 1·0 to 0·8. State which reading is at the top of the scale and find the relative density of the liquid which gives a reading halfway between the two marks.

SOLUTION: The less dense the liquid the greater the volume of stem to be immersed to give the same upthrust to support the same floating hydrometer. Hence the 0·8 mark is above the 1·0 mark. (In fact the readings always decrease towards the top of the instrument regardless of the range of liquids for which the instrument is used.) When the hydrometer is placed in water the volume immersed is 20 cm³.

When the hydrometer is placed in liquid ($d = 0.8$) the volume immersed is $20 \div 0.8 = 25$ cm³.

The difference in volume is 5 cm³ between the stem marks 1·0 and 0·8. The mark halfway between 1·0 and 0·8 corresponds to an immersed volume of $20 + 2.5$ cm³, i.e. 22·5 cm³ of the new displaced liquid has a mass of 20 g.

The relative density of the new liquid $= \dfrac{20}{22 \cdot 5} = \dfrac{8}{9}$ (*Answer*).

Exercise 83

1. A body floats in water with 0·75 of its volume immersed, find its relative density.

2. A wooden cube floats two-thirds immersed in water with four edges vertical, find its relative density.

3. If a body floats with two-thirds of its volume immersed in a liquid of relative density 0·8, find the relative density of the body.

4. A common hydrometer has a mass of 20 g and floats in water with the base of its stem at the water level. If the stem has a cross-sectional area of 0·4 cm² and a length of 10 cm, find the range of liquids for which the instrument is suitable.

5. A common hydrometer has a mass of 20 g and a stem calibrated from 1·0 to 1·2. Find the relative density of the liquid which gives a reading halfway up the scale.

6. Some lead shot is placed in a flat-based test tube which is then sealed with a stopper. The tube has a uniform cross-sectional area of 1·5 cm² and floats exactly half immersed in water. If the complete apparatus has a mass of 6 g calculate the depth immersed.

7. Using the test tube of Question 6, what extra force is needed to push the tube 2 cm farther down?

8. If the test tube in Question 6 is floated in a liquid of relative density 0·8, what depth of the tube is immersed?

9. Take a jug half filled with water and place an egg in the water: the egg should rest on the bottom. Remove the egg and pour salt into the water and mix the salt and water. Provided enough salt has been added the egg should now float. Now place the fingers of one hand in the liquid and slowly pour fresh water over your hand into the jug until it is full. What happens to the egg? Why?

(6) Buoyancy and Equilibrium

We have already seen that a body will always float in a liquid provided the density of the body is less than the density of the liquid. We know from experience that it is easier to float in salt water than fresh water—as the experiment in Question 9 above confirmed. But we have only dealt with small bodies, some of which have probably been unstable, so how does an iron ship or a submarine manage to float? How does a submarine fire a torpedo without springing a leak or disturbing the aim? A very simplified explanation can be offered as follows.

When a ship floats at rest the resultant force on the ship is zero, that is, the upthrust is equal and opposite to the weight of the ship. But whereas the weight of the ship acts through its centre of gravity, the upthrust acts through the *centre of gravity of the displaced water*. This latter point, labelled H in Fig. 227, is called the **centre of buoyancy**.

If W is the magnitude of the total weight of the boat and contents acting through G, then the upthrust is W acting through H. When the boat rolls into the position shown in Fig. 227 (*b*) the shape of the displaced volume changes and the centre of buoyancy moves to H_1. So now the line of action of the

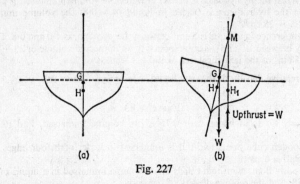

(*a*) (*b*)

Fig. 227

upthrust intersects the original line of symmetry in the point M, called the **metacentre**. If M is above G, then, taking moments about G, we see that the upthrust will tend to restore the boat to the symmetrical position shown in Fig. 227 (*a*): in other words, the boat is in stable equilibrium. Clearly the lower we get G the greater the moment of the upthrust about G tending to right the boat.

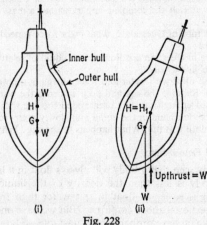

(i) (ii)

Fig. 228

A submarine, when completely immersed, has its centre of buoyancy at the metacentre. So it is essential to have the centre of gravity G of the submarine below H as in Fig. 228. This condition is easily fulfilled by adding ballast to the lower part of the vessel.

The double-hull type of submarine has one hull inside another and the space between is divided into tanks holding fuel or water. In order to surface the submarine, compressed air forces the water out of the tanks and keeps it out by the use of valves. This decreases the total weight of the submarine while keeping the same outer displacement, so the vessel rises. To dive, valves are opened and the tanks between the hulls are flooded to make the submarine sink to the required level.

Torpedoes are fired from tubes by pressurized air, and there is one difficulty to be overcome. When the tube is opened by firing the torpedo, water will rush in and the buoyancy or upthrust on the torpedo will upset the aim. This is overcome by a method of 'water round the tube' in which the tube is flooded before firing so that when the tube is opened there will be no change in upthrust on the torpedo.

Clearly the equilibrium of a boat will depend on the loading of its cargo. A shifting cargo means a shifting centre of gravity, which could be disastrous in a storm. In the early part of the nineteenth century there was often notorious disregard for safety and many ships were lost through overloading. In many cases the insurance to be claimed encouraged a deliberate overloading of the older ships. Samuel Plimsoll (1824–98) was a Member of Parliament who managed to check this practice as a result of the passing of the Merchant Shipping Act of 1876 whereby ships were compelled to have an official load-line drawn on the side of the hull. This load-line became known as the Plimsoll mark, and is illustrated in Fig. 229.

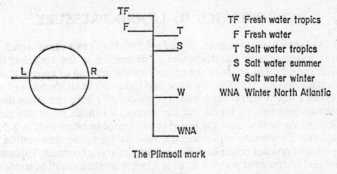

The Plimsoll mark

Fig. 229

The deadweight of a ship is the total weight of fuel, cargo, stores, etc., and the water-line will be in a different position according to the relative density of the water in which the ship floats. Thus the same deadweight will have different water-lines in fresh water and sea-water corresponding to summer and winter, hence the need for the many load-lines as shown in the Plimsoll mark. The meaning of the letters is as follows: $TF=$ fresh water, tropics; $F=$ fresh water; $T=$ salt water, tropics; $S=$ salt water, summer; $W=$ salt water, winter; $WNA=$ winter, North Atlantic. The position of the lines are calculated from a table of ships' lengths and design compiled by Lloyd's Register of Shipping; the letters LR indicate that the vessel complies with the requirements of the Register.

Exercise 84

1. A small boat has 100 000 kg of iron in its construction. What is the minimum volume of the hull if it is to float in fresh water? (Ignore any parts not made of iron.)

2. A syrup tin has a depth of 7·5 cm and a uniform cross-sectional area of 50 cm². The mass is 150 g including syrup. What depth of the tin is immersed when it floats in water? How much lead shot can be loaded into the tin before it sinks?

3. A cylindrical tin with a cross-sectional area of 40 cm² and depth 8 cm has a mass of 100 g. How much water has to be poured into the tin in order to immerse it in water to within 1 cm of the top?

4. The tin of Question 3 floats in water after 200 g of water are poured into it. If a hydrometer of mass 30 g is now floated inside the tin, what will happen?

5. A hollow cube with an internal edge of 2·5 cm has a wooden cube of relative density 0·75 and mass 6 g placed inside. Find the minimum volume of water to be poured into the hollow cube in order to make the small cube float.

6. A ship whose sides may be considered vertical has a total mass of 1000 t and sinks a depth of 0·1 m when it takes on a load of 100 t in fresh water ($d = 1$). Find the distance it rises when it sails into salt water ($d = 1·024$).

CHAPTER FIFTEEN

HYDROSTATICS II: LIQUID PRESSURE

We have seen how the contact of two solid bodies at rest causes equal and opposite forces of action and reaction to be applied to the bodies at their points of contact. The result of the contact for the purpose of applied mathematics is adequately described by these two forces at the single point of contact. When a liquid is in contact with a container or a body the forces due to the contact between the liquid and the surface are much more difficult to identify. We know that since the liquid is considered to be smooth, any force arising from the contact with a surface must be normal to that surface but the number of contact points have merged into an area of contact. This means that instead of speaking about a force at a point we shall speak in terms of a force applied over an area. This distribution of force over area is called **pressure**.

DEFINITION: *Pressure is the distribution of force over a surface.*

Note that (*a*) pressure is measured by the force per unit area. The basic SI unit of pressure is the newton per square metre. (N m⁻².)

(*b*) The pressure over a plane area is said to be **uniform** if equal forces act on all equal areas.

(*c*) The **average** pressure of a fluid over a plane area is measured by the resultant thrust of the fluid divided by the area. For example, if a total force or thrust of 40 newtons acts over an area of 0·02 m² the average pressure on the area is $\frac{40}{0·02} = 2000$ N m⁻². Alternatively, imagine that a tin holds 50 newtons of paint. If the tin has a base area of 0·04 m², then the weight of the paint is supported by the base area of the tin. Therefore the average pressure on the base of the tin is $\frac{50}{0.04} = 1250$ N m^{-2}. Intuitively we know that the pressure over the base of the tin is a *uniform* pressure of 1250 N m^{-2}.

(1) Fundamental Principles

The existence of liquid pressure is easily demonstrated by loosely corking a half-filled bottle of water. By turning the bottle over at different angles we see the cork ejected or we have to exert a force to prevent it from being ejected, and the need for this force must be due to the liquid pressure on the cork.

Two other demonstrations of liquid pressure are illustrated in Fig. 230. In Fig. 230 (*a*) a tin has two sets of equal holes symmetrically drilled along two vertical straight lines on its side. The tin is then placed under a tap which keeps the tin full throughout the demonstration. The jets will be observed as shown. Intuitively we know that the jet with the greatest horizontal range must be associated with the greatest liquid pressure or, to put it another way, the greater the depth of the hole the greater the pressure.

Another idea suggested by the approximate symmetry of the jets is that the magnitude of the pressure is the same in both directions, and we confirm this by drilling a tin with sets of holes on the same level. As we see in Fig. 230 (*b*) the jets from holes at the same level form approximately the same shaped curve.

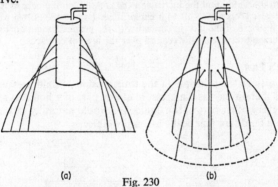

(a)   Fig. 230   (b)

These two demonstrations suggest that the pressure in a liquid not only has the same magnitude in any direction at the same depth, but it also depends upon the depth. But what do we mean by the pressure at a certain depth? For example, if the pressure is over an area how can all the area be at the same depth? In order to overcome this difficulty we shall introduce the idea of 'pressure at a point', but first let us consider one or two examples.

Example: A cylindrical boiler has an inspection plate of area 0·3 m² held in place by 32 bolts. Calculate the force on each of the bolts when the pressure inside the boiler is 50 000 N m^{-2}.

SOLUTION: We assume that the pressure within the boiler is uniform. The thrust on the inspection plate is

$$50\,000 \text{ N m}^{-2} \times 0.3 \text{ m}^2 = 15\,000 \text{ N}$$

Since this thrust is shared equally by the 32 bolts, the force on one of the bolts is

$$15\,000 \div 32 = 468.75 \text{ N} \quad (Answer).$$

Example: A piston has a flat circular head of diameter 0·1 m which is subjected to a thrust of 600 N. Calculate the average pressure on the piston-head.

SOLUTION: Since pressure is force per unit area, the average pressure on the piston-head is

$$\frac{600}{0·05\,^2\pi} = 76\,300 \text{ N m}^{-2}$$

This means that a pressure of 76 300 N m^{-2} on the piston-head will produce a thrust of 600 N (*Answer*).

We must note that the area stated when giving the magnitude of the pressure is not necessarily the same as the area upon which such pressure is applied. For example, the basic unit of pressure is the newton per square metre but the actual area involved in the two examples above was much less than a square metre. Changing the area will change the thrust on the area but the pressure will remain the same.

Exercise 85

1. A cylindrical boiler in the shape of a right-circular cylinder contains steam at a uniform pressure of 75 000 N m^{-2}. If the circular ends of the boiler have a radius of 1 m, calculate the thrust on one end. ($\pi = 3\cdot14$)

2. A woman of weight 450 N balances on the heel of one shoe. If the heel has an area of 400 mm^2, calculate in N m^{-2} the average pressure on the floor covered by the heel.

3. A cylinder with a base area of 0·005 m^2 contains water. If a body of mass 40 g is floated on the water, find the increase in the pressure on the base of the beaker.

4. A car jack gives a side lift to a car of mass 1 t and raises two wheels off the ground. If the base of the jack in contact with the ground has an area of 800 mm^2, find the average pressure on the ground over the area of contact.

(2) Pascal's Law

Known also as the principle of the transmission of fluid pressure, Pascal's Law states that an increase in pressure at any part of a liquid at rest is transmitted to every other part of the liquid. We would naturally expect this from the fact that fluids are practically incompressible.

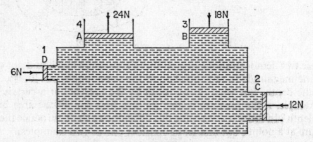

Fig. 231

The apparatus which can be used to demonstrate the truth of this law is shown in Fig. 231 and consists of any large vessel on to which are bolted outlet tubes which carry tight-fitting pistons. With the vessel completely filled with water (it is extremely important to ensure that all air has been forced out), we apply a force to one of the pistons and measure the force which has to be applied on each of the other pistons in order to keep the liquid in the same position of equilibrium. But as soon as an *additional* force of magnitude F is applied to any one of the pistons we discover the following:

(*a*) If all the pistons have the same cross-sectional area, then an *additional* force of magnitude F has to be applied to every one of the pistons to maintain equilibrium.

(*b*) If the pistons have different cross-sectional areas, then the *additional* force to be applied to any piston to maintain equilibrium is directly proportional to its cross sectional area.

To illustrate this numerically, let the areas of pistons A, B, C, and D in Fig. 231 be in the ratio $4:3:2:1$. If an additional force of 24 N is applied to A, then equilibrium will only be maintained by applying additional forces of 18 N at B, 12 N at C, and 6 N at D. We notice that the force per unit area is of magnitude 6 N in each case, so the pressures on each of the pistons are equal. We therefore deduce that a pressure applied anywhere to a fluid will be transmitted throughout the fluid. This is often referred to as the fundamental law of hydrostatics.

If we consider the fluid pressure to be transmitted to every part or point of the fluid we shall need to define what we mean by the **pressure at a point.** Provided we have a small enough area surrounding the point, we can suggest that the pressure over the area may be assumed to be both uniform and equal to the average pressure—so every part or point of this area has the same pressure.

DEFINITION: *The pressure at a point of a fluid is the average pressure found by taking any small plane area containing the point, provided that the area is small enough to ensure that the pressure over it is uniform.*

(3) Pressure due to Weight of Liquid

Having defined pressure at a point we now find the pressure that is due to the weight of a liquid. From the demonstrations in Fig. 230 we have already deduced that the pressure at a point in a liquid must depend on the depth of the point, and it seems reasonable to expect that it will also have something to do with the density of the liquid.

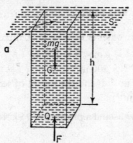

Fig. 232

Let us try to find the pressure at the point Q in Fig. 232. We can consider Q to lie in a horizontal square of area a which is the base of a vertical rectangular column of liquid of height h from Q to the free surface S, i.e. Q is at a depth h below the surface. Now all the liquid is in equilibrium and so is this column.

The vertical forces which keep it in equilibrium are the weight mg of the column (acting through the centre of gravity G) and the upward vertical thrust F on the base (acting through Q). Since the base through Q is horizontal the pressure p on the base area is uniform, so

$$F = pa$$

In equilibrium we have

$$F = mg$$

But the weight of the column of liquid is given by the weight of a unit volume of the liquid multiplied by the volume of the column. Therefore

$$mg = \rho\,gha$$

where ρ is the density of the liquid, $g = 9{\cdot}80$ or $9{\cdot}81$ m s^{-2}

$$F = mg = pa = \rho gha$$

Hence

$$p = \rho gh$$

To extend this result consider the type of problem we have answered already about the pressure at the base of a beaker or tin. Thus 5 kg of water in a cubical tin of edge 0·2 m exerts a thrust on the base of $5g$ N. Since the base has an area of 0·04 m² the uniform pressure on the base is

$$5g \text{ N} \div 0{\cdot}04\text{ m}^2 = 125g \text{ N m}^{-2}.$$

Now the 5 kg of water has a volume of 0·005 m³ since the density of water is 1000 kg m⁻³. With the base area of the tin being 0·04 m², we see that the depth of the water in the tin will be

$$0{\cdot}005 \text{ m}^3 \div 0{\cdot}04 \text{ m}^2 = 0{\cdot}125 \text{ m}$$

Using the result $p = \rho gh$ with $\rho = 1000$ kg m⁻³ and $h = 0{\cdot}125$ m we see that the pressure at any point on the base is given by

$$p = 1000 \text{ kg m}^{-3} \times g \text{ m s}^{-2} \times 0{\cdot}125 \text{ m}.$$
$$p = 1000\,g \text{ N m}^{-3} \times 0{\cdot}125 \text{ m}.$$
$$p = 125\,g \text{ N m}^{-2} \quad \text{(the same result as before).}$$

Note that this is the pressure due to the weight of the liquid.

Example: Find the pressure at a depth of 100 m in sea-water with relative density 1·024.

SOLUTION: The density of sea-water is 1024 kg m⁻³, therefore

$$p = 1024 \text{ kg m}^{-3} \times g \text{ m s}^{-2} \times 100 \text{ m}$$
$$= 1024g \text{ N m}^{-3} \times 100 \text{ m}$$
$$p = 102\,400g \text{ N m}^{-2} \text{ due to the weight of sea-water} \quad (Answer).$$
$$(\text{or } p = 102{\cdot}4g \text{ kN m}^{-2})$$

Example: A container can withstand a pressure of 75 kN m⁻². At what depth in sea-water will it collapse?

SOLUTION: Let h be the depth at which the container collapses.

$$75\,000 \text{ N m}^{-2} = 1024 \text{ kg m}^{-3} \times g \text{ m s}^{-2} \times h$$
$$= 1024g \text{ N m}^{-3} \times h$$
$$\frac{75\,000 \text{ N m}^{-2}}{1024g \text{ N m}^{-3}} = h$$

Hence

$$h = 7{\cdot}45 \text{ m} \quad (Answer).$$

Where the units of measurement have been confused the expression of the pressure in basic units is still a simple process, as the following problem shows.

Example: Find the pressure due to the weight of liquid at a depth of 9 cm if the relative density of the liquid is 0·8.

SOLUTION: We need to express all the measurements in terms of kilogrammes and metres. Thus the depth of 9 cm becomes 0·09 m and the density of the liquid is expressed as 800 kg m^{-3}

$$\therefore\ p = 800 \text{ kg m}^{-3} \times g \text{ m s}^{-2} \times 0.09 \text{ m}$$
$$= 800g \text{ N m}^{-3} \times 0.09 \text{ m}$$
$$= 72g \text{ N m}^{-2} \quad (Answer).$$

Exercise 86

1. Using the apparatus shown in Fig. 231 find the additional forces on each of the other three pistons when the following additional forces are applied:

(i) 48 N to A, (ii) 9 N to B, (iii) 60 N to C, (iv) 1 N to D.

2. A bottle is filled with water and the cork is thrust into the neck of the bottle with a force of 100 N. If the base areas of the bottle and cork are circles of diameter 10 cm and 2 cm respectively, calculate the thrust and pressure on the base of the bottle due to the action of the cork. ($\pi = 3.14$)

3. If piston A in Fig. 231 is depressed a distance a, how far will piston D travel if B and C are held fixed?

4. Find the pressure in kN m^{-2} due to the weight of sea-water ($d = 1.024$) at a depth of 500 m.

5. A cylindrical tin with a vertical axis and base area 0·0001 m^2 contains 1500 g of water. If the tin is lifted vertically with an acceleration of 2 m s^{-2} find the increase in pressure on the base due to the motion. ($g = 9.80$ m s^{-2})

6. Find the pressure due to the weight of methylated spirit ($d = 0.83$) at a depth of 6 cm.

7. Find the increase in pressure in the last of the worked examples above if the depth of the liquid is increased by 2 cm.

8. A corked bottle rests at a depth of 10 m on the bottom of a salt-water lake. Find the force applied to the cork due to the weight of water if the cork has an area of 4 cm^2.

(4) Hydraulic Machines

In 1795 Joseph Bramah, a British inventor, applied Pascal's Law to make the Bramah or hydraulic press. This consists of two interconnected cylinders of different cross-sectional areas such that a piston plunger in one cylinder can be used to transmit a force to a piston ram in the other cylinder. A simplified picture of the apparatus is shown in Fig 233, the arithmetical details are the same as for Fig. 231.

The effort E is applied at the end of a handle in the form of a lever of the second type (page 239) which in turn applies a downward thrust D on the smaller piston. The increase in pressure is transmitted through the liquid so as to give a small lift to the load L. Suppose the ratio of the areas of the piston heads is 10:1.

An effort of $E = 100$ N applied to the lever is converted into a downward thrust of 500 N at D. The pressure increase is transmitted to the other piston which, being ten times greater in area, receives an upward thrust of 5000 N so that the load supported is $L = 5000$ N. The mechanical advantage of the complete machine is given by

$$\text{M.A.} = \frac{5000 \text{ N}}{100 \text{ N}} = 50$$

We see, therefore, that increasing the ratio of the contact areas will increase the mechanical advantage of the machine.

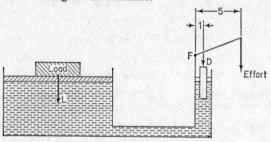

Fig. 233

Example: The ratio of the area of liquid contact of the ram piston to the plunger piston in a Bramah press is 30:1. If an effort of 400 N is applied to a lever with a mechanical advantage of 5, calculate the load which can be lifted and the mechanical advantage of the complete machine.

SOLUTION: Using Fig. 233 the force given to the plunger is $400 \times 5 = 2000$ N. Since the ratio of piston contact areas is 30:1, the load which can be lifted is

$$2000 \times 30 = 60\,000 \text{ N}$$

The mechanical advantage of the complete machine is given by

$$\text{M.A.} = \frac{60\,000}{400} = 150 \quad (\textit{Answer}).$$

In everyday terms this means that a small woman applying her whole weight to the lever could lift a 6-tonne lorry. The basic principle is used in public weigh-bridges which determine the weight of a lorry or truck leaving goods yards, etc.

The difficulty with the apparatus in Fig. 233 is that as soon as the effort is removed the load will descend. When used as a press we really need to be able to build up the pressure like a screw press (page 250) so that the operator may leave the machine and attend to other tasks. For this purpose we employ a pump action with the aid of spring-loaded valves, as shown in Fig. 234 where the apparatus is shown as a simplified book press.

. The operation of apparatus is as follows: downward movement of the plunger forces the valve A to open and the valve B to close. Liquid passes from P to Q and transmits the pressure to the load which is then lifted. When the plunger is raised the pressure in P becomes less than in Q and R, with the result that valve A closes and valve B opens to flood the chamber P with liquid again. The pumping action is repeated to force more liquid into chamber Q, thereby raising the load farther. The machine may be left at any time with the pressure maintained as long as valve A is closed. To release the pressure and return to the original position, tap C is unscrewed and the excess liquid in Q returns to R.

If we combine the complete machine into one housing we have the well-known hydraulic or bottle jack used for lifting cars from underneath the axles, etc. Naturally with such great pressures being built up the machine must incorporate safety valves.

Valve

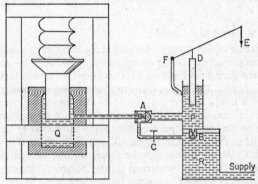

Fig. 234

Example: Using the details of Fig. 234, if the ratio of the areas of liquid contact of the plunger to the ram is 1:25 and the stroke of the plunger is 0·1 m find:

(i) the distance moved by the load in 10 strokes of the plunger,

(ii) the work done by an effort of 100 N applied at the end of the handle with a mechanical advantage of 3,

(iii) the load which the effort of 100 N can move assuming that no frictional forces occur.

SOLUTION: (i) volume of liquid which passes from chamber P to chamber Q is the same for each stroke of the plunger. The distance moved by the load is therefore

$$0·1 \div 25 = 0·004 \text{ m per stroke}$$

(ii) The velocity ratio of the handle is the same as the mechanical advantage, and the distance moved by the effort is therefore $0·1 \times 3 = 0·3$ m vertically downwards. So the work done per stroke is $100 \text{ N} \times 0·3 \text{ m} = 30 \text{ J}$ *(Answer)*.

(iii) The mechanical advantage of the complete machine is 75, so the load supported by an effort of 100 N is 7500 N *(Answer)*.

Exercise 87

1. A hydraulic press, assumed to be an ideal machine, is operated with a handle which has a mechanical advantage of 4. If the ratio of the areas of ram to plunger is 30:1 find:

(i) the load lifted by an effort of 100 N,

(ii) the effort needed to lift a load of 1200 N,

(iii) the distance moved by the load for 10 strokes of the plunger if one stroke is 0·1 m long.

2. A safety valve like Fig. 208 (*f*) is operated by a lever of M.A. = 6 and carries a weight of 8 N at the end. If the liquid contact area of the valve is 2 mm², calculate the pressure necessary to 'blow' the valve.

3. A garden syringe has a plunger whose liquid contact area is 250 times as great as the ejection hole at the end of the syringe. If a force of 4 N is applied to the plunger, what force is applied across the hole?

4. A steam-hammer has a ram piston of area 0·03 m² operated by steam at a uniform pressure of 200 kN m⁻². Calculate the work done by the force on the piston as it moves a distance of 0·1 m.

5. The hammer in Question 4, after being pushed for 0·1 m, moves freely with the steam shut off. If in stamping out a metal disc the hammer is brought to rest in 0·002 m, calculate the resistance of the metal.

6. Using the diagram of Fig. 234, where on the machine would you fit a safety valve?

(5) Horizontal Thrust in a Liquid

The calculation of the thrust on an area is made easier by assuming that the pressure is uniform; we have ensured uniform pressure by restricting our inquiries to the thrust on the base of a container or to the thrust produced by a hydraulic press. But if we return to the problem of finding the pressure at a point in a liquid due to the weight of the liquid, we observe that the pressure over a non-horizontal plane area will not be uniform. Take for example, the vertical rectangle with sides *a* and *b* shown in Fig. 235. The upper edge is parallel to the surface of the liquid, and the centroid *G* of the rectangle is at a depth *h*. We shall consider the forces on one side of the area only. Since *Q* is lower than *T* the pressure at *Q* must be greater than the pressure at *T*. With the pressure varying according to depth we see that the pressure over the whole rectangle is not uniform.

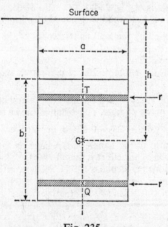

Fig. 235

We can overcome this difficulty by considering the thrust on narrow strips of equal width *r* and parallel to the surface of the liquid. We choose these strips an equal distance above and below *G* so that *TG* = *GQ*. Now since the strips are horizontal and have a very narrow width *r*, we may assume that the pressure is uniform over the whole strip. Thus the pressure on the strip through T is the pressure at the point T, and the pressure on the strip through Q is the pressure at Q.

Pressure on strip T is $p = \rho g(h - TG)$; thrust on strip is $ar\rho g(h - TG)$
Pressure on strip Q is $p = \rho g(h + GQ)$; thrust on strip is $ar\rho g(h + GQ)$

Hence the sum of the thrusts is $ar\rho g(h - TG + h + GQ) = 2ar\rho gh$. But $2ar$ is the sum of the areas and ρgh is the pressure at the centroid G of the rectangle.

It follows that if we continue to take pairs of strips like this, we shall eventually cover the complete rectangle; moreover for each pair of strips, the sum of the thrusts will be the sum of the areas multiplied by the pressure at G. We therefore deduce that the thrust on the whole rectangle is equal to the product,

Area × Pressure at centroid of the area

In the case of the rectangle in Fig. 235 the thrust is $ab\rho gh$.

Example: A reservoir holds water to a depth of 12·5 m and contains an emergency rectangular drain-off door placed vertically in one side of the reservoir. If the door measures 0·5 m × 1 m and its shorter edge is horizontal and 10·5 m from the surface of the water, find the thrust on the door due to the water.

SOLUTION: We have just proved that the thrust on a plane area due to the weight of liquid in which it is immersed is given by the product

Area × Pressure at centroid of the area

Here, the centroid of the door is at a depth of 11 m and $\rho = 1000$ kg m^{-3}, hence the pressure at the centroid of the door is

$$1000 \text{ kg m}^{-3} \times g \text{ m s}^{-2} \times 11 \text{ m} = 11\,000g \text{ N m}^{-2}$$

The thrust on the door is $0 \cdot 5 \times 1 \times 11\,000g = 5500g$ N (*Answer*).
Substituting $g = 9 \cdot 80$ m s^{-2} this result becomes 53 900 N, which is a considerable force.

We have just calculated the force on a door, but this has no special significance because an equal area similarly placed in the sides of a reservoir takes exactly the same thrust. With the considerable force on only small areas such as these it becomes almost impossible to appreciate the enormous forces which are applied to the dams of artificial lakes. For example, the Hoover dam at Boulder, Nevada, U.S.A., has a height of approximately 240 m and holds back about 50 000 million litres of water. An area of 0·5 m² at the base of the dam such as the door in the worked example above would receive a thrust of approximately

$$0 \cdot 5 \times 1 \cdot 0 \times 1000 \times 9 \cdot 80 \times 240 = 1\,176\,000 \text{ N.}$$

Since the average car has a weight of approximately 10 000 N, this means that the thrust on the area is sufficient to support over 100 cars. Creating artificial lakes also increases the thrust on the floor of the lake, so the dam has to be founded well below the lake bottom to prevent the lake breaking out from *under* the dam.

Example: The Grand Coulee dam in Washington, U.S.A., has a height of approximately 170 m and holds approximately 14 000 million litres of water. Find the thrust on a plane rectangular area 1 m × 2 m drawn on the side of the dam with centroid 150 m below the surface.

SOLUTION: It makes no difference what the angle of inclination of the area is to the vertical, the thrust will be given by the product of the area and the pressure at the centroid.

$$\text{Thrust} = 1 \text{ m} \times 2 \text{ m} \times 1000 \text{ kg m}^{-3} \times g \text{ m s}^{-2} \times 150 \text{ m}$$
$$= 300\,000g \text{ N}$$
$$= 2\,940\,000 \text{ N} \quad (Answer).$$

This is roughly equivalent to the weight of 300 cars.

To find the thrust on an immersed area of any shape is beyond the scope of the present work, but it should be understood that not only does the basic method outlined above still apply, but the final result is still the same: thrust is the product of the area and pressure at the centroid of the area. For example, a circle of radius 2 m with its centre at a depth of 100 m in sea-water experiences a thrust on one side equal to

$$T = \pi \times 2 \text{ m} \times 2 \text{ m} \times 1024 \text{ kg m}^{-3} \times g \text{ m s}^{-2} \times 100 \text{ m}$$
$$= \pi g \ 409\ 600 \text{ N}$$

A triangle of area A m² whose centroid is at a depth of h m in sea-water experiences a thrust on one side due to the weight of sea-water given by

$$T = A \times 1024 \times g \times h \text{ N}$$

Notice, however, that although we know the magnitude of the thrust we do not know its point of application.

Exercise 88

1. A sluice gate in the vertical side wall of a reservoir is in the form of a rectangle 1 m × 1·5 m with its longer side at the base of the wall. If the water is at a height of 10·5 m, find the thrust on the door due to the weight of the water. ($g = 9·80$ m s⁻²)

2. A lock gate has a width of 4 m and is immersed on one side in 10 m of water. Find the thrust on the lock gate.

3. If the other side of the lock gate in Question 3 is only immersed in 3 m of water, what is the resultant thrust on the door?

4. A cube of edge 0·25 m rests on the horizontal floor of a reservoir which contains water to a depth of 10·25 m. Find the thrust on each face of the cube.

5. *ABCD* is a rectangle drawn on the side of a vertical lock gate, with $AB = 2$ m and $BC = 6$ m. If the upper side AB is in the surface of the water, find the thrust on the area (i) △*ABCD*, (ii) △*ABD*, (iii) △*DBC*.

CHAPTER SIXTEEN

HYDROSTATICS III: BAROMETERS AND BOYLE'S LAW

Liquids and gases are both classified as fluids, but gases have a certain elusive quality which makes an investigation of their properties more difficult. To begin with they are not usually visible to the naked eye and they have the ability to always fill a container of any shape. We recall that ice, water, and steam were used in an earlier chapter as examples of three states of the same mass. To get some idea of the volume which can be occupied by the same mass in a different state we should remember that any volume of water when boiled under ordinary kitchen conditions will give a volume of steam which is about 1700 times as great as the volume of water. Since the relative density of water is 1, this means that the relative density of steam is approximately 0·0006.

The relative density of the atmosphere at sea-level is 0·0012, that is the atmosphere has a density of 1·2 kg m ⁹. Since the atmosphere has weight it must exert a force on any surface of contact, and yet, as you reach forward to turn this page, you do not experience any resistance to the motion as you do in moving your arm through water when swimming (indeed it is the resistance of the water which enables us to swim). Does this lack of resistance, together with the lack of feeling of any force on our bodies, mean that the atmosphere exerts so small a pressure that we are not sensitive enough to feel it?

(1) Atmospheric Pressure

The first experiment to obtain a measurement of the pressure of the atmosphere was carried out in 1643 by the Italian physicist Evangelista Torricelli. A straight glass tube about 1 m long and sealed at one end was completely filled with mercury. The open end was then closed with the finger and the tube inverted into a dish of mercury. (Before reading on, what would you expect to happen?)

It was discovered that instead of all the mercury pouring out into the dish the head of mercury merely fell to a particular height above the level of the mercury in the dish. As the tube was moved into different positions as shown in Fig. 236, the difference in level between the mercury in the tube and the mercury in the dish remained equal to approximately 0·760 m, the height varying slightly from day to day. The open space above the mercury is still known as a Torricellian vacuum: it is not a perfect vacuum because it contains a minute quantity of mercury vapour.

Because the mercury is at rest the pressure over the horizontal surface of the mercury in the dish must be uniform; but at the point X the pressure is due to the weight of the mercury column only, and we know from Chapter Fifteen that this is given by the relation

$$p = \rho g h$$

where ρ is the density of mercury and h is the height of the column. At points such as Y on the same level as X the pressure is due to the atmosphere only. It follows therefore that the atmospheric pressure A is given by the equation,

$$A = \rho g h$$

From the experiment we have $A = 13\,600\,\text{kg m}^{-3} \times g\,\text{m s}^{-2} \times 0{\cdot}76\text{m}$
$$= 13\,600g\,\text{N m}^{-3} \times 0\,{\cdot}76\,\text{m}$$
$$= 101\,292{\cdot}8\,\text{N m}^{-2}$$
so the atmospheric pressure $\quad A = 101{\cdot}293\,\text{kN m}^{-2}$

Expressing this result in everyday terms, the pressure of the atmosphere has approximately the same effect as a mass of 1 kg placed on every square centimetre of a surface; a full bottle of wine placed upside down on its cork produces roughly the same pressure over the area of contact.

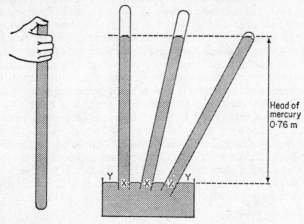

Fig. 236

Now pause and look at the palm of your hand. It probably has an area of approximately $0{\cdot}1\,\text{m} \times 0{\cdot}1\,\text{m} = 0{\cdot}01\,\text{m}^2$, in which case the thrust on your palm due to the same atmospheric pressure as above is

$$101\,292{\cdot}8\,\text{N m}^{-2} \times 0{\cdot}01\,\text{m} = 1012{\cdot}928\,\text{N}$$

This is a force equivalent to the weight of approximately 1000 copies of this book which you are reading. But do not worry: since the pressure at a point is the same in all directions, there is an equal and opposite force on the back of your hand so the resultant force on your hand due to the atmosphere is zero. Our bodies do not collapse because we are built to withstand these forces.

To convince yourself of the existence of these forces, take an empty tin with a well-fitting screw cap (such as the tins in which car brake fluid is sold) and notice how difficult it is to crush. Pour in a little water and boil, then screw the cap back on tightly, using a cloth to protect your hands. Remove the source of heat and await the collapse of the tin due to atmospheric pressure. A less spectacular demonstration is to fill a tumbler with water, place a postcard firmly on the top, and invert the glass.

The barometer. Torricelli's experiment provides us with an instrument for measuring the pressure of the atmosphere. Any instrument which measures atmospheric pressure is called a barometer. The reader may think that when the atmospheric pressure has been measured it has been found once and for all, so why do we need a barometer? The fact is that the atmospheric pressure varies according to the height at which the measurement is taken and also according to the type of weather which occurs in the locality of the measurement. To get some idea of the possible variations in atmospheric pressure we should note that

(a) the highest sea-level barometric pressure ever recorded was 107·8 kN m^{-2} (in the U.S.S.R., Jan. 1900),

(b) the lowest sea-level barometric pressure ever recorded was 87·7 kN m^{-2} (in the Pacific, north west of Guam, Sept. 1958).

In terms of the height of the mercury column, the greatest height was approximately 0·81 m and the smallest height was approximately 0·66 m.

When used in weather-forecasting it is not the barometer reading which is important but the *trend* of the reading, i.e. whether the pressure is rising or falling. The reason for this is that dull rainy weather is associated with regions of low pressure (cyclones) and fine weather is associated with regions of high pressure (anticyclones). Consequently you know that if the pressure is rising you are due for a period of fine weather; if, on the other hand, the pressure is falling you are due for a period of rainy weather. The technical reason in this case is that when the atmosphere receives an increase in the concentration of water vapour, it becomes less dense, and, as we can see from the equation $p = \rho g h$, as soon as ρ decreases so does the pressure. Thus a falling pressure indicates concentration of water vapour in the air and therefore a possibility of rain.

In the Northern Hemisphere, centres of low pressure normally drift East to North East and the centres of high pressure drift South East to East. Before the advent of weather satellites, forecasters had to rely on plotting the movement of the cyclones and anticyclones and estimating where they would be in 24 h. time—a prediction all too frequently upset by the occurrence of unexpected surface winds. While very few winds approach anywhere near the world-record wind speed of 332 km h^{-1} (registered in 1934 at Mt. Washington, U.S.A.) they can nevertheless upset the most carefully considered forecasts.

The Torricelli tube is too crude for accurate readings to be taken, but its commercial counterpart the **Fortin barometer** (Fig. 237 (a)) is very efficient. Instead of carrying the mercury in an open dish, the Fortin barometer has a reservoir of mercury in a chamois-leather bag within a metal cylinder which can be adjusted so that the height of the column is always measured from the same point. A screw operating from underneath forces the reservoir level up to the zero mark given by the tip of the pointer from which the height of the column is measured.

As long as we have a body which responds uniformly to the atmospheric pressure acting on it then we have an instrument with which to measure the pressure. When the body is calibrated we have our barometer. Such a body could be any sealed container of sufficiently elastic material that would respond to small changes in pressure. A partially evacuated metal box is what is used in the **aneroid barometer** shown in Fig. 237 (b). As the air pressure changes so the box expands or contracts, operating an elementary linkage to register the pressure over a scale. The main advantage of this instrument is its size.

While the Fortin barometer is over a metre in length and is usually fixed to a wall, the aneroid barometer can be very small and can be fitted to aircraft as an altitude meter. Another advantage of the aneroid barometer is that no liquid is used, indeed the meaning of the word aneroid is 'without liquid'.

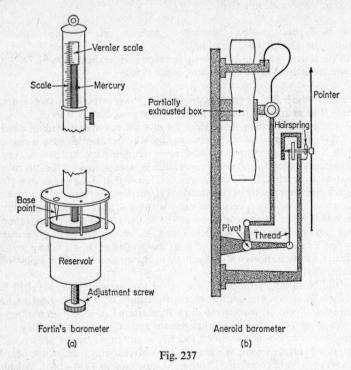

Fortin's barometer   Aneroid barometer

(a)   (b)

Fig. 237

Atmospheric pressure varies according to height above sea-level, which is why a barometer is used as an altimeter in an aeroplane. The following table shows the approximate pressure according to height.

Pressure (kN m^{-2}):	100	75	50	30	20	10	5
Height (km):	Sea-level	4	9	14·5	18·5	25	32

Since jet aircraft operate most efficiently at altitudes between 8000 m and 14 000 m (8–14 km) they have to be pressurized in order to ensure passenger comfort. The pressure adopted is usually sufficient to simulate the conditions at a height of about 2·5 km.

Example: Given that the atmospheric pressure supports a column of mercury 0·76 m high, find the height of a column of water which could be supported by the same pressure. ($g = 9·80$ m s^{-2})

SOLUTION: Taking the density of mercury as 13 600 kg m^{-3}, the atmospheric pressure is

$$A = 13\,600 \text{ kg m}^{-3} \times g \text{ m s}^{-2} \times 0·76 \text{ m}$$
$$= 101\,292·8 \text{ N m}^{-2}$$

Let h be the height of the column of water supported by the same pressure. Then

$$A = 1000 \text{ kg m}^{-3} \times g \text{ m s}^{-2} \times h$$
$$101\,292 \cdot 8 = 9800h$$
$$\therefore h = 10 \cdot 336 \text{ m} \quad (Answer).$$

The arithmetic of this solution possibly obscures the simplicity of the idea, so let us consider the problem.

The pressure due to the column of mercury is

Density of mercury $\times g \times$ Height of column of mercury

The pressure due to the column of water is

Density of water $\times g \times$ Height of column of water

Since these are both the same pressure, we have

$$13\,600 \times g \times 0 \cdot 76 = 1000 \times g \times h$$
$$13 \cdot 6 \times 0 \cdot 76 = h$$
$$= 10 \cdot 336 \text{ as before}$$

This result shows that if Torricelli had experimented with water instead of mercury he would have required a tube of length greater than 10·336 m in order to obtain his vacuum. It is reported that Pascal carried out the experiment in 1644 with a tube about 13 m long filled with wine. The result carries an important implication for pumps and siphons because it shows that atmospheric pressure will not lift water through a height greater than 10·336 m.

Example: Find the pressure at a depth of 10 m in sea-water when the barometric height is 0·75 m of mercury.

SOLUTION: Here the pressure will be the atmospheric pressure plus the pressure due to the weight of the sea-water. We could also describe this as being the pressure due to the weight of the sea-water and the weight of the atmosphere.

Pressure due to the weight of the sea-water is $p = 1024g \times 10 \text{ N m}^{-2}$
Atmospheric pressure is $\qquad\qquad\qquad A = 13\,600g \times 0 \cdot 75 \text{ N m}^{-2}$
$$= 10\,200g \text{ N m}^{-2}$$
Sum of the pressures is $p + A$ $\qquad\qquad = 20440g \text{ N m}^{-2} \quad (Answer).$

Exercise 89

1. If Toricelli had used a tube which was only 0·6 m long, what would have happened?

2. How is it possible to tell if any air has entered the space at the top of the Torricelli tube?

3. If the barometric height is 0·75 m of mercury, find the height of a similar barometer using glycerine with a relative density of 1·25. What possible advantage could there be in using such a barometer?

4. Some air is introduced into the space at the top of a Torricelli barometer (bubbles are blown into the base of the tube) and the level of the mercury falls 20 mm. Find the pressure of the air.

5. Find the pressure at a depth of 20 m in water ($d = 1$) due to the atmosphere and the weight of liquid, when the mercury barometer reads 0·75 m.

6. Fill a tumbler full of water, place a postcard over the water and invert the tumbler holding the card in place. Explain what happens when you remove your hand from the card.

(2) Pressure of a Gas

With the atmosphere exerting pressure it is obvious that air must have density, i.e. mass per unit volume. But a given mass of air (or any other gas) can have its volume made almost as large or as small as we please, and the density will alter accordingly. It follows that whenever we quote the density of a gas we should also give the pressure at which that density was determined. We shall see very shortly how to relate pressure, volume, and density with the aid of Boyle's Law, but for the moment let us consider how we can find the mass or weight of a given volume of air. One method is to evacuate a container by pumping out the air. A tap on the container is used to prevent any inflow as the weight of the evacuated container is found. The tap is then opened, and the weight of the container is found again; the difference in readings gives the weight of the air which has passed into the vessel. By knowing the volume we are able to calculate the density of the air.

The experiment is a delicate one since the density of air is taken at 1·3 kg m^{-3} at an atmospheric pressure represented by 0·76 m of mercury.

For example the volume of a football is approximately 0·005 m^3, when it contains air at atmospheric pressure (which is not enough for playing the game) the mass of air in the ball is only 0·0065 kg, i.e. a weight of 0·0637 N or approximately $\frac{1}{16}$ of the weight of this book.

Example: A balloon which has a weight of 0·1 N when not inflated is filled with hydrogen ($\rho = 0·09$ kg m^{-3}) until its volume is approximately 0·01 m^3. Given that the density of air is 1·3 kg m^{-3}, determine whether or not the balloon will rise when released.

SOLUTION: The problem is exactly the same as some of those in Chapter Fourteen where we inquired whether or not a body would float. We consider the balloon immersed in air so that the buoyancy effect is to give the balloon an upthrust equivalent to the weight of fluid (air) displaced.

Upthrust = 0·01 m^3 × 1·3 kg m^{-3} × 9·80 m s^{-2} = 0·1274 kg m s^{-2} = 0·1274 N
Weight of hydrogen is 0·01 × 0·09 × g = 0·008 82 N
Resultant vertical force on the balloon is 0·1274 − 0·1 − 0·008 82 = 0·018 58 N
Since the resultant force on the balloon is vertically upwards it will rise. (*Answer*).

This example illustrates the mechanical principle of ballooning and early airship flight. Helium is another light gas which can be used and is preferred since unlike hydrogen it is non-inflammable. In the more common hot-air balloon a flame heater is suspended beneath the balloon cover to heat the air within. Since hot air is less dense than cold air, eventually the upthrust on the balloon takes the complete apparatus off the ground. A balloon stops ascending because the density of the atmosphere decreases with height and thereby decreases the upthrust on the balloon.

(3) Hare's Apparatus

The atmosphere has been shown to be able to support a column of liquid whose height is given by the relation $A = \rho g h$. Rearranging this formula we see that

$$\rho = \frac{A}{gh}$$

which suggests that with a knowledge of A from a mercury barometer and g from the locality of the experiment we can find the density ρ of a liquid by observing the height h of a column of the liquid. We note that the density of the liquid is inversely proportional to the height of the column, or to put it another way the greater the density the shorter the column. This gives us an idea for finding the density of a liquid by simply making a Torricellian barometer filled with the liquid concerned and then measuring the height of the column. The drawback, of course, is that we shall require tubes up to 12 m long in some cases. This difficulty is neatly overcome by using the following method.

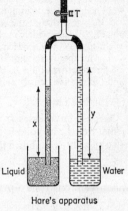

Hare's apparatus consists of two glass tubes interconnected with rubber tubing to a glass T-piece, as shown in Fig. 238. Some air is sucked out of the tubes and in so doing raises a column of liquid in each of the glass tubes which have been immersed in two dishes, one containing water and one the liquid of unknown density. When a column of readable height has been raised, the tap is closed and the height of each column noted. Let the

Hare's apparatus

Fig. 238

density of the liquid be ρ and the density of the water 1000 kg m⁻³. If the pressure in the space above the columns is S, then the atmospheric pressure on the surface of each liquid is equal to the pressure due to the weight of the liquid column plus the pressure in the space. Thus

$$S + \rho g x = S + 1000 g y$$

$$\therefore \quad \rho = 1000 \frac{y}{x}$$

Alternatively $\dfrac{\rho}{1000} = \dfrac{y}{x} = d$ the relative density of the liquid.

Example: Using Hare's apparatus a column of water of height 0·15 m balances a column of liquid of height 0·2 m. Find the relative density of the liquid.

SOLUTION: The relative density $d = \dfrac{0 \cdot 15}{0 \cdot 20} = 0 \cdot 75$ (*Answer*).

(The density of the liquid is 750 kg m⁻³.)

The apparatus is obviously useful for liquids which must be kept separated, and the method is very convenient because the readings are so easy to take. We can vary the pressure S to take several readings so as to yield a more reliable average result, in a comparatively short period of time.

Exercise 90

1. The weight of a balloon is found when it is flat and again when it is inflated. Assuming that your equipment is sufficiently sensitive, will the difference in weight give the weight of the air in the balloon?

2. What weight can be lifted by a balloon which holds 500 m³ of gas with a density of 0·1 kg m⁻³ when the density of the atmosphere is 1·3 kg m⁻³?

3. If the balloon in Question 2 carries a load of 400g N, find the density of the

layer of air where the balloon stops rising assuming that the volume of the balloon remains unchanged.

4. Using Hare's apparatus it is found that a 0·21-m column of water balances a 0·15-m column of liquid. (i) Find the relative density of the liquid. (ii) If the pressure within the apparatus is changed so that the water column is lowered to 0·1 m, to what height is the column of liquid lowered?

5. Suppose that at the end of the experiment in Question 4 the liquid tube is replaced by a tube which has twice the diameter of the previous one. What is the height of the liquid column which will balance the 0·21-m column of water?

6. An experiment with Hare's apparatus used methylated spirit of relative density 0·8 and another liquid *A*. If 0·2 m of methylated spirit balanced 0·12 m of liquid *A*, find the relative density and density of *A*.

(4) Boyle's Law

Boyle's Law arises from studying the behaviour of a given mass of gas when varying its volume and pressure while keeping its temperature constant. The apparatus as shown in Fig. 239 (*a*) consists of an open glass tube joined by

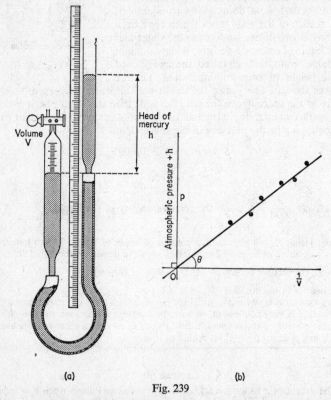

(a)   (b)

Fig. 239

rubber tubing to a measuring tube which is graduated to read off the volume contained. Mercury is poured into the apparatus and the tap in the volume tube closed so that air is trapped. When the level of the mercury is the same in both tubes the pressure of the gas must be atmospheric pressure, which we note at the time of the experiment. The reservoir is now raised so that the pressure of the gas is atmospheric pressure + the pressure due to the head of mercury given by h. We then note the corresponding reading for the volume v. By varying h we obtain all the necesssary readings for plotting a graph of pressure p against $\dfrac{1}{v}$ as shown in Fig. 239 (*b*). The graph is found to be a straight line which, when produced back, is seen to pass through O the origin of co-ordinates. This tells us that

$$p \div \frac{1}{v} = \tan \theta$$

$$p \div \frac{1}{v} = \tan \theta = k \quad \text{(a constant)}$$

$$pv = k \quad \text{(a constant) is the result of the experiment.}$$

An alternative statement of this relation is that the volume is inversely proportional to the pressure. The conclusion from this experiment, which is carried out with other gases for confirmation, is known as Boyle's Law:

When the temperature of a given mass gas is kept constant the volume of the gas is inversely proportional to its pressure.

We summarize the results of this law in the statement $pv =$ a constant.

We said earlier that we would obtain a relation between pressure, volume and density. So far we only have a relation involving pressure and volume, but we can easily introduce the density of the gas by remembering that, for Boyle's Law to apply, the mass m of gas is kept constant as well as its temperature.

The relation $pv = k$ may be rearranged as

$$p = \frac{k}{v} = \frac{m}{v} \times \frac{k}{m}, \text{ where } m \text{ is the mass of the gas.}$$

Now k and m are constants, therefore $\dfrac{k}{m}$ is a constant equal to c (say); $\dfrac{m}{v}$ is the density of the gas at that particular pressure p. Therefore

$$p = \rho c$$

In other words, the density of the gas is directly proportional to the pressure: doubling the pressure doubles the density, dividing the pressure by 3 also divides the density by 3, and so on. We now see why it is so important when quoting the density of a gas to quote its pressure also. In addition we should quote the temperature because this, too, affects the density.

Example: The tube of an imperfect mercury barometer contains some air above the mercury. The length of the air space was 0·06 m when the length of the mercury column was 0·72 m and 0·084 m when the mercury column was 0·73 m. Find the true barometric height.

SOLUTION: Let the cross-sectional area of the barometer tube be a m², and let A, the true atmospheric pressure, be measured in metres of mercury. The atmospheric pressure is equal to the sum of the pressures due to the weight of the mercury column and the pressure in the space above the column. This is exactly the same situation as in the use of Hare's apparatus.

In the first case $p_1 = A - 0.72$; $v_1 = 0.06a$

second case $p_2 = A - 0.73$; $v_2 = 0.084a$

If $pv = k$, then $p_1 v_1 = k$ and $p_2 v_2 = k$

$$\therefore p_1 v_1 = p_2 v_2$$
$$(A - 0.72)\,0.06 = (A - 0.73)\,0.084$$

Multiplying by 100

$$6A - 4.32 = 8.4A - 6.132$$
$$1.812 = 2.4A$$
$$A = 0.755$$

The true barometric height is 0.755 m of mercury   (*Answer*).

Example: The air in a plastic bag trebles in volume as it rises from the bottom of a fresh-water lake. Find the depth of the lake if the barometric height is 0.76 m of mercury.

SOLUTION: Let h be the depth of the lake and V the volume of the air in the bag when it is at the bottom of the lake.

$$p_1 = (13\,600 \times g \times 0.76 + 1000gh)\,\text{N m}^{-2}, v_1 = V$$
$$p_2 = 13\,600 \times g \times 0.76\,\text{N m}^{-2}, v_2 = 3V$$

Assuming that there is no change in temperature we may apply Boyle's Law:

$$p_1 v_1 = p_2 v_2$$
$$(13\,600g \times 0.76 + 1000gh)\,V = (13\,600g \times 0.76)\,3V$$

Dividing throughout by Vg we have

$$1000h = 13\,600 \times 0.76 \times 2$$
$$h = 13.6 \times 1.52 = 20.672$$

The depth of the lake is 20.672 m   (*Answer*).

Exercise 91

1. A gas occupies a volume of 0.01 m³ at a pressure of 100 kN m⁻². What will its volume be when the pressure is 150 kN m⁻²?

2. A gas has a volume of 4 m³ at a pressure of 300 kN m⁻². What is the numerical value of the constant in Boyle's Law for this gas?

3. What pressure is needed to compress the gas in Question 2 into a volume of 1.5 m³? Find the ratio of the densities of the gas at these two pressures.

4. A bubble of air ascends from the bottom of a fresh-water lake at a time when the barometric height is 0.75 m. If the volume of the bubble is doubled by the time it surfaces, what is the depth of the lake?

5. A plastic bag containing air of volume 1 m³ at atmospheric pressure is caught on the mast of a submarine which dives in fresh water. If the barometric height is 0.76 m of mercury of relative density 13.6, find the volume of the air in the bag at a depth of 30 m.

6. Some air is injected into the vacuum space in a mercury barometer. The result of the injection pushes the mercury column down 0.02 m. What is the pressure of the air in the space in N m⁻² if $g = 9.80$ m s⁻²?

7. An imperfect barometer contains some air in the vacuum space. The length of the air space was 0.06 m when the mercury column was 0.72 m high and 0.08 m when the mercury column was 0.73 m high. Find the true barometric height.

8. The following table gives a set of results from a Boyle's Law experiment:

p (metres):	0·24	0·35	0·49	0·67	1·24	1·84
l (metres):	1	0·9	0·8	0·7	0·5	0·4

where p is the head of mercury and l is the length of gas column. Draw a graph of these results plotting p against $\frac{1}{l}$ in the manner of Fig. 239, and from the graph obtain the atmospheric pressure.

(5) Pumps

The Bramah press is probably the most important self-contained machine based on hydrostatic principles. The most common extension of this idea is the hydraulic braking system in motor-cars, where a foot-operated plunger has its thrust converted (in keeping with Pascal's Law) into the braking force applied at the discs.

Less spectacular machines like the air or water pump can be found in places where the motor-car and the Bramah press are seldom seen. The common **lift pump** is essentially a cylinder, two valves, and a piston. As its name suggests, it moves water by lifting it in a manner shown by Fig 240 (a).

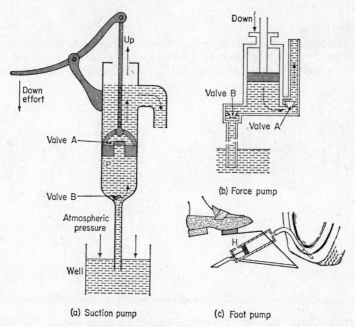

(a) Suction pump   (c) Foot pump

Fig. 240

On the up stroke of the piston, valve A closes and valve B opens. Water is sucked into space P and at the same time water is lifted out of space Q to be discharged through the outlet S. On the downstroke, valve B closes and valve A opens as the piston passes down the cylinder ready to begin the lifting operation again.

When we referred just now to the water being 'sucked into space P' this really meant that the upward movement of the piston decreases the pressure in P so that atmospheric pressure exerted over the surface of the water in the well forces a column of water up the pipe T, somewhat in the manner of Hare's apparatus. With a barometric pressure of 0·76 m of mercury, the height of the water column which can be supported by the atmosphere is $13·6 \times 0·76 = 10·336$ m; so the piston must not rise above this distance from the surface of the well water if the pump is to work. In practice the inevitable leakage at the valves means that the pump cannot operate at a height of more than 9 m above the level of the water in the well.

The **force pump** in Fig. 240 (*b*) needs no explanation; like the lift pump it depends on the atmospheric pressure for its function, and its height of operation is similarly limited. Heavy-duty plunger pumps can be used to force water to much greater heights.

The simple air foot-pump of Fig. 240 (*c*) has no valve of its own, but relies on the valve fitted to the inner-tube of the tyre. As the plunger passes the hole H the air is compressed and forced into the tube. On the return stroke the inner-tube valve closes, the plunger is forced back by a spring, and fresh air leaks in past the plunger washer and then through the hole H for a repeat of the operation.

(6) The Siphon

The siphon is another device which relies on atmospheric pressure for its function. To make a siphon, take a piece of rubber tubing and immerse one end in a beaker of water, as shown in Fig. 241 (*a*); keep the other end held

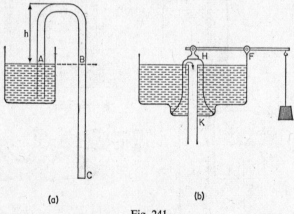

(a)   (b)

Fig. 241

below this level. The siphon is started by sucking water through the tube; once started it will empty the beaker. We can best explain the operation of the siphon by considering the pressure situation when the tube is filled with water. The pressure at A and B is the same as atmospheric pressure, and as long as h is less than 10·336 m no vacuum will form in the bend of the tube. The smaller the height h the less effort needed to suck the water through. The pressure inside the tube at C is atmospheric pressure $+$ the pressure due to the weight of the liquid in *BC*. But the pressure outside the tube at *C* is only atmospheric pressure. Consequently the water will flow out of the tube until the beaker is emptied and air enters the tubing. The common lavatory flush in Fig. 241 (*b*) operates on the same principle.

Example: The diving-bell shown in Fig. 242 enables men to work on a river bed when repairing the foundations of a bridge. In its original form the device was a bell-shaped vessel without a bottom. As the bell is lowered into the river the water level rises inside it according to the depth of immersion, so air has to be pumped in to force out the water.

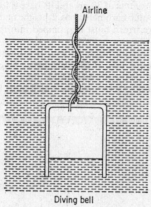

Airline

Diving bell

Fig. 242

The bell in this example has an interior height of 4 m. It is lowered into fresh water until the water rises 1·5 m inside the bell. If the barometric height is 0·75 m of mercury (relative density 13·6) find the pressure inside the bell and the depth of the bell.

SOLUTION: We shall assume that there is no change in temperature, and apply Boyle's Law to the air trapped in the diving bell. Let *a* be the cross-sectional area of the interior of the bell:

At the surface $\qquad p_1 = 13\,600g \times 0\cdot75 \text{ N m}^{-2}, v_1 = 4a \text{ m}^3$

When lowered $\qquad p_2 = \qquad ? \qquad v_2 = 2\cdot5a \text{ m}^3$

By Boyle's Law,

$$p_1v_1 = p_2v_2$$
$$13\,600g \times 0\cdot75 \times 4a = p_2 \times 2\cdot5a$$
$$p_2 = 16\,320g \text{ N m}^{-2} \quad (Answer).$$

Now p_2 must be due to the atmospheric pressure plus the pressure due to the weight of the water. If *h* is the depth of the water level inside the bell then

$$p_2 = 16\,320g = 13\,600g \times 0\cdot75 + 1000gh$$
$$h = \frac{16\,320 - 10\,200}{1000}$$
$$= 6\cdot12 \text{ m} \quad (Answer).$$

Thus, the level of the water in the diving bell is 6·12 m below the surface of the river

Exercise 92

1. Examining the siphon drawn in Fig. 241 (*a*) explain what would happen if a hole was made in the tubing at (i) a point between *B* and *C*, (ii) a point at the top of the bend.

2. Explain the action of the lavatory flush in Fig. 241 (*b*).

3. At a time when the water barometer has a height of 10 m, a small tin is sunk in the manner of a diving-bell to a depth of 20 m in fresh water. Find the volume of the air in the tin at this depth as a fraction of the volume of the tin. (Ignore the height of the water in the tin since it is small compared with the other distances.)

4. If a diving-bell has a volume of 30 m³ and its base is lowered to a depth of 10 m in fresh water when the water barometer has a height of 10 m, find the volume of air at atmospheric pressure which must be pumped into the bell to keep the water out.

5. A diving-bell has an interior height of 5 m. It is lowered into fresh water until the water rises 1 m inside the bell. If the barometric height is 0·75 m of mercury (relative density 13·6), find (i) the pressure inside the bell, (ii) the depth of the bell, (iii) the volume of air at atmospheric pressure which must be pumped in to keep the water out.

TABLES

N	0	1	2	3	4	5	6	7	8	9	1	2	3	4	5	6	7	8	9
10	·0000	0043	0086	0128	0170	0212	0253	0294	0334	0374	4	8	12	17	21	25	29	33	37
11	·0414	0453	0492	0531	0569	0607	0645	0682	0719	0755	4	8	11	15	19	23	26	30	34
12	·0792	0828	0864	0899	0934	0969	1004	1038	1072	1106	3	7	10	14	17	21	24	28	31
13	·1139	1173	1206	1239	1271	1303	1335	1367	1399	1430	3	6	10	13	16	19	23	26	29
14	·1461	1492	1523	1553	1584	1614	1644	1673	1703	1732	3	6	9	12	15	18	21	24	27
15	·1761	1790	1818	1847	1875	1903	1931	1959	1987	2014	3	6	8	11	14	17	20	22	25
16	·2041	2068	2095	2122	2148	2175	2201	2227	2253	2279	3	5	8	11	13	16	18	21	24
17	·2304	2330	2355	2380	2405	2430	2455	2480	2504	2529	2	5	7	10	12	15	17	20	22
18	·2553	2577	2601	2625	2648	2672	2695	2718	2742	2765	2	5	7	9	12	14	16	19	21
19	·2788	2810	2833	2856	2878	2900	2923	2945	2967	2989	2	4	7	9	11	13	16	18	20
20	·3010	3032	3054	3075	3096	3118	3139	3160	3181	3201	2	4	6	8	11	13	15	17	19
21	·3222	3243	3263	3284	3304	3324	3345	3365	3385	3404	2	4	6	8	10	12	14	16	18
22	·3424	3444	3464	3483	3502	3522	3541	3560	3579	3598	2	4	6	8	10	12	14	15	17
23	·3617	3636	3655	3674	3692	3711	3729	3747	3766	3784	2	4	6	7	9	11	13	15	17
24	·3802	3820	3838	3856	3874	3892	3909	3927	3945	3962	2	4	5	7	9	11	12	14	16
25	·3979	3997	4014	4031	4048	4065	4082	4099	4116	4133	2	3	5	7	9	10	12	14	15
26	·4150	4166	4183	4200	4216	4232	4249	4265	4281	4298	2	3	5	7	8	10	11	13	15
27	·4314	4330	4346	4362	4378	4393	4409	4425	4440	4456	2	3	5	6	8	9	11	13	14
28	·4472	4487	4502	4518	4533	4548	4564	4579	4594	4609	2	3	5	6	8	9	11	12	14
29	·4624	4639	4654	4669	4683	4698	4713	4728	4742	4757	1	3	4	6	7	9	10	12	13
30	·4771	4786	4800	4814	4829	4843	4857	4871	4886	4900	1	3	4	6	7	9	10	11	13
31	·4914	4928	4942	4955	4969	4983	4997	5011	5024	5038	1	3	4	6	7	8	10	11	12
32	·5051	5065	5079	5092	5105	5119	5132	5145	5159	5172	1	3	4	5	7	8	9	11	12
33	·5185	5198'	5211	5224	5237	5250	5263	5276	5289	5302	1	3	4	5	6	8	9	10	12
34	·5315	5328	5340	5353	5366	5378	5391	5403	5416	5428	1	3	4	5	6	8	9	10	11
35	·5441	5453	5465	5478	5490	5502	5514	5527	5539	5551	1	2	4	5	6	7	9	10	11
36	·5563	5575	5587	5599	5611	5623	5635	5647	5658	5670	1	2	4	5	6	7	8	10	11
37	·5682	5694	5705	5717	5729	5740	5752	5763	5775	5786	1	2	3	5	6	7	8	9	10
38	·5798	5809	5821	5832	5843	5855	5866	5877	5888	5899	1	2	3	5	6	7	8	9	10
39	·5911	5922	5933	5944	5955	5966	5977	5988	5999	6010	1	2	3	4	5	7	8	9	10
40	·6021	6031	6042	6053	6064	6075	6085	6096	6107	6117	1	2	3	4	5	6	8	9	10
41	·6128	6138	6149	6160	6170	6180	6191	6201	6212	6222	1	2	3	4	5	6	7	8	9
42	·6232	6243	6253	6263	6274	6284	6294	6304	6314	6325	1	2	3	4	5	6	7	8	9
43	·6335	6345	6355	6365	6375	6385	6395	6405	6415	6425	1	2	3	4	5	6	7	8	9
44	·6435	6444	6454	6464	6474	6484	6493	6503	6513	6522	1	2	3	4	5	6	7	8	9
45	·6532	6542	6551	6561	6571	6580	6590	6599	6609	6618	1	2	3	4	5	6	7	8	9
46	·6628	6637	6646	6656	6665	6675	6684	6693	6702	6712	1	2	3	4	5	6	7	7	8
47	·6721	6730	6739	6749	6758	6767	6776	6785	6794	6803	1	2	3	4	5	5	6	7	8
48	·6812	6821	6830	6839	6848	6857	6866	6875	6884	6893	1	2	3	4	4	5	6	7	8
49	·6902	6911	6920	6928	6937	6946	6955	6964	6972	6981	1	2	3	4	4	5	6	7	8
50	·6990	6998	7007	7016	7024	7033	7042	7050	7059	7067	1	2	3	3	4	5	6	7	8
51	·7076	7084	7093	7101	7110	7118	7126	7135	7143	7152	1	2	3	3	4	5	6	7	8
52	·7160	7168	7177	7185	7193	7202	7210	7218	7226	7235	1	2	2	3	4	5	6	7	7
53	·7243	7251	7259	7267	7275	7284	7292	7300	7308	7316	1	2	2	3	4	5	6	6	7
54	·7324	7332	7340	7348	7356	7364	7372	7380	7388	7396	1	2	2	3	4	5	6	6	7
	0	1	2	3	4	5	6	7	8	9	1	2	3	4	5	6	7	8	9

N	0	1	2	3	4	5	6	7	8	9	1	2	3	4	5	6	7	8	9
55	·7404	7412	7419	7427	7435	7443	7451	7459	7466	7474	1	2	2	3	4	5	5	6	7
56	·7482	7490	7497	7505	7513	7520	7528	7536	7543	7551	1	2	2	3	4	5	5	6	7
57	·7559	7566	7574	7582	7589	7597	7604	7612	7619	7627	1	2	2	3	4	5	5	6	7
58	·7634	7642	7649	7657	7664	7672	7679	7686	7694	7701	1	1	2	3	4	4	5	6	7
59	·7709	7716	7723	7731	7738	7745	7752	7760	7767	7774	1	1	2	3	4	4	5	6	7
60	·7782	7789	7796	7803	7810	7818	7825	7832	7839	7846	1	1	2	3	4	4	5	6	6
61	·7853	7860	7868	7875	7882	7889	7896	7903	7910	7917	1	1	2	3	4	4	5	6	6
62	·7924	7931	7938	7945	7952	7959	7966	7973	7980	7987	1	1	2	3	3	4	5	6	6
63	·7993	8000	8007	8014	8021	8028	8035	8041	8048	8055	1	1	2	3	3	4	5	5	6
64	·8062	8069	8075	8082	8089	8096	8102	8109	8116	8122	1	1	2	3	3	4	5	5	6
65	·8129	8136	8142	8149	8156	8162	8169	8176	8182	8189	1	1	2	3	3	4	5	5	6
66	·8195	8202	8209	8215	8222	8228	8235	8241	8248	8254	1	1	2	3	3	4	5	5	6
67	·8261	8267	8274	8280	8287	8293	8299	8306	8312	8319	1	1	2	3	3	4	5	5	6
68	·8325	8331	8338	8344	8351	8357	8363	8370	8376	8382	1	1	2	3	3	4	4	5	6
69	·8388	8395	8401	8407	8414	8420	8426	8432	8439	8445	1	1	2	2	3	4	4	5	6
70	·8451	8457	8463	8470	8476	8482	8488	8494	8500	8506	1	1	2	2	3	4	4	5	6
71	·8513	8519	8525	8531	8537	8543	8549	8555	8561	8567	1	1	2	2	3	4	4	5	5
72	·8573	8579	8585	8591	8597	8603	8609	8615	8621	8627	1	1	2	2	3	4	4	5	5
73	·8633	8639	8645	8651	8657	8663	8669	8675	8681	8686	1	1	2	2	3	4	4	5	5
74	·8692	8698	8704	8710	8716	8722	8727	8733	8739	8745	1	1	2	2	3	4	4	5	5
75	·8751	8756	8762	8768	8774	8779	8785	8791	8797	8802	1	1	2	2	3	3	4	5	5
76	·8808	8814	8820	8825	8831	8837	8842	8848	8854	8859	1	1	2	2	3	3	4	5	5
77	·8865	8871	8876	8882	8887	8893	8899	8904	8910	8915	1	1	2	2	3	3	4	4	5
78	·8921	8927	8932	8938	8943	8949	8954	8960	8965	8971	1	1	2	2	3	3	4	4	5
79	·8976	8982	8987	8993	8998	9004	9009	9015	9020	9025	1	1	2	2	3	3	4	4	5
80	·9031	9036	9042	9047	9053	9058	9063	9069	9074	9079	1	1	2	2	3	3	4	4	5
81	·9085	9090	9096	9101	9106	9112	9117	9122	9128	9133	1	1	2	2	3	3	4	4	5
82	·9138	9143	9149	9154	9159	9165	9170	9175	9180	9186	1	1	2	2	3	3	4	4	5
83	·9191	9196	9201	9206	9212	9217	9222	9227	9232	9238	1	1	2	2	3	3	4	4	5
84	·9243	9248	9253	9258	9263	9269	9274	9279	9284	9289	1	1	2	2	3	3	4	4	5
85	·9294	9299	9304	9309	9315	9320	9325	9330	9335	9340	1	1	2	2	3	3	4	4	5
86	·9345	9350	9355	9360	9365	9370	9375	9380	9385	9390	1	1	1	2	3	3	4	4	5
87	·9395	9400	9405	9410	9415	9420	9425	9430	9435	9440	0	1	1	2	2	3	3	4	4
88	·9445	9450	9455	9460	9465	9469	9474	9479	9484	9489	0	1	1	2	2	3	3	4	4
89	·9494	9499	9504	9509	9513	9518	9523	9528	9533	9538	0	1	1	2	2	3	3	4	4
90	·9542	9547	9552	9557	9562	9566	9571	9576	9581	9586	0	1	1	2	2	3	3	4	4
91	·9590	9595	9600	9605	9609	9614	9619	9624	9628	9633	0	1	1	2	2	3	3	4	4
92	·9638	9643	9647	9652	9657	9661	9666	9671	9675	9680	0	1	1	2	2	3	3	4	4
93	·9685	9689	9694	9699	9703	9708	9713	9717	9722	9727	0	1	1	2	2	3	3	4	4
94	·9731	9736	9741	9745	9750	9754	9759	9763	9768	9773	0	1	1	2	2	3	3	4	4
95	·9777	9782	9786	9791	9795	9800	9805	9809	9814	9818	0	1	1	2	2	3	3	4	4
96	·9823	9827	9832	9836	9841	9845	9850	9854	9859	9863	0	1	1	2	2	3	3	4	4
97	·9868	9872	9877	9881	9886	9890	9894	9899	9903	9908	0	1	1	2	2	3	3	4	4
98	·9912	9917	9921	9926	9930	9934	9939	9943	9948	9952	0	1	1	2	2	3	3	4	4
99	·9956	9961	9965	9969	9974	9978	9983	9987	9991	9996	0	1	1	2	2	3	3	3	4
N	0	1	2	3	4	5	6	7	8	9	1	2	3	4	5	6	7	8	9

N	0	1	2	3	4	5	6	7	8	9	1	2	3	4	5	6	7	8	9
·00	1000	1002	1005	1007	1009	1012	1014	1016	1019	1021	0	0	1	1	1	1	2	2	2
·01	1023	1026	1028	1030	1033	1035	1038	1040	1042	1045	0	0	1	1	1	1	2	2	2
·02	1047	1050	1052	1054	1057	1059	1062	1064	1067	1069	0	0	1	1	1	1	2	2	2
·03	1072	1074	1076	1079	1081	1084	1086	1089	1091	1094	0	0	1	1	1	1	2	2	2
·04	1096	1099	1102	1104	1107	1109	1112	1114	1117	1119	0	1	1	1	1	2	2	2	2
·05	1122	1125	1127	1130	1132	1135	1138	1140	1143	1146	0	1	1	1	1	2	2	2	2
·06	1148	1151	1153	1156	1159	1161	1164	1167	1169	1172	0	1	1	1	1	2	2	2	2
·07	1175	1178	1180	1183	1186	1189	1191	1194	1197	1199	0	1	1	1	1	2	2	2	2
·08	1202	1205	1208	1211	1213	1216	1219	1222	1225	1227	0	1	1	1	1	2	2	2	3
·09	1230	1233	1236	1239	1242	1245	1247	1250	1253	1256	0	1	1	1	1	2	2	2	3
·10	1259	1262	1265	1268	1271	1274	1276	1279	1282	1285	0	1	1	1	1	2	2	2	3
·11	1288	1291	1294	1297	1300	1303	1306	1309	1312	1315	0	1	1	1	2	2	2	2	3
·12	1318	1321	1324	1327	1330	1334	1337	1340	1343	1346	0	1	1	1	2	2	2	3	3
·13	1349	1352	1355	1358	1361	1365	1368	1371	1374	1377	0	1	1	1	2	2	2	3	3
·14	1380	1384	1387	1390	1393	1396	1400	1403	1406	1409	0	1	1	1	2	2	2	3	3
·15	1413	1416	1419	1422	1426	1429	1432	1435	1439	1442	0	1	1	1	2	2	2	3	3
·16	1445	1449	1452	1455	1459	1462	1466	1469	1472	1476	0	1	1	1	2	2	2	3	3
·17	1479	1483	1486	1489	1493	1496	1500	1503	1507	1510	0	1	1	1	2	2	2	3	3
·18	1514	1517	1521	1524	1528	1531	1535	1538	1542	1545	0	1	1	1	2	2	2	3	3
·19	1549	1552	1556	1560	1563	1567	1570	1574	1578	1581	0	1	1	1	2	2	3	3	3
·20	1585	1589	1592	1596	1600	1603	1607	1611	1614	1618	0	1	1	1	2	2	3	3	3
·21	1622	1626	1629	1633	1637	1641	1644	1648	1652	1656	0	1	1	2	2	2	3	3	3
·22	1660	1663	1667	1671	1675	1679	1683	1687	1690	1694	0	1	1	2	2	2	3	3	3
·23	1698	1702	1706	1710	1714	1718	1722	1726	1730	1734	0	1	1	2	2	2	3	3	4
·24	1738	1742	1746	1750	1754	1758	1762	1766	1770	1774	0	1	1	2	2	2	3	3	4
·25	1778	1782	1786	1791	1795	1799	1803	1807	1811	1816	0	1	1	2	2	2	3	3	4
·26	1820	1824	1828	1832	1837	1841	1845	1849	1854	1858	0	1	1	2	2	3	3	3	4
·27	1862	1866	1871	1875	1879	1884	1888	1892	1897	1901	0	1	1	2	2	3	3	3	4
·28	1905	1910	1914	1919	1923	1928	1932	1936	1941	1945	0	1	1	2	2	3	3	4	4
·29	1950	1954	1959	1963	1968	1972	1977	1982	1986	1991	0	1	1	2	2	3	3	4	4
·30	1995	2000	2004	2009	2014	2018	2023	2028	2032	2037	0	1	1	2	2	3	3	4	4
·31	2042	2046	2051	2056	2061	2065	2070	2075	2080	2084	0	1	1	2	2	3	3	4	4
·32	2089	2094	2099	2104	2109	2113	2118	2123	2128	2133	0	1	1	2	2	3	3	4	4
·33	2138	2143	2148	2153	2158	2163	2168	2173	2178	2183	0	1	1	2	2	3	3	4	4
·34	2188	2193	2198	2203	2208	2213	2218	2223	2228	2234	1	1	2	2	3	3	4	4	5
·35	2239	2244	2249	2254	2259	2265	2270	2275	2280	2286	1	1	2	2	3	3	4	4	5
·36	2291	2296	2301	2307	2312	2317	2323	2328	2333	2339	1	1	2	2	3	3	4	4	5
·37	2344	2350	2355	2360	2366	2371	2377	2382	2388	2393	1	1	2	2	3	3	4	4	5
·38	2399	2404	2410	2415	2421	2427	2432	2438	2443	2449	1	1	2	2	3	3	4	4	5
·39	2455	2460	2466	2472	2477	2483	2489	2495	2500	2506	1	1	2	2	3	3	4	5	5
·40	2512	2518	2523	2529	2535	2541	2547	2553	2559	2564	1	1	2	2	3	4	4	5	5
·41	2570	2576	2582	2588	2594	2600	2606	2612	2618	2624	1	1	2	2	3	4	4	5	5
·42	2630	2636	2642	2649	2655	2661	2667	2673	2679	2685	1	1	2	2	3	4	4	5	6
·43	2692	2698	2704	2710	2716	2723	2729	2735	2742	2748	1	1	2	3	3	4	4	5	6
·44	2754	2761	2767	2773	2780	2786	2793	2799	2805	2812	1	1	2	3	3	4	4	5	6
·45	2818	2825	2831	2838	2844	2851	2858	2864	2871	2877	1	1	2	3	3	4	5	5	6
·46	2884	2891	2897	2904	2911	2917	2924	2931	2938	2944	1	1	2	3	3	4	5	5	6
·47	2951	2958	2965	2972	2979	2985	2992	2999	3006	3013	1	1	2	3	3	4	5	5	6
·48	3020	3027	3034	3041	3048	3055	3062	3069	3076	3083	1	1	2	3	4	4	5	6	6
·49	3090	3097	3105	3112	3119	3126	3133	3141	3148	3155	1	1	2	3	4	4	5	6	6
N	0	1	2	3	4	5	6	7	8	9	1	2	3	4	5	6	7	8	9

N	0	1	2	3	4	5	6	7	8	9	1	2	3	4	5	6	7	8	9
·50	3162	3170	3177	3184	3192	3199	3206	3214	3221	3228	1	1	2	3	4	4	5	6	7
·51	3236	3243	3251	3258	3266	3273	3281	3289	3296	3304	1	2	2	3	4	5	5	6	7
·52	3311	3319	3327	3334	3342	3350	3357	3365	3373	3381	1	2	2	3	4	5	5	6	7
·53	3388	3396	3404	3412	3420	3428	3436	3443	3451	3459	1	2	2	3	4	5	6	6	7
·54	3467	3475	3483	3491	3499	3508	3516	3524	3532	3540	1	2	2	3	4	5	6	6	7
·55	3548	3556	3565	3573	3581	3589	3597	3606	3614	3622	1	2	2	3	4	5	6	7	7
·56	3631	3639	3648	3656	3664	3673	3681	3690	3698	3707	1	2	3	3	4	5	6	7	8
·57	3715	3724	3733	3741	3750	3758	3767	3776	3784	3793	1	2	3	3	4	5	6	7	8
·58	3802	3811	3819	3828	3837	3846	3855	3864	3873	3882	1	2	3	4	4	5	6	7	8
·59	3890	3899	3908	3917	3926	3936	3945	3954	3963	3972	1	2	3	4	5	5	6	7	8
·60	3981	3990	3999	4009	4018	4027	4036	4046	4055	4064	1	2	3	4	5	6	6	7	8
·61	4074	4083	4093	4102	4111	4121	4130	4140	4150	4159	1	2	3	4	5	6	7	8	9
·62	4169	4178	4188	4198	4207	4217	4227	4236	4246	4256	1	2	3	4	5	6	7	8	9
·63	4266	4276	4285	4295	4305	4315	4325	4335	4345	4355	1	2	3	4	5	6	7	8	9
·64	4365	4375	4385	4395	4406	4416	4426	4436	4446	4457	1	2	3	4	5	6	7	8	9
·65	4467	4477	4487	4498	4508	4519	4529	4539	4550	4560	1	2	3	4	5	6	7	8	9
·66	4571	4581	4592	4603	4613	4624	4634	4645	4656	4667	1	2	3	4	5	6	7	9	10
·67	4677	4688	4699	4710	4721	4732	4742	4753	4764	4775	1	2	3	4	5	7	8	9	10
·68	4786	4797	4808	4819	4831	4842	4853	4864	4875	4887	1	2	3	4	6	7	8	9	10
·69	4898	4909	4920	4932	4943	4955	4966	4977	4989	5000	1	2	3	5	6	7	8	9	10
·70	5012	5023	5035	5047	5058	5070	5082	5093	5105	5117	1	2	4	5	6	7	8	9	11
·71	5129	5140	5152	5164	5176	5188	5200	5212	5224	5236	1	2	4	5	6	7	8	10	11
·72	5248	5260	5272	5284	5297	5309	5321	5333	5346	5358	1	2	4	5	6	7	9	10	11
·73	5370	5383	5395	5408	5420	5433	5445	5458	5470	5483	1	3	4	5	6	8	9	10	11
·74	5495	5508	5521	5534	5546	5559	5572	5585	5598	5610	1	3	4	5	6	8	9	10	12
·75	5623	5636	5649	5662	5675	5689	5702	5715	5728	5741	1	3	4	5	7	8	9	10	12
·76	5754	5768	5781	5794	5808	5821	5834	5848	5861	5875	1	3	4	5	7	8	9	11	12
·77	5888	5902	5916	5929	5943	5957	5970	5984	5998	6012	1	3	4	5	7	8	10	11	12
·78	6026	6039	6053	6067	6081	6095	6109	6124	6138	6152	1	3	4	6	7	8	10	11	13
·79	6166	6180	6194	6209	6223	6237	6252	6266	6281	6295	1	3	4	6	7	9	10	11	13
·80	6310	6324	6339	6353	6368	6383	6397	6412	6427	6442	1	3	4	6	7	9	10	12	13
·81	6457	6471	6486	6501	6516	6531	6546	6561	6577	6592	2	3	5	6	8	9	11	12	14
·82	6607	6622	6637	6653	6668	6683	6699	6714	6730	6745	2	3	5	6	8	9	11	12	14
·83	6761	6776	6792	6808	6823	6839	6855	6871	6887	6902	2	3	5	6	8	9	11	13	14
·84	6918	6934	6950	6966	6982	6998	7015	7031	7047	7063	2	3	5	6	8	10	11	13	15
·85	7079	7096	7112	7129	7145	7161	7178	7194	7211	7228	2	3	5	7	8	10	12	13	15
·86	7244	7261	7278	7295	7311	7328	7345	7362	7379	7396	2	3	5	7	8	10	12	13	15
·87	7413	7430	7447	7464	7482	7499	7516	7534	7551	7568	2	3	5	7	9	10	12	14	16
·88	7586	7603	7621	7638	7656	7674	7691	7709	7727	7745	2	4	5	7	9	11	12	14	16
·89	7762	7780	7798	7816	7834	7852	7870	7889	7907	7925	2	4	5	7	9	11	13	14	16
·90	7943	7962	7980	7998	8017	8035	8054	8072	8091	8110	2	4	6	7	9	11	13	15	17
·91	8128	8147	8166	8185	8204	8222	8241	8260	8279	8299	2	4	6	8	9	11	13	15	17
·92	8318	8337	8356	8375	8395	8414	8433	8453	8472	8492	2	4	6	8	10	12	14	15	17
·93	8511	8531	8551	8570	8590	8610	8630	8650	8670	8690	2	4	6	8	10	12	14	16	18
·94	8710	8730	8750	8770	8790	8810	8831	8851	8872	8892	2	4	6	8	10	12	14	16	18
·95	8913	8933	8954	8974	8995	9016	9036	9057	9078	9099	2	4	6	8	10	12	15	17	19
·96	9120	9141	9162	9183	9204	9226	9247	9268	9290	9311	2	4	6	8	11	13	15	17	19
·97	9333	9354	9376	9397	9419	9441	9462	9484	9506	9528	2	4	7	9	11	13	15	17	20
·98	9550	9572	9594	9616	9638	9661	9683	9705	9727	9750	2	4	7	9	11	13	16	18	20
·99	9772	9795	9817	9840	9863	9886	9908	9931	9954	9977	2	5	7	9	11	14	16	18	20
N	0	1	2	3	4	5	6	7	8	9	1	2	3	4	5	6	7	8	9

Natural Sines

°	0′	6′	12′	18′	24′	30′	36′	42′	48′	54′	1′	2′	3′	4′	5′
0	·0000	0017	0035	0052	0070	0087	0105	0122	0140	0157	3	6	9	12	15
1	·0175	0192	0209	0227	0244	0262	0279	0297	0314	0332	3	6	9	12	15
2	·0349	0366	0384	0401	0419	0436	0454	0471	0488	0506	3	6	9	12	15
3	·0523	0541	0558	0576	0593	0610	0628	0645	0663	0680	3	6	9	12	15
4	·0698	0715	0732	0750	0767	0785	0802	0819	0837	0854	3	6	9	12	14
5	·0872	0889	0906	0924	0941	0958	0976	0993	1011	1028	3	6	9	12	14
6	·1045	1063	1080	1097	1115	1132	1149	1167	1184	1201	3	6	9	12	14
7	·1219	1236	1253	1271	1288	1305	1323	1340	1357	1374	3	6	9	12	14
8	·1392	1409	1426	1444	1461	1478	1495	1513	1530	1547	3	6	9	12	14
9	·1564	1582	1599	1616	1633	1650	1668	1685	1702	1719	3	6	9	11	14
10	·1736	1754	1771	1788	1805	1822	1840	1857	1874	1891	3	6	9	11	14
11	·1908	1925	1942	1959	1977	1994	2011	2028	2045	2062	3	6	9	11	14
12	·2079	2096	2113	2130	2147	2164	2181	2198	2215	2233	3	6	9	11	14
13	·2250	2267	2284	2300	2317	2334	2351	2368	2385	2402	3	6	8	11	14
14	·2419	2436	2453	2470	2487	2504	2521	2538	2554	2571	3	6	8	11	14
15	·2588	2605	2622	2639	2656	2672	2689	2706	2723	2740	3	6	8	11	14
16	·2756	2773	2790	2807	2823	2840	2857	2874	2890	2907	3	6	8	11	14
17	·2924	2940	2957	2974	2990	3007	3024	3040	3057	3074	3	6	8	11	14
18	·3090	3107	3123	3140	3156	3173	3190	3206	3223	3239	3	6	8	11	14
19	·3256	3272	3289	3305	3322	3338	3355	3371	3387	3404	3	5	8	11	14
20	·3420	3437	3453	3469	3486	3502	3518	3535	3551	3567	3	5	8	11	14
21	·3584	3600	3616	3633	3649	3665	3681	3697	3714	3730	3	5	8	11	14
22	·3746	3762	3778	3795	3811	3827	3843	3859	3875	3891	3	5	8	11	13
23	·3907	3923	3939	3955	3971	3987	4003	4019	4035	4051	3	5	8	11	13
24	·4067	4083	4099	4115	4131	4147	4163	4179	4195	4210	3	5	8	11	13
25	·4226	4242	4258	4274	4289	4305	4321	4337	4352	4368	3	5	8	11	13
26	·4384	4399	4415	4431	4446	4462	4478	4493	4509	4524	3	5	8	10	13
27	·4540	4555	4571	4586	4602	4617	4633	4648	4664	4679	3	5	8	10	13
28	·4695	4710	4726	4741	4756	4772	4787	4802	4818	4833	3	5	8	10	13
29	·4848	4863	4879	4894	4909	4924	4939	4955	4970	4985	3	5	8	10	13
30	·5000	5015	5030	5045	5060	5075	5090	5105	5120	5135	3	5	8	10	13
31	·5150	5165	5180	5195	5210	5225	5240	5255	5270	5284	2	5	7	10	12
32	·5299	5314	5329	5344	5358	5373	5388	5402	5417	5432	2	5	7	10	12
33	·5446	5461	5476	5490	5505	5519	5534	5548	5563	5577	2	5	7	10	12
34	·5592	5606	5621	5635	5650	5664	5678	5693	5707	5721	2	5	7	10	12
35	·5736	5750	5764	5779	5793	5807	5821	5835	5850	5864	2	5	7	9	12
36	·5878	5892	5906	5920	5934	5948	5962	5976	5990	6004	2	5	7	9	12
37	·6018	6032	6046	6060	6074	6088	6101	6115	6129	6143	2	5	7	9	12
38	·6157	6170	6184	6198	6211	6225	6239	6252	6266	6280	2	5	7	9	11
39	·6293	6307	6320	6334	6347	6361	6374	6388	6401	6414	2	4	7	9	11
40	·6428	6441	6455	6468	6481	6494	6508	6521	6534	6547	2	4	7	9	11
41	·6561	6574	6587	6600	6613	6626	6639	6652	6665	6678	2	4	7	9	11
42	·6691	6704	6717	6730	6743	6756	6769	6782	6794	6807	2	4	6	9	11
43	·6820	6833	6845	6858	6871	6884	6896	6909	6921	6934	2	4	6	8	11
44	·6947	6959	6972	6984	6997	7009	7022	7034	7046	7059	2	4	6	8	10
	0′	6′	12′	18′	24′	30′	36′	42′	48′	54′	1′	2′	3′	4′	5′

	0'	6'	12'	18'	24'	30'	36'	42'	48'	54'	1'	2'	3'	4'	5'
45°	·7071	7083	7096	7108	7120	7133	7145	7157	7169	7181	2	4	6	8	10
46	·7193	7206	7218	7230	7242	7254	7266	7278	7290	7302	2	4	6	8	10
47	·7314	7325	7337	7349	7361	7373	7385	7396	7408	7420	2	4	6	8	10
48	·7431	7443	7455	7466	7478	7490	7501	7513	7524	7536	2	4	6	8	10
49	·7547	7559	7570	7581	7593	7604	7615	7627	7638	7649	2	4	6	8	9
50	·7660	7672	7683	7694	7705	7716	7727	7738	7749	7760	2	4	6	7	9
51	·7771	7782	7793	7804	7815	7826	7837	7848	7859	7869	2	4	5	7	9
52	·7880	7891	7902	7912	7923	7934	7944	7955	7965	7976	2	4	5	7	9
53	·7986	7997	8007	8018	8028	8039	8049	8059	8070	8080	2	3	5	7	9
54	·8090	8100	8111	8121	8131	8141	8151	8161	8171	8181	2	3	5	7	8
55	·8192	8202	8211	8221	8231	8241	8251	8261	8271	8281	2	3	5	7	8
56	·8290	8300	8310	8320	8329	8339	8348	8358	8368	8377	2	3	5	6	8
57	·8387	8396	8406	8415	8425	8434	8443	8453	8462	8471	2	3	5	6	8
58	·8480	8490	8499	8508	8517	8526	8536	8545	8554	8563	2	3	5	6	8
59	·8572	8581	8590	8599	8607	8616	8625	8634	8643	8652	1	3	4	6	7
60	·8660	8669	8678	8686	8695	8704	8712	8721	8729	8738	1	3	4	6	7
61	·8746	8755	8763	8771	8780	8788	8796	8805	8813	8821	1	3	4	6	7
62	·8829	8838	8846	8854	8862	8870	8878	8886	8894	8902	1	3	4	5	7
63	·8910	8918	8926	8934	8942	8949	8957	8965	8973	8980	1	3	4	5	6
64	·8988	8996	9003	9011	9018	9026	9033	9041	9048	9056	1	3	4	5	6
65	·9063	9070	9078	9085	9092	9100	9107	9114	9121	9128	1	2	4	5	6
66	·9135	9143	9150	9157	9164	9171	9178	9184	9191	9198	1	2	3	5	6
67	·9205	9212	9219	9225	9232	9239	9245	9252	9259	9265	1	2	3	4	6
68	·9272	9278	9285	9291	9298	9304	9311	9317	9323	9330	1	2	3	4	5
69	·9336	9342	9348	9354	9361	9367	9373	9379	9385	9391	1	2	3	4	5
70	·9397	9403	9409	9415	9421	9426	9432	9438	9444	9449	1	2	3	4	5
71	·9455	9461	9466	9472	9478	9483	9489	9494	9500	9505	1	2	3	4	5
72	·9511	9516	9521	9527	9532	9537	9542	9548	9553	9558	1	2	3	4	4
73	·9563	9568	9573	9578	9583	9588	9593	9598	9603	9608	1	2	2	3	4
74	·9613	9617	9622	9627	9632	9636	9641	9646	9650	9655	1	2	2	3	4
75	·9659	9664	9668	9673	9677	9681	9686	9690	9694	9699	1	1	2	3	4
76	·9703	9707	9711	9715	9720	9724	9728	9732	9736	9740	1	1	2	3	3
77	·9744	9748	9751	9755	9759	9763	9767	9770	9774	9778	1	1	2	3	3
78	·9781	9785	9789	9792	9796	9799	9803	9806	9810	9813	1	1	2	2	3
79	·9816	9820	9823	9826	9829	9833	9836	9839	9842	9845	1	1	2	2	3
80	·9848	9851	9854	9857	9860	9863	9866	9869	9871	9874	0	1	1	2	2
81	·9877	9880	9882	9885	9888	9890	9893	9895	9898	9900	0	1	1	2	2
82	·9903	9905	9907	9910	9912	9914	9917	9919	9921	9923	0	1	1	2	2
83	·9925	9928	9930	9932	9934	9936	9938	9940	9942	9943	0	1	1	1	2
84	·9945	9947	9949	9951	9952	9954	9956	9957	9959	9960	0	1	1	1	1
85	·9962	9963	9965	9966	9968	9969	9971	9972	9973	9974	0	0	1	1	1
86	·9976	9977	9978	9979	9980	9981	9982	9983	9984	9985	0	0	1	1	1
87	·9986	9987	9988	9989	9990	9990	9991	9992	9993	9993	0	0	0	1	1
88	·9994	9995	9995	9996	9996	9997	9997	9997	9998	9998					
89	·9998	9999	9999	9999	9999	1·000	1·000	1·000	1·000	1·000					
	0'	6'	12'	18'	24'	30'	36'	42'	48'	54'	1'	2'	3'	4'	5'

	0′	6′	12′	18′	24′	30′	36′	42′	48′	54′	1′	2′	3′	4′	5′
0°	1·0000	1·000	1·000	1·000	1·000	1·000	9999	9999	9999	9999					
1	·9998	9998	9998	9997	9997	9997	9996	9996	9995	9995					
2	·9994	9993	9993	9992	9991	9990	9990	9989	9988	9987					
3	·9986	9985	9984	9983	9982	9981	9980	9979	9978	9977					
4	·9976	9974	9973	9972	9971	9969	9968	9966	9965	9963	1′	2′	3′	4′	5′
5	·9962	9960	9959	9957	9956	9954	9952	9951	9949	9947					
6	·9945	9943	9942	9940	9938	9936	9934	9932	9930	9928	0	1	1	1	2
7	·9925	9923	9921	9919	9917	9914	9912	9910	9907	9905	0	1	1	2	2
8	·9903	9900	9898	9895	9893	9890	9888	9885	9882	9880	0	1	1	2	2
9	·9877	9874	9871	9869	9866	9863	9860	9857	9854	9851	0	1	1	2	2
10	·9848	9845	9842	9839	9836	9833	9829	9826	9823	9820	1	1	2	2	3
11	·9816	9813	9810	9806	9803	9799	9796	9792	9789	9785	1	1	2	2	3
12	·9781	9778	9774	9770	9767	9763	9759	9755	9751	9748	1	1	2	3	3
13	·9744	9740	9736	9732	9728	9724	9720	9715	9711	9707	1	1	2	3	3
14	·9703	9699	9694	9690	9686	9681	9677	9673	9668	9664	1	1	2	3	4
15	·9659	9655	9650	9646	9641	9636	9632	9627	9622	9617	1	2	2	3	4
16	·9613	9608	9603	9598	9593	9588	9583	9578	9573	9568	1	2	2	3	4
17	·9563	9558	9553	9548	9542	9537	9532	9527	9521	9516	1	2	3	3	4
18	·9511	9505	9500	9494	9489	9483	9478	9472	9466	9461	1	2	3	4	5
19	·9455	9449	9444	9438	9432	9426	9421	9415	9409	9403	1	2	3	4	5
20	·9397	9391	9385	9379	9373	9367	9361	9354	9348	9342	1	2	3	4	5
21	·9336	9330	9323	9317	9311	9304	9298	9291	9285	9278	1	2	3	4	5
22	·9272	9265	9259	9252	9245	9239	9232	9225	9219	9212	1	2	3	4	6
23	·9205	9198	9191	9184	9178	9171	9164	9157	9150	9143	1	2	3	5	6
24	·9135	9128	9121	9114	9107	9100	9092	9085	9078	9070	1	2	4	5	6
25	·9063	9056	9048	9041	9033	9026	9018	9011	9003	8996	1	3	4	5	6
26	·8988	8980	8973	8965	8957	8949	8942	8934	8926	8918	1	3	4	5	6
27	·8910	8902	8894	8886	8878	8870	8862	8854	8846	8838	1	3	4	5	7
28	·8829	8821	8813	8805	8796	8788	8780	8771	8763	8755	1	3	4	6	7
29	·8746	8738	8729	8721	8712	8704	8695	8686	8678	8669	1	3	4	6	7
30	·8660	8652	8643	8634	8625	8616	8607	8599	8590	8581	1	3	4	6	7
31	·8572	8563	8554	8545	8536	8526	8517	8508	8499	8490	2	3	5	6	8
32	·8480	8471	8462	8453	8443	8434	8425	8415	8406	8396	2	3	5	6	8
33	·8387	8377	8368	8358	8348	8339	8329	8320	8310	8300	2	3	5	6	8
34	·8290	8281	8271	8261	8251	8241	8231	8221	8211	8202	2	3	5	7	8
35	·8192	8181	8171	8161	8151	8141	8131	8121	8111	8100	2	3	5	7	8
36	·8090	8080	8070	8059	8049	8039	8028	8018	8007	7997	2	3	5	7	9
37	·7986	7976	7965	7955	7944	7934	7923	7912	7902	7891	2	4	5	7	9
38	·7880	7869	7859	7848	7837	7826	7815	7804	7793	7782	2	4	5	7	9
39	·7771	7760	7749	7738	7727	7716	7705	7694	7683	7672	2	4	6	7	9
40	·7660	7649	7638	7627	7615	7604	7593	7581	7570	7559	2	4	6	8	9
41	·7547	7536	7524	7513	7501	7490	7478	7466	7455	7443	2	4	6	8	10
42	·7431	7420	7408	7396	7385	7373	7361	7349	7337	7325	2	4	6	8	10
43	·7314	7302	7290	7278	7266	7254	7242	7230	7218	7206	2	4	6	8	10
44	·7193	7181	7169	7157	7145	7133	7120	7108	7096	7083	2	4	6	8	10

SUBTRACT

	0′	6′	12′	18′	24′	30′	36′	42′	48′	54′	1′	2′	3′	4′	5′
45°	·7071	7059	7046	7034	7022	7009	6997	6984	6972	6959	2	4	6	8	10
46	·6947	6934	6921	6909	6896	6884	6871	6858	6845	6833	2	4	6	8	11
47	·6820	6807	6794	6782	6769	6756	6743	6730	6717	6704	2	4	6	9	11
48	·6691	6678	6665	6652	6639	6626	6613	6600	6587	6574	2	4	7	9	11
49	·6561	6547	6534	6521	6508	6494	6481	6468	6455	6441	2	4	7	9	11
50	·6428	6414	6401	6388	6374	6361	6347	6334	6320	6307	2	4	7	9	11
51	·6293	6280	6266	6252	6239	6225	6211	6198	6184	6170	2	5	7	9	11
52	·6157	6143	6129	6115	6101	6088	6074	6060	6046	6032	2	5	7	9	12
53	·6018	6004	5990	5976	5962	5948	5934	5920	5906	5892	2	5	7	9	12
54	·5878	5864	5850	5835	5821	5807	5793	5779	5764	5750	2	5	7	9	12
55	·5736	5721	5707	5693	5678	5664	5650	5635	5621	5606	2	5	7	10	12
56	·5592	5577	5563	5548	5534	5519	5505	5490	5476	5461	2	5	7	10	12
57	·5446	5432	5417	5402	5388	5373	5358	5344	5329	5314	2	5	7	10	12
58	·5299	5284	5270	5255	5240	5225	5210	5195	5180	5165	2	5	7	10	12
59	·5150	5135	5120	5105	5090	5075	5060	5045	5030	5015	3	5	8	10	13
60	·5000	4985	4970	4955	4939	4924	4909	4894	4879	4863	3	5	8	10	13
61	·4848	4833	4818	4802	4787	4772	4756	4741	4726	4710	3	5	8	10	13
62	·4695	4679	4664	4648	4633	4617	4602	4586	4571	4555	3	5	8	10	13
63	·4540	4524	4509	4493	4478	4462	4446	4431	4415	4399	3	5	8	10	13
64	·4384	4368	4352	4337	4321	4305	4289	4274	4258	4242	3	5	8	11	13
65	·4226	4210	4195	4179	4163	4147	4131	4115	4099	4083	3	5	8	11	13
66	·4067	4051	4035	4019	4003	3987	3971	3955	3939	3923	3	5	8	11	13
67	·3907	3891	3875	3859	3843	3827	3811	3795	3778	3762	3	5	8	11	13
68	·3746	3730	3714	3697	3681	3665	3649	3633	3616	3600	3	5	8	11	14
69	·3584	3567	3551	3535	3518	3502	3486	3469	3453	3437	3	5	8	11	14
70	·3420	3404	3387	3371	3355	3338	3322	3305	3289	3272	3	5	8	11	14
71	·3256	3239	3223	3206	3190	3173	3156	3140	3123	3107	3	6	8	11	14
72	·3090	3074	3057	3040	3024	3007	2990	2974	2957	2940	3	6	8	11	14
73	·2924	2907	2890	2874	2857	2840	2823	2807	2790	2773	3	6	8	11	14
74	·2756	2740	2723	2706	2689	2672	2656	2639	2622	2605	3	6	8	11	14
75	·2588	2571	2554	2538	2521	2504	2487	2470	2453	2436	3	6	8	11	14
76	·2419	2402	2385	2368	2351	2334	2317	2300	2284	2267	3	6	8	11	14
77	·2250	2233	2215	2198	2181	2164	2147	2130	2113	2096	3	6	9	11	14
78	·2079	2062	2045	2028	2011	1994	1977	1959	1942	1925	3	6	9	11	14
79	·1908	1891	1874	1857	1840	1822	1805	1788	1771	1754	3	6	9	11	14
80	·1736	1719	1702	1685	1668	1650	1633	1616	1599	1582	3	6	9	11	14
81	·1564	1547	1530	1513	1495	1478	1461	1444	1426	1409	3	6	9	12	14
82	·1392	1374	1357	1340	1323	1305	1288	1271	1253	1236	3	6	9	12	14
83	·1219	1201	1184	1167	1149	1132	1115	1097	1080	1063	3	6	9	12	14
84	·1045	1028	1011	0993	0976	0958	0941	0924	0906	0889	3	6	9	12	14
85	·0872	0854	0837	0819	0802	0785	0767	0750	0732	0715	3	6	9	12	14
86	·0698	0680	0663	0645	0628	0610	0593	0576	0558	0541	3	6	9	12	15
87	·0523	0506	0488	0471	0454	0436	0419	0401	0384	0366	3	6	9	12	15
88	·0349	0332	0314	0297	0279	0262	0244	0227	0209	0192	3	6	9	12	15
89	·0175	0157	0140	0122	0105	0087	0070	0052	0035	0017	3	6	9	12	15

SUBTRACT

	0'	6'	12'	18'	24'	30'	36'	42'	48'	54'	1'	2'	3'	4'	5'
0	0·0000	0017	0035	0052	0070	0087	0105	0122	0140	0157	3	6	9	12	15
1	0·0175	0192	0209	0227	0244	0262	0279	0297	0314	0332	3	6	9	12	15
2	0·0349	0367	0384	0402	0419	0437	0454	0472	0489	0507	3	6	9	12	15
3	0·0524	0542	0559	0577	0594	0612	0629	0647	0664	0682	3	6	9	12	15
4	0·0699	0717	0734	0752	0769	0787	0805	0822	0840	0857	3	6	9	12	15
5	0·0875	0892	0910	0928	0945	0963	0981	0998	1016	1033	3	6	9	12	15
6	0·1051	1069	1086	1104	1122	1139	1157	1175	1192	1210	3	6	9	12	15
7	0·1228	1246	1263	1281	1299	1317	1334	1352	1370	1388	3	6	9	12	15
8	0·1405	1423	1441	1459	1477	1495	1512	1530	1548	1566	3	6	9	12	15
9	0·1584	1602	1620	1638	1655	1673	1691	1709	1727	1745	3	6	9	12	15
10	0·1763	1781	1799	1817	1835	1853	1871	1890	1908	1926	3	6	9	12	15
11	0·1944	1962	1980	1998	2016	2035	2053	2071	2089	2107	3	6	9	12	15
12	0·2126	2144	2162	2180	2199	2217	2235	2254	2272	2290	3	6	9	12	15
13	0·2309	2327	2345	2364	2382	2401	2419	2438	2456	2475	3	6	9	12	15
14	0·2493	2512	2530	2549	2568	2586	2605	2623	2642	2661	3	6	9	12	16
15	0·2679	2698	2717	2736	2754	2773	2792	2811	2830	2849	3	6	9	13	16
16	0·2867	2886	2905	2924	2943	2962	2981	3000	3019	3038	3	6	9	13	16
17	0·3057	3076	3096	3115	3134	3153	3172	3191	3211	3230	3	6	10	13	16
18	0·3249	3269	3288	3307	3327	3346	3365	3385	3404	3424	3	6	10	13	16
19	0·3443	3463	3482	3502	3522	3541	3561	3581	3600	3620	3	7	10	13	16
20	0·3640	3659	3679	3699	3719	3739	3759	3779	3799	3819	3	7	10	13	17
21	0·3839	3859	3879	3899	3919	3939	3959	3979	4000	4020	3	7	10	13	17
22	0·4040	4061	4081	4101	4122	4142	4163	4183	4204	4224	3	7	10	14	17
23	0·4245	4265	4286	4307	4327	4348	4369	4390	4411	4431	3	7	10	14	17
24	0·4452	4473	4494	4515	4536	4557	4578	4599	4621	4642	4	7	11	14	18
25	0·4663	4684	4706	4727	4748	4770	4791	4813	4834	4856	4	7	11	14	18
26	0·4877	4899	4921	4942	4964	4986	5008	5029	5051	5073	4	7	11	15	18
27	0·5095	5117	5139	5161	5184	5206	5228	5250	5272	5295	4	7	11	15	18
28	0·5317	5340	5362	5384	5407	5430	5452	5475	5498	5520	4	8	11	15	19
29	0·5543	5566	5589	5612	5635	5658	5681	5704	5727	5750	4	8	12	15	19
30	0·5774	5797	5820	5844	5867	5890	5914	5938	5961	5985	4	8	12	16	20
31	0·6009	6032	6056	6080	6104	6128	6152	6176	6200	6224	4	8	12	16	20
32	0·6249	6273	6297	6322	6346	6371	6395	6420	6445	6469	4	8	12	16	20
33	0·6494	6519	6544	6569	6594	6619	6644	6669	6694	6720	4	8	13	17	21
34	0·6745	6771	6796	6822	6847	6873	6899	6924	6950	6976	4	9	13	17	21
35	0·7002	7028	7054	7080	7107	7133	7159	7186	7212	7239	4	9	13	18	22
36	0·7265	7292	7319	7346	7373	7400	7427	7454	7481	7508	5	9	14	18	23
37	0·7536	7563	7590	7618	7646	7673	7701	7729	7757	7785	5	9	14	18	23
38	0·7813	7841	7869	7898	7926	7954	7983	8012	8040	8069	5	9	14	19	24
39	0·8098	8127	8156	8185	8214	8243	8273	8302	8332	8361	5	10	15	20	24
40	0·8391	8421	8451	8481	8511	8541	8571	8601	8632	8662	5	10	15	20	25
41	0·8693	8724	8754	8785	8816	8847	8878	8910	8941	8972	5	10	16	21	26
42	0·9004	9036	9067	9099	9131	9163	9195	9228	9260	9293	5	11	16	21	27
43	0·9325	9358	9391	9424	9457	9490	9523	9556	9590	9623	6	11	17	22	28
44	0·9657	9691	9725	9759	9793	9827	9861	9896	9930	9965	6	11	17	23	29
	0'	6'	12'	18'	24'	30'	36'	42'	48'	54'	1'	2'	3'	4'	5'

	0'	6'	12'	18'	24'	30'	36'	42'	48'	54'	1'	2'	3'	4'	5'
45	1·0000	0035	0070	0105	0141	0176	0212	0247	0283	0319	6	12	18	24	30
46	1·0355	0392	0428	0464	0501	0538	0575	0612	0649	0686	6	12	18	25	31
47	1·0724	0761	0799	0837	0875	0913	0951	0990	1028	1067	6	13	19	25	32
48	1·1106	1145	1184	1224	1263	1303	·1343	1383	1423	1463	7	13	20	26	33
49	1·1504	1544	1585	1626	1667	1708	1750	1792	1833	1875	7	14	21	28	34
50	1·1918	1960	2002	2045	2088	2131	2174	2218	2261	2305	7	14	22	29	36
51	1·2349	2393	2437	2482	2527	2572	2617	2662	2708	2753	8	15	23	30	38
52	1·2799	2846	2892	2938	2985	3032	3079	3127	3175	3222	8	16	24	31	39
53	1·3270	3319	3367	3416	3465	3514	3564	3613	3663	3713	8	16	25	33	41
54	1·3764	3814	3865	3916	3968	4019	4071	4124	4176	4229	9	17	26	34	43
55	1·4281	4335	4388	4442	4496	4550	4605	4659	4715	4770	9	18	27	36	45
56	1·4826	4882	4938	4994	5051	5108	5166	5224	5282	5340	10	19	29	38	48
57	1·5399	5458	5517	5577	5637	5697	5757	5818	5880	5941	10	20	30	40	50
58	1·6003	6066	6128	6191	6255	6319	6383	6447	6512	6577	11	21	32	43	53
59	1·6643	6709	6775	6842	6909	6977	7045	7113	7182	7251	11	23	34	45	56
60	1·7321	7391	7461	7532	7603	7675	7747	7820	7893	7966	12	24	36	48	60
61	1·8040	8115	8190	8265	8341	8418	8495	8572	8650	8728	13	26	38	51	64
62	1·8807	8887	8967	9047	9128	9210	9292	9375	9458	9542	14	27	41	55	68
63	1·9626	9711	9797	9883	9970	0057	0145	0233	0323	0413	15	29	44	58	73
64	2·0503	0594	0686	0778	0872	0965	1060	1155	1251	1348	16	31	47	63	78
65	2·1445	1543	1642	1742	1842	1943	2045	2148	2251	2355	17	34	51	68	85
66	2·2460	2566	2673	2781	2889	2998	3109	3220	3332	3445	18	37	55	73	91
67	2·3559	3673	3789	3906	4023	4142	4262	4383	4504	4627	20	40	60	79	99
68	2·4751	4876	5002	5129	5257	5386	5517	5649	5782	5916	22	43	65	87	108
69	2·6051	6187	6325	6464	6605	6746	6889	7034	7179	7326	24	47	71	95	119
70	2·7475	7625	7776	7929	8083	8239	8397	8556	8716	8878	26	52	78	104	130
71	2·9042	9208	9375	9544	9714	9887	0061	0237	0415	0595	29	58	87	116	144
72	3·0777	0961	1146	1334	1524	1716	1910	2106	2305	2506	32	64	97	129	161
73	3·2709	2914	3122	3332	3544	3759	3977	4197	4420	4646	36	72	108	144	180
74	3·4874	5105	5339	5576	5816	6059	6305	6554	6806	7062	41	81	122	163	203
75	3·7321	7583	7848	8118	8391	8667	8947	9232	9520	9812	46	93	139	186	232
76	4·0108	0408	0713	1022	1335	1653	1976	2303	2635	2972	53	107	160	214	267
77	4·3315	3662	4015	4373	4737	5107	5483	5864	6252	6646	62	124	186	248	310
78	4·7046	7453	7867	8288	8716	9152	9594	0045	0504	0970	73	146	220	293	366
79	5·1446	1929	2422	2924	3435	3955	4486	5026	5578	6140	87	175	263	350	438
80	5·671	5·730	5·789	5·850	5·912	5·976	6·041	6·107	6·174	6·243					
81	6·314	6·386	6·460	6·535	6·612	6·691	6·772	6·855	6·940	7·026					
82	7·115	7·207	7·300	7·396	7·495	7·596	7·700	7·806	7·916	8·028					
83	8·144	8·264	8·386	8·513	8·643	8·777	8·915	9·058	9·205	9·357					
84	9·51	9·68	9·84	10·02	10·20	10·39	10·58	10·78	10·99	11·20					
85	11·43	11·66	11·91	12·16	12·43	12·71	13·00	13·30	13·62	13·95		Differences untrustworthy here			
86	14·30	14·67	15·06	15·46	15·89	16·35	16·83	17·34	17·89	18·46					
87	19·08	19·74	20·45	21·20	22·02	22·90	23·86	24·90	26·03	27·27					
88	28·64	30·14	31·82	33·69	35·80	38·19	40·92	44·07	47·74	52·08					
89	57·29	63·66	71·62	81·85	95·49	114·6	143·2	191·0	286·5	573·0					
	0'	6'	12'	18'	24'	30'	36'	42'	48'	54'	1'	2'	3'	4'	5'

The black type indicates that the integer changes.

	0	1	2	3	4	5	6	7	8	9	1	2	3	4	5	6	7	8	9
10	1000	1020	1040	1061	1082	1103	1124	1145	1166	1188	2	4	6	8	10	13	15	17	19
11	1210	1232	1254	1277	1300	1323	1346	1369	1392	1416	2	5	7	9	11	14	16	18	21
12	1440	1464	1488	1513	1538	1563	1588	1613	1638	1664	2	5	7	10	12	15	17	20	22
13	1690	1716	1742	1769	1796	1823	1850	1877	1904	1932	3	5	8	11	13	16	19	22	24
14	1960	1988	2016	2045	2074	2103	2132	2161	2190	2220	3	6	9	12	14	17	20	23	26
15	2250	2280	2310	2341	2372	2403	2434	2465	2496	2528	3	6	9	12	15	19	22	25	28
16	2560	2592	2624	2657	2690	2723	2756	2789	2822	2856	3	7	10	13	16	20	23	26	30
17	2890	2924	2958	2993	3028	3063	3098	3133	3168	3204	3	7	10	14	17	21	24	28	31
18	3240	3276	3312	3349	3386	3423	3460	3497	3534	3572	4	7	11	15	18	22	26	30	33
19	3610	3648	3686	3725	3764	3803	3842	3881	3920	3960	4	8	12	16	19	23	27	31	35
20	4000	4040	4080	4121	4162	4203	4244	4285	4326	4368	4	8	12	16	20	25	29	33	37
21	4410	4452	4494	4537	4580	4623	4666	4709	4752	4796	4	9	13	17	21	26	30	34	39
22	4840	4884	4928	4973	5018	5063	5108	5153	5198	5244	4	9	13	18	22	27	31	36	40
23	5290	5336	5382	5429	5476	5523	5570	5617	5664	5712	5	9	14	19	23	28	33	38	42
24	5760	5808	5856	5905	5954	6003	6052	6101	6150	6200	5	10	15	20	24	29	34	39	44
25	6250	6300	6350	6401	6452	6503	6554	6605	6656	6708	5	10	15	20	25	31	36	41	46
26	6760	6812	6864	6917	6970	7023	7076	7129	7182	7236	5	11	16	21	26	32	37	42	48
27	7290	7344	7398	7453	7508	7563	7618	7673	7728	7784	5	11	16	22	28	33	38	44	49
28	7840	7896	7952	8009	8066	8123	8180	8237	8294	8352	6	11	17	23	28	34	40	46	51
29	8410	8468	8526	8585	8644	8703	8762	8821	8880	8940	6	12	18	24	30	35	41	47	53
30	9000	9060	9120	9181	9242	9303	9364	9425	9486	9548	6	12	18	24	31	37	43	49	55
31	9610	9672	9734	9797	9860	9923	9986				6	13	19	25	31	38	44	50	56
31								1005	1011	1018	1	1	2	3	3	4	5	5	6
32	1024	1030	1037	1043	1050	1056	1063	1069	1076	1082	1	1	2	3	3	4	5	5	6
33	1089	1096	1102	1109	1116	1122	1129	1136	1142	1149	1	1	2	3	3	4	5	5	6
34	1156	1163	1170	1176	1183	1190	1197	1204	1211	1218	1	1	2	3	3	4	5	6	6
35	1225	1232	1239	1246	1253	1260	1267	1274	1282	1289	1	1	2	3	4	4	5	6	6
36	1296	1303	1310	1318	1325	1332	1340	1347	1354	1362	1	1	2	3	4	4	5	6	7
37	1369	1376	1384	1391	1399	1406	1414	1421	1429	1436	1	2	2	3	4	5	5	6	7
38	1444	1452	1459	1467	1475	1482	1490	1498	1505	1513	1	2	2	3	4	5	5	6	7
39	1521	1529	1537	1544	1552	1560	1568	1576	1584	1592	1	2	2	3	4	5	6	6	7
40	1600	1608	1616	1624	1632	1640	1648	1656	1665	1673	1	2	2	3	4	5	6	6	7
41	1681	1689	1697	1706	1714	1722	1731	1739	1747	1756	1	2	2	3	4	5	6	7	7
42	1764	1772	1781	1789	1798	1806	1815	1823	1832	1840	1	2	3	3	4	5	6	7	8
43	1849	1858	1866	1875	1884	1892	1901	1910	1918	1927	1	2	3	3	4	5	6	7	8
44	1936	1945	1954	1962	1971	1980	1989	1998	2007	2016	1	2	3	4	5	5	6	7	8
45	2025	2034	2043	2052	2061	2070	2079	2088	2098	2107	1	2	3	4	5	5	6	7	8
46	2116	2125	2134	2144	2153	2162	2172	2181	2190	2200	1	2	3	4	5	6	7	7	8
47	2209	2218	2228	2237	2247	2256	2266	2275	2285	2294	1	2	3	4	5	6	7	8	9
48	2304	2314	2323	2333	2343	2352	2362	2372	2381	2391	1	2	3	4	5	6	7	8	9
49	2401	2411	2421	2430	2440	2450	2460	2470	2480	2490	1	2	3	4	5	6	7	8	9
50	2500	2510	2520	2530	2540	2550	2560	2570	2581	2591	1	2	3	4	5	6	7	8	9
51	2601	2611	2621	2632	2642	2652	2663	2673	2683	2694	1	2	3	4	5	6	7	8	9
52	2704	2714	2725	2735	2746	2756	2767	2777	2788	2798	1	2	3	4	5	6	7	8	9
53	2809	2820	2830	2841	2852	2862	2873	2884	2894	2905	1	2	3	4	5	6	7	9	10
54	2916	2927	2938	2948	2959	2970	2981	2992	3003	3014	1	2	3	4	6	7	8	9	10
	0	1	2	3	4	5	6	7	8	9	1	2	3	4	5	6	7	8	9

The position of the decimal point must be determined by inspection.

	0	1	2	3	4	5	6	7	8	9	1	2	3	4	5	6	7	8	9
55	3025	3036	3047	3058	3069	3080	3091	3102	3114	3125	1	2	3	4	6	7	8	9	10
56	3136	3147	3158	3170	3181	3192	3204	3215	3226	3238	1	2	3	5	6	7	8	9	10
57	3249	3260	3272	3283	3295	3306	3318	3329	3341	3352	1	2	3	5	6	7	8	9	10
58	3364	3376	3387	3399	3411	3422	3434	3446	3457	3469	1	2	4	5	6	7	8	9	11
59	3481	3493	3505	3516	3528	3540	3552	3564	3576	3588	1	2	4	5	6	7	8	10	11
60	3600	3612	3624	3636	3648	3660	3672	3684	3697	3709	1	2	4	5	6	7	8	10	11
61	3721	3733	3745	3758	3770	3782	3795	3807	3819	3832	1	2	4	5	6	7	9	10	11
62	3844	3856	3869	3881	3894	3906	3919	3931	3944	3956	1	3	4	5	6	8	9	10	11
63	3969	3982	3994	4007	4020	4032	4045	4058	4070	4083	1	3	4	5	6	8	9	10	11
64	4096	4109	4122	4134	4147	4160	4173	4186	4199	4212	1	3	4	5	6	8	9	10	12
65	4225	4238	4251	4264	4277	4290	4303	4316	4330	4343	1	3	4	5	7	8	9	10	12
66	4356	4369	4382	4396	4409	4422	4436	4449	4462	4476	1	3	4	5	7	8	9	11	12
67	4489	4502	4516	4529	4543	4556	4570	4583	4597	4610	1	3	4	5	7	8	9	11	12
68	4624	4638	4651	4665	4679	4692	4706	4720	4733	4747	1	3	4	5	7	8	10	11	12
69	4761	4775	4789	4802	4816	4830	4844	4858	4872	4886	1	3	4	6	7	8	10	11	13
70	4900	4914	4928	4942	4956	4970	4984	4998	5013	5027	1	3	4	6	7	8	10	11	13
71	5041	5055	5069	5084	5098	5112	5127	5141	5155	5170	1	3	4	6	7	9	10	11	13
72	5184	5198	5213	5227	5242	5256	5271	5285	5300	5314	1	3	4	6	7	9	10	11	13
73	5329	5344	5358	5373	5388	5402	5417	5432	5446	5461	1	3	4	6	7	9	10	12	13
74	5476	5491	5506	5520	5535	5550	5565	5580	5595	5610	1	3	4	6	7	9	10	12	13
75	5625	5640	5655	5670	5685	5700	5715	5730	5746	5761	2	3	5	6	8	9	11	12	14
76	5776	5791	5806	5822	5837	5852	5868	5883	5898	5914	2	3	5	6	8	9	11	12	14
77	5929	5944	5960	5975	5991	6006	6022	6037	6053	6068	2	3	5	6	8	9	11	12	14
78	6084	6100	6115	6131	6147	6162	6178	6194	6209	6225	2	3	5	6	8	9	11	13	14
79	6241	6257	6273	6288	6304	6320	6336	6352	6368	6384	2	3	5	6	8	10	11	13	14
80	6400	6416	6432	6448	6464	6480	6496	6512	6529	6545	2	3	5	6	8	10	11	13	14
81	6561	6577	6593	6610	6626	6642	6659	6675	6691	6708	2	3	5	7	8	10	11	13	15
82	6724	6740	6757	6773	6790	6806	6823	6839	6856	6872	2	3	5	7	8	10	12	13	15
83	6889	6906	6922	6939	6956	6972	6989	7006	7022	7039	2	3	5	7	8	10	12	13	15
84	7056	7073	7090	7106	7123	7140	7157	7174	7191	7208	2	3	5	7	8	10	12	14	15
85	7225	7242	7259	7276	7293	7310	7327	7344	7362	7379	2	3	5	7	9	10	12	14	15
86	7396	7413	7430	7448	7465	7482	7500	7517	7534	7552	2	3	5	7	9	10	12	14	16
87	7569	7586	7604	7621	7639	7656	7674	7691	7709	7726	2	4	5	7	9	11	12	14	16
88	7744	7762	7779	7797	7815	7832	7850	7868	7885	7903	2	4	5	7	9	11	12	14	16
89	7921	7939	7957	7974	7992	8010	8028	8046	8064	8082	2	4	5	7	9	11	13	14	16
90	8100	8118	8136	8154	8172	8190	8208	8226	8245	8263	2	4	5	7	9	11	13	14	16
91	8281	8299	8317	8336	8354	8372	8391	8409	8427	8446	2	4	5	7	9	11	13	15	16
92	8464	8482	8501	8519	8538	8556	8575	8593	8612	8630	2	4	6	7	9	11	13	15	17
93	8649	8668	8686	8705	8724	8742	8761	8780	8798	8817	2	4	6	7	9	11	13	15	17
94	8836	8855	8874	8892	8911	8930	8949	8968	8987	9006	2	4	6	8	9	11	13	15	17
95	9025	9044	9063	9082	9101	9120	9139	9158	9178	9197	2	4	6	8	10	11	13	15	17
96	9216	9235	9254	9274	9293	9312	9332	9351	9370	9390	2	4	6	8	10	12	14	15	17
97	9409	9428	9448	9467	9487	9506	9526	9545	9565	9584	2	4	6	8	10	12	14	16	18
98	9604	9624	9643	9663	9683	9702	9722	9742	9761	9781	2	4	6	8	10	12	14	16	18
99	9801	9821	9841	9860	9880	9900	9920	9940	9960	9980	2	4	6	8	10	12	14	16	18
	0	**1**	**2**	**3**	**4**	**5**	**6**	**7**	**8**	**9**	**1**	**2**	**3**	**4**	**5**	**6**	**7**	**8**	**9**

The position of the decimal point must be determined by inspection.

	0	1	2	3	4	5	6	7	8	9	1	2	3	4	5	6	7	8	9
10	1000	1005	1010	1015	1020	1025	1030	1034	1039	1044	0	1	1	2	2	3	3	4	4
	3162	3178	3194	3209	3225	3240	3256	3271	3286	3302	2	3	5	6	8	9	11	12	14
11	1049	1054	1058	1063	1068	1072	1077	1082	1086	1091	0	1	1	2	2	3	3	4	4
	3317	3332	3347	3362	3376	3391	3406	3421	3435	3450	1	3	4	6	7	9	10	12	13
12	1095	1100	1105	1109	1114	1118	1122	1127	1131	1136	0	1	1	2	2	3	3	4	4
	3464	3479	3493	3507	3521	3536	3550	3564	3578	3592	1	3	4	6	7	8	10	11	13
13	1140	1145	1149	1153	1158	1162	1166	1170	1175	1179	0	1	1	2	2	3	3	3	4
	3606	3619	3633	3647	3661	3674	3688	3701	3715	3728	1	3	4	5	7	8	10	11	12
14	1183	1187	1192	1196	1200	1204	1208	1212	1217	1221	0	1	1	2	2	3	3	3	4
	3742	3755	3768	3782	3795	3808	3821	3834	3847	3860	1	3	4	5	7	8	9	11	12
15	1225	1229	1233	1237	1241	1245	1249	1253	1257	1261	0	1	1	2	2	3	3	3	4
	3873	3886	3899	3912	3924	3937	3950	3962	3975	3987	1	3	4	5	6	8	9	10	11
16	1265	1269	1273	1277	1281	1285	1288	1292	1296	1300	0	1	1	2	2	3	3	3	4
	4000	4012	4025	4037	4050	4062	4074	4087	4099	4111	1	2	4	5	6	7	9	10	11
17	1304	1308	1311	1315	1319	1323	1327	1330	1334	1338	0	1	1	2	2	2	3	3	3
	4123	4135	4147	4159	4171	4183	4195	4207	4219	4231	1	2	4	5	6	7	8	10	11
18	1342	1345	1349	1353	1356	1360	1364	1367	1371	1375	0	1	1	1	2	2	3	3	3
	4243	4254	4266	4278	4290	4301	4313	4324	4336	4347	1	2	3	5	6	7	8	9	10
19	1378	1382	1386	1389	1393	1396	1400	1404	1407	1411	0	1	1	1	2	2	3	3	3
	4359	4370	4382	4393	4405	4416	4427	4438	4450	4461	1	2	3	5	6	7	8	9	10
20	1414	1418	1421	1425	1428	1432	1435	1439	1442	1446	0	1	1	1	2	2	2	3	3
	4472	4483	4494	4506	4517	4528	4539	4550	4561	4572	1	2	3	4	5	7	8	9	10
21	1449	1453	1456	1459	1463	1466	1470	1473	1476	1480	0	1	1	1	2	2	2	3	3
	4583	4593	4604	4615	4626	4637	4648	4658	4669	4680	1	2	3	4	5	6	8	9	10
22	1483	1487	1490	1493	1497	1500	1503	1507	1510	1513	0	1	1	1	2	2	2	3	3
	4690	4701	4712	4722	4733	4743	4754	4764	4775	4785	1	2	3	4	5	6	7	8	9
23	1517	1520	1523	1526	1530	1533	1536	1539	1543	1546	0	1	1	1	2	2	2	3	3
	4796	4806	4817	4827	4837	4848	4858	4868	4879	4889	1	2	3	4	5	6	7	8	9
24	1549	1552	1556	1559	1562	1565	1568	1572	1575	1578	0	1	1	1	2	2	2	3	3
	4899	4909	4919	4930	4940	4950	4960	4970	4980	4990	1	2	3	4	5	6	7	8	9
25	1581	1584	1587	1591	1594	1597	1600	1603	1606	1609	0	1	1	1	2	2	2	3	3
	5000	5010	5020	5030	5040	5050	5060	5070	5079	5089	1	2	3	4	5	6	7	8	9
26	1612	1616	1619	1622	1625	1628	1631	1634	1637	1640	0	1	1	1	2	2	2	3	3
	5099	5109	5119	5128	5138	5148	5158	5167	5177	5187	1	2	3	4	5	6	7	8	9
27	1643	1646	1649	1652	1655	1658	1661	1664	1667	1670	0	1	1	1	2	2	2	3	3
	5196	5206	5215	5225	5235	5244	5254	5263	5273	5282	1	2	3	4	5	6	7	8	9
28	1673	1676	1679	1682	1685	1688	1691	1694	1697	1700	0	1	1	1	1	2	2	2	3
	5292	5301	5310	5320	5329	5339	5348	5357	5367	5376	1	2	3	4	5	6	7	7	8
29	1703	1706	1709	1712	1715	1718	1720	1723	1726	1729	0	1	1	1	1	2	2	2	3
	5385	5394	5404	5413	5422	5431	5441	5450	5459	5468	1	2	3	4	5	5	6	7	8
30	1732	1735	1738	1741	1744	1746	1749	1752	1755	1758	0	1	1	1	1	2	2	2	3
	5477	5486	5495	5505	5514	5523	5532	5541	5550	5559	1	2	3	4	4	5	6	7	8
31	1761	1764	1766	1769	1772	1775	1778	1780	1783	1786	0	1	1	1	1	2	2	2	3
	5568	5577	5586	5595	5604	5612	5621	5630	5639	5648	1	2	3	3	4	5	6	7	8
32	1789	1792	1794	1797	1800	1803	1806	1808	1811	1814	0	1	1	1	1	2	2	2	2
	5657	5666	5675	5683	5692	5701	5710	5718	5727	5736	1	2	3	3	4	5	6	7	8
	0	1	2	3	4	5	6	7	8	9	1	2	3	4	5	6	7	8	9

The first significant figure and the position of the decimal point must be determined by inspection.

	0	1	2	3	4	5	6	7	8	9	1	2	3	4	5	6	7	8	9
33	1817	1819	1822	1825	1828	1830	1833	1836	1838	1841	0	1	1	1	1	2	2	2	2
	5745	5753	5762	5771	5779	5788	5797	5805	5814	5822	1	2	3	3	4	5	6	7	8
34	1844	1847	1849	1852	1855	1857	1860	1863	1865	1868	0	1	1	1	1	2	2	2	2
	5831	5840	5848	5857	5865	5874	5882	5891	5899	5908	1	2	3	3	4	5	6	7	8
35	1871	1873	1876	1879	1881	1884	1887	1889	1892	1895	0	1	1	1	1	2	2	2	2
	5916	5925	5933	5941	5950	5958	5967	5975	5983	5992	1	2	2	3	4	5	6	7	8
36	1897	1900	1903	1905	1908	1910	1913	1916	1918	1921	0	1	1	1	1	2	2	2	2
	6000	6008	6017	6025	6033	6042	6050	6058	6066	6075	1	2	2	3	4	5	6	7	7
37	1924	1926	1929	1931	1934	1936	1939	1942	1944	1947	0	1	1	1	1	2	2	2	2
	6083	6091	6099	6107	6116	6124	6132	6140	6148	6156	1	2	2	3	4	5	6	7	7
38	1949	1952	1954	1957	1960	1962	1965	1967	1970	1972	0	1	1	1	1	2	2	2	2
	6164	6173	6181	6189	6197	6205	6213	6221	6229	6237	1	2	2	3	4	5	6	6	7
39	1975	1977	1980	1982	1985	1987	1990	1992	1995	1997	0	1	1	1	1	2	2	2	2
	6245	6253	6261	6269	6277	6285	6293	6301	6309	6317	1	2	2	3	4	5	6	6	7
40	2000	2002	2005	2007	2010	2012	2015	2017	2020	2022	0	0	1	1	1	1	2	2	2
	6325	6332	6340	6348	6356	6364	6372	6380	6387	6395	1	2	2	3	4	5	6	6	7
41	2025	2027	2030	2032	2035	2037	2040	2042	2045	2047	0	0	1	1	1	1	2	2	2
	6403	6411	6419	6427	6434	6442	6450	6458	6465	6473	1	2	2	3	4	5	5	6	7
42	2049	2052	2054	2057	2059	2062	2064	2066	2069	2071	0	0	1	1	1	1	2	2	2
	6481	6488	6496	6504	6512	6519	6527	6535	6542	6550	1	2	2	3	4	5	5	6	7
43	2074	2076	2078	2081	2083	2086	2088	2090	2093	2095	0	0	1	1	1	1	2	2	2
	6557	6565	6573	6580	6588	6595	6603	6611	6618	6626	1	2	2	3	4	5	5	6	7
44	2098	2100	2102	2105	2107	2110	2112	2114	2117	2119	0	0	1	1	1	1	2	2	2
	6633	6641	6648	6656	6663	6671	6678	6686	6693	6701	1	2	2	3	4	5	5	6	7
45	2121	2124	2126	2128	2131	2133	2135	2138	2140	2142	0	0	1	1	1	1	2	2	2
	6708	6716	6723	6731	6738	6745	6753	6760	6768	6775	1	1	2	3	4	5	5	6	7
46	2145	2147	2149	2152	2154	2156	2159	2161	2163	2166	0	0	1	1	1	1	2	2	2
	6782	6790	6797	6804	6812	6819	6826	6834	6841	6848	1	1	2	3	4	4	5	6	7
47	2168	2170	2173	2175	2177	2179	2182	2184	2186	2189	0	0	1	1	1	1	2	2	2
	6856	6863	6870	6877	6885	6892	6899	6907	6914	6921	1	1	2	3	4	4	5	6	7
48	2191	2193	2195	2198	2200	2202	2205	2207	2209	2211	0	0	1	1	1	1	2	2	2
	6928	6935	6943	6950	6957	6964	6971	6979	6986	6993	1	1	2	3	4	4	5	6	6
49	2214	2216	2218	2220	2223	2225	2227	2229	2232	2234	0	0	1	1	1	1	2	2	2
	7000	7007	7014	7021	7029	7036	7043	7050	7057	7064	1	1	2	3	4	4	5	6	6
50	2236	2238	2241	2243	2245	2247	2249	2252	2254	2256	0	0	1	1	1	1	2	2	2
	7071	7078	7085	7092	7099	7106	7113	7120	7127	7134	1	1	2	3	4	4	5	6	6
51	2258	2261	2263	2265	2267	2269	2272	2274	2276	2278	0	0	1	1	1	1	2	2	2
	7141	7148	7155	7162	7169	7176	7183	7190	7197	7204	1	1	2	3	4	4	5	6	6
52	2280	2283	2285	2287	2289	2291	2293	2296	2298	2300	0	0	1	1	1	1	2	2	2
	7211	7218	7225	7232	7239	7246	7253	7259	7266	7273	1	1	2	3	3	4	5	6	6
53	2302	2304	2307	2309	2311	2313	2315	2317	2319	2322	0	0	1	1	1	1	2	2	2
	7280	7287	7294	7301	7308	7314	7321	7328	7335	7342	1	1	2	3	3	4	5	5	6
54	2324	2326	2328	2330	2332	2335	2337	2339	2341	2343	0	0	1	1	1	1	1	2	2
	7348	7355	7362	7369	7376	7382	7389	7396	7403	7409	1	1	2	3	3	4	5	5	6
	0	1	2	3	4	5	6	7	8	9	1	2	3	4	5	6	7	8	9

The first significant figure and the position of the decimal point must be determined by inspection.

Square Roots

	0	1	2	3	4	5	6	7	8	9	1 2 3	4 5 6	7 8 9
55	2345	2347	2349	2352	2354	2356	2358	2360	2362	2364	0 0 1	1 1 1	1 2 2
	7416	7423	7430	7436	7443	7450	7457	7463	7470	7477	1 1 2	3 3 4	5 5 6
56	2366	2369	2371	2373	2375	2377	2379	2381	2383	2385	0 0 1	1 1 1	1 2 2
	7483	7490	7497	7503	7510	7517	7523	7530	7537	7543	1 1 2	3 3 4	5 5 6
57	2387	2390	2392	2394	2396	2398	2400	2402	2404	2406	0 0 1	1 1 1	1 2 2
	7550	7556	7563	7570	7576	7583	7589	7596	7603	7609	1 1 2	3 3 4	5 5 6
58	2408	2410	2412	2415	2417	2419	2421	2423	2425	2427	0 0 1	1 1 1	1 2 2
	7616	7622	7629	7635	7642	7649	7655	7662	7668	7675	1 1 2	3 3 4	5 5 6
59	2429	2431	2433	2435	2437	2439	2441	2443	2445	2447	0 0 1	1 1 1	1 2 2
	7681	7688	7694	7701	7707	7714	7720	7727	7733	7740	1 1 2	3 3 4	5 5 6
60	2449	2452	2454	2456	2458	2460	2462	2464	2466	2468	0 0 1	1 1 1	1 2 2
	7746	7752	7759	7765	7772	7778	7785	7791	7797	7804	1 1 2	3 3 4	4 5 6
61	2470	2472	2474	2476	2478	2480	2482	2484	2486	2488	0 0 1	1 1 1	1 2 2
	7810	7817	7823	7829	7836	7842	7849	7855	7861	7868	1 1 2	3 3 4	4 5 6
62	2490	2492	2494	2496	2498	2500	2502	2504	2506	2508	0 0 1	1 1 1	1 2 2
	7874	7880	7887	7893	7899	7906	7912	7918	7925	7931	1 1 2	3 3 4	4 5 6
63	2510	2512	2514	2516	2518	2520	2522	2524	2526	2528	0 0 1	1 1 1	1 2 2
	7937	7944	7950	7956	7962	7969	7975	7981	7987	7994	1 1 2	3 3 4	4 5 6
64	2530	2532	2534	2536	2538	2540	2542	2544	2546	2548	0 0 1	1 1 1	1 2 2
	8000	8006	8012	8019	8025	8031	8037	8044	8050	8056	1 1 2	2 3 4	4 5 6
65	2550	2551	2553	2555	2557	2559	2561	2563	2565	2567	0 0 1	1 1 1	1 2 2
	8062	8068	8075	8081	8087	8093	8099	8106	8112	8118	1 1 2	2 3 4	4 5 5
66	2569	2571	2573	2575	2577	2579	2581	2583	2585	2587	0 0 1	1 1 1	1 2 2
	8124	8130	8136	8142	8149	8155	8161	8167	8173	8179	1 1 2	2 3 4	4 5 5
67	2588	2590	2592	2594	2596	2598	2600	2602	2604	2606	0 0 1	1 1 1	1 2 2
	8185	8191	8198	8204	8210	8216	8222	8228	8234	8240	1 1 2	2 3 4	4 5 5
68	2608	2610	2612	2613	2615	2617	2619	2621	2623	2625	0 0 1	1 1 1	1 2 2
	8246	8252	8258	8264	8270	8276	8283	8289	8295	8301	1 1 2	2 3 4	4 5 5
69	2627	2629	2631	2632	2634	2636	2638	2640	2642	2644	0 0 1	1 1 1	1 2 2
	8307	8313	8319	8325	8331	8337	8343	8349	8355	8361	1 1 2	2 3 4	4 5 5
70	2646	2648	2650	2651	2653	2655	2657	2659	2661	2663	0 0 1	1 1 1	1 2 2
	8367	8373	8379	8385	8390	8396	8402	8408	8414	8420	1 1 2	2 3 4	4 5 5
71	2665	2666	2668	2670	2672	2674	2676	2678	2680	2681	0 0 1	1 1 1	1 1 2
	8426	8432	8438	8444	8450	8456	8462	8468	8473	8479	1 1 2	2 3 3	4 5 5
72	2683	2685	2687	2689	2691	2693	2694	2696	2698	2700	0 0 1	1 1 1	1 1 2
	8485	8491	8497	8503	8509	8515	8521	8526	8532	8538	1 1 2	2 3 3	4 5 5
73	2702	2704	2706	2707	2709	2711	2713	2715	2717	2718	0 0 1	1 1 1	1 1 2
	8544	8550	8556	8562	8567	8573	8579	8585	8591	8597	1 1 2	2 3 3	4 5 5
74	2720	2722	2724	2726	2728	2729	2731	2733	2735	2737	0 0 1	1 1 1	1 1 2
	8602	8608	8614	8620	8626	8631	8637	8643	8649	8654	1 1 2	2 3 3	4 5 5
75	2739	2740	2742	2744	2746	2748	2750	2751	2753	2755	0 0 1	1 1 1	1 1 2
	8660	8666	8672	8678	8683	8689	8695	8701	8706	8712	1 1 2	2 3 3	4 5 5
76	2757	2759	2760	2762	2764	2766	2768	2769	2771	2773	0 0 1	1 1 1	1 1 2
	8718	8724	8729	8735	8741	8746	8752	8758	8764	8769	1 1 2	2 3 3	4 5 5
77	2775	2777	2778	2780	2782	2784	2786	2787	2789	2791	0 0 1	1 1 1	1 1 2
	8775	8781	8786	8792	8798	8803	8809	8815	8820	8826	1 1 2	2 3 3	4 4 5
	0	1	2	3	4	5	6	7	8	9	1 2 3	4 5 6	7 8 9

The first significant figure and the position of the decimal point must be determined by inspection.

	0	1	2	3	4	5	6	7	8	9	1 2 3	4 5 6	7 8 9
78	2793 8832	2795 8837	2796 8843	2798 8849	2800 8854	2802 8860	2804 8866	2805 8871	2807 8877	2809 8883	0 0 1 1 1 2	1 1 1 2 3 3	1 1 2 4 4 5
79	2811 8888	2812 8894	2814 8899	2816 8905	2818 8911	2820 8916	2821 8922	2823 8927	2825 8933	2827 8939	0 0 1 1 1 2	1 1 1 2 3 3	1 1 2 4 4 5
80	2828 8944	2830 8950	2832 8955	2834 8961	2835 8967	2837 8972	2839 8978	2841 8983	2843 8989	2844 8994	0 0 1 1 1 2	1 1 1 2 3 3	1 1 2 4 4 5
81	2846 9000	2848 9006	2850 9011	2851 9017	2853 9022	2855 9028	2857 9033	2858 9039	2860 9044	2862 9050	0 0 1 1 1 2	1 1 1 2 3 3	1 1 2 4 4 5
82	2864 9055	2865 9061	2867 9066	2869 9072	2871 9077	2872 9083	2874 9088	2876 9094	2877 9099	2879 9105	0 0 1 1 1 2	1 1 1 2 3 3	1 1 2 4 4 5
83	2881 9110	2883 9116	2884 9121	2886 9127	2888 9132	2890 9138	2891 9143	2893 9149	2895 9154	2897 9160	0 0 1 1 1 2	1 1 1 2 3 3	1 1 2 4 4 5
84	2898 9165	2900 9171	2902 9176	2903 9182	2905 9187	2907 9192	2909 9198	2910 9203	2912 9209	2914 9214	0 0 1 1 1 2	1 1 1 2 3 3	1 1 2 4 4 5
85	2915 9220	2917 9225	2919 9230	2921 9236	2922 9241	2924 9247	2926 9252	2927 9257	2929 9263	2931 9268	0 0 1 1 1 2	1 1 1 2 3 3	1 1 2 4 4 5
86	2933 9274	2934 9279	2936 9284	2938 9290	2939 9295	2941 9301	2943 9306	2944 9311	2946 9317	2948 9322	0 0 1 1 1 2	1 1 1 2 3 3	1 1 2 4 4 5
87	2950 9327	2951 9333	2953 9338	2955 9343	2956 9349	2958 9354	2960 9359	2961 9365	2963 9370	2965 9375	0 0 1 1 1 2	1 1 1 2 3 3	1 1 2 4 4 5
88	2966 9381	2968 9386	2970 9391	2972 9397	2973 9402	2975 9407	2977 9413	2978 9418	2980 9423	2982 9429	0 0 1 1 1 2	1 1 1 2 3 3	1 1 2 4 4 5
89	2983 9434	2985 9439	2987 9445	2988 9450	2990 9455	2992 9460	2993 9466	2995 9471	2997 9476	2998 9482	0 0 1 1 1 2	1 1 1 2 3 3	1 1 2 4 4 5
90	3000 9487	3002 9492	3003 9497	3005 9503	3007 9508	3008 9513	3010 9518	3012 9524	3013 9529	3015 9534	0 0 0 1 1 2	1 1 1 2 3 3	1 1 1 4 4 5
91	3017 9539	3018 9545	3020 9550	3022 9555	3023 9560	3025 9566	3027 9571	3028 9576	3030 9581	3032 9586	0 0 0 1 1 2	1 1 1 2 3 3	1 1 1 4 4 5
92	3033 9592	3035 9597	3036 9602	3038 9607	3040 9612	3041 9618	3043 9623	3045 9628	3046 9633	3048 9638	0 0 0 1 1 2	1 1 1 2 3 3	1 1 1 4 4 5
93	3050 9644	3051 9649	3053 9654	3055 9659	3056 9664	3058 9670	3059 9675	3061 9680	3063 9685	3064 9690	0 0 0 1 1 2	1 1 1 2 3 3	1 1 1 4 4 5
94	3066 9695	3068 9701	3069 9706	3071 9711	3072 9716	3074 9721	3076 9726	3077 9731	3079 9737	3081 9742	0 0 0 1 1 2	1 1 1 2 3 3	1 1 1 4 4 5
95	3082 9747	3084 9752	3085 9757	3087 9762	3089 9767	3090 9772	3092 9778	3094 9783	3095 9788	3097 9793	0 0 0 1 1 2	1 1 1 2 3 3	1 1 1 4 4 5
96	3098 9798	3100 9803	3102 9808	3103 9813	3105 9818	3106 9823	3108 9829	3110 9834	3111 9839	3113 9844	0 0 0 1 1 2	1 1 1 2 3 3	1 1 1 4 4 5
97	3114 9849	3116 9854	3118 9859	3119 9864	3121 9869	3122 9874	3124 9879	3126 9884	3127 9889	3129 9894	0 0 0 1 1 2	1 1 1 2 3 3	1 1 1 4 4 5
98	3130 9899	3132 9905	3134 9910	3135 9915	3137 9920	3138 9925	3140 9930	3142 9935	3143 9940	3145 9945	0 0 0 0 1 1	1 1 1 2 2 3	1 1 1 3 4 4
99	3146 9950	3148 9955	3150 9960	3151 9965	3153 9970	3154 9975	3156 9980	3158 9985	3159 9990	3161 9995	0 0 0 0 1 1	1 1 1 2 2 3	1 1 1 3 4 4
	0	**1**	**2**	**3**	**4**	**5**	**6**	**7**	**8**	**9**	**1 2 3**	**4 5 6**	**7 8 9**

The first significant figure and the position of the decimal point must
be determined by inspection.

SUBTRACT

	0	1	2	3	4	5	6	7	8	9	1	2	3	4	5	6	7	8	9
1·0	1·0000	·9901	·9804	·9709	·9615	·9524	·9434	·9346	·9259	·9174	9	18	28	37	46	55	64	73	83
1·1	·9091	·9009	·8929	·8850	·8772	·8696	·8621	·8547	·8475	·8403	8	15	23	31	38	46	54	61	69
1·2	·8333	·8264	·8197	·8130	·8065	·8000	·7937	·7874	·7813	·7752	6	13	19	26	32	38	45	51	58
1·3	·7692	·7634	·7576	·7519	·7463	·7407	·7353	·7299	·7246	·7194	5	11	16	22	27	33	38	44	49
1·4	·7143	·7092	·7042	·6993	·6944	·6897	·6849	·6803	·6757	·6711	5	10	14	19	24	29	33	38	43
1·5	·6667	·6623	·6579	·6536	·6494	·6452	·6410	·6369	·6329	·6289	4	8	13	17	21	25	29	33	38
1·6	·6250	·6211	·6173	·6135	·6098	·6061	·6024	·5988	·5952	·5917	4	7	11	15	18	22	26	29	33
1·7	·5882	·5848	·5814	·5780	·5747	·5714	·5682	·5650	·5618	·5587	3	7	10	13	16	20	23	26	30
1·8	·5556	·5525	·5495	·5464	·5435	·5405	·5376	·5348	·5319	·5291	3	6	9	12	15	18	20	23	26
1·9	·5263	·5236	·5208	·5181	·5155	·5128	·5102	·5076	·5051	·5025	3	5	8	11	13	16	18	21	24
2·0	·5000	·4975	·4950	·4926	·4902	·4878	·4854	·4831	·4808	·4785	2	5	7	10	12	14	17	19	21
2·1	·4762	·4739	·4717	·4695	·4673	·4651	·4630	·4608	·4587	·4566	2	4	7	9	11	13	15	17	20
2·2	·4545	·4525	·4505	·4484	·4464	·4444	·4425	·4405	·4386	·4367	2	4	6	8	10	12	14	16	18
2·3	·4348	·4329	·4310	·4292	·4274	·4255	·4237	·4219	·4202	·4184	2	4	5	7	9	11	13	14	16
2·4	·4167	·4149	·4132	·4115	·4098	·4082	·4065	·4049	·4032	·4016	2	3	5	7	8	10	12	13	15
2·5	·4000	·3984	·3968	·3953	·3937	·3922	·3906	·3891	·3876	·3861	2	3	5	6	8	9	11	12	14
2·6	·3846	·3831	·3817	·3802	·3788	·3774	·3759	·3745	·3731	·3717	1	3	4	6	7	8	10	11	13
2·7	·3704	·3690	·3676	·3663	·3650	·3636	·3623	·3610	·3597	·3584	1	3	4	5	7	8	9	11	12
2·8	·3571	·3559	·3546	·3534	·3521	·3509	·3497	·3484	·3472	·3460	1	2	4	5	6	7	9	10	11
2·9	·3448	·3436	·3425	·3413	·3401	·3390	·3378	·3367	·3356	·3344	1	2	3	5	6	7	8	9	10
3·0	·3333	·3322	·3311	·3300	·3289	·3279	·3268	·3257	·3247	·3236	1	2	·3	4	5	6	7	9	10
3·1	·3226	·3215	·3205	·3195	·3185	·3175	·3165	·3155	·3145	·3135	1	2	3	4	5	6	7	8	9
3·2	·3125	·3115	·3106	·3096	·3086	·3077	·3067	·3058	·3049	·3040	1	2	3	4	5	6	7	8	9
3·3	·3030	·3021	·3012	·3003	·2994	·2985	·2976	·2967	·2959	·2950	1	2	3	4	4	5	6	7	8
3·4	·2941	·2933	·2924	·2915	·2907	·2899	·2890	·2882	·2874	·2865	1	2	3	3	4	5	6	7	8
3·5	·2857	·2849	·2841	·2833	·2825	·2817	·2809	·2801	·2793	·2786	1	2	2	3	4	5	6	6	7
3·6	·2778	·2770	·2762	·2755	·2747	·2740	·2732	·2725	·2717	·2710	1	2	2	3	4	5	5	6	7
3·7	·2703	·2695	·2688	·2681	·2674	·2667	·2660	·2653	·2646	·2639	1	1	2	3	4	4	5	6	6
3·8	·2632	·2625	·2618	·2611	·2604	·2597	·2591	·2584	·2577	·2571	1	1	2	3	3	4	5	5	6
3·9	·2564	·2558	·2551	·2545	·2538	·2532	·2525	·2519	·2513	·2506	1	1	2	3	3	4	4	5	6
4·0	·2500	·2494	·2488	·2481	·2475	·2469	·2463	·2457	·2451	·2445	1	1	2	2	3	4	4	5	5
4·1	·2439	·2433	·2427	·2421	·2415	·2410	·2404	·2398	·2392	·2387	1	1	2	2	3	3	4	5	5
4·2	·2381	·2375	·2370	·2364	·2358	·2353	·2347	·2342	·2336	·2331	1	1	2	2	3	3	4	4	5
4·3	·2326	·2320	·2315	·2309	·2304	·2299	·2294	·2288	·2283	·2278	1	1	2	2	3	3	4	4	5
4·4	·2273	·2268	·2262	·2257	·2252	·2247	·2242	·2237	·2232	·2227	1	1	2	2	3	3	4	4	4
4·5	·2222	·2217	·2212	·2208	·2203	·2198	·2193	·2188	·2183	·2179	0	1	1	2	2	3	3	4	4
4·6	·2174	·2169	·2165	·2160	·2155	·2151	·2146	·2141	·2137	·2132	0	1	1	2	2	3	3	4	4
4·7	·2128	·2123	·2119	·2114	·2110	·2105	·2101	·2096	·2092	·2088	0	1	1	2	2	3	3	4	4
4·8	·2083	·2079	·2075	·2070	·2066	·2062	·2058	·2053	·2049	·2045	0	1	1	2	2	3	3	3	4
4·9	·2041	·2037	·2033	·2028	·2024	·2020	·2016	·2012	·2008	·2004	0	1	1	2	2	2	3	3	4
5·0	·2000	·1996	·1992	·1988	·1984	·1980	·1976	·1972	·1969	·1965	0	1	1	2	2	2	3	3	3
5·1	·1961	·1957	·1953	·1949	·1946	·1942	·1938	·1934	·1931	·1927	0	1	1	2	2	2	3	3	3
5·2	·1923	·1919	·1916	·1912	·1908	·1905	·1901	·1898	·1894	·1890	0	1	1	1	2	2	3	3	3
5·3	·1887	·1883	·1880	·1876	·1873	·1869	·1866	·1862	·1859	·1855	0	1	1	1	2	2	3	3	3
5·4	·1852	·1848	·1845	·1842	·1838	·1835	·1832	·1828	·1825	·1821	0	1	1	1	2	2	2	3	3
	0	**1**	**2**	**3**	**4**	**5**	**6**	**7**	**8**	**9**	**1**	**2**	**3**	**4**	**5**	**6**	**7**	**8**	

SUBTRACT

	0	1	2	3	4	5	6	7	8	9	1	2	3	4	5	6	7	8	9
5·5	·1818	·1815	·1812	·1808	·1805	·1802	·1799	·1795	·1792	·1789	0	1	1	1	2	2	2	3	3
5·6	·1786	·1783	·1779	·1776	·1773	·1770	·1767	·1764	·1761	·1757	0	1	1	1	2	2	2	3	3
5·7	·1754	·1751	·1748	·1745	·1742	·1739	·1736	·1733	·1730	·1727	0	1	1	1	2	2	2	2	3
5·8	·1724	·1721	·1718	·1715	·1712	·1709	·1706	·1704	·1701	·1698	0	1	1	1	1	2	2	2	3
5·9	·1695	·1692	·1689	·1686	·1684	·1681	·1678	·1675	·1672	·1669	0	1	1	1	1	2	2	2	3
6·0	·1667	·1664	·1661	·1658	·1656	·1653	·1650	·1647	·1645	·1642	0	1	1	1	1	2	2	2	3
6·1	·1639	·1637	·1634	·1631	·1629	·1626	·1623	·1621	·1618	·1616	0	1	1	1	1	2	2	2	2
6·2	·1613	·1610	·1608	·1605	·1603	·1600	·1597	·1595	·1592	·1590	0	1	1	1	1	2	2	2	2
6·3	·1587	·1585	·1582	·1580	·1577	·1575	·1572	·1570	·1567	·1565	0	0	1	1	1	1	2	2	2
6·4	·1563	·1560	·1558	·1555	·1553	·1550	·1548	·1546	·1543	·1541	0	0	1	1	1	1	2	2	2
6·5	·1538	·1536	·1534	·1531	·1529	·1527	·1524	·1522	·1520	·1517	0	0	1	1	1	1	2	2	2
6·6	·1515	·1513	·1511	·1508	·1506	·1504	·1502	·1499	·1497	·1495	0	0	1	1	1	1	2	2	2
6·7	·1493	·1490	·1488	·1486	·1484	·1481	·1479	·1477	·1475	·1473	0	0	1	1	1	1	2	2	2
6·8	·1471	·1468	·1466	·1464	·1462	·1460	·1458	·1456	·1453	·1451	0	0	1	1	1	1	2	2	2
6·9	·1449	·1447	·1445	·1443	·1441	·1439	·1437	·1435	·1433	·1431	0	0	1	1	1	1	1	2	2
7·0	·1429	·1427	·1425	·1422	·1420	·1418	·1416	·1414	·1412	·1410	0	0	1	1	1	1	1	2	2
7·1	·1408	·1406	·1404	·1403	·1401	·1399	·1397	·1395	·1393	·1391	0	0	1	1	1	1	1	2	2
7·2	·1389	·1387	·1385	·1383	·1381	·1379	·1377	·1376	·1374	·1372	0	0	1	1	1	1	1	2	2
7·3	·1370	·1368	·1366	·1364	·1362	·1361	·1359	·1357	·1355	·1353	0	0	1	1	1	1	1	2	2
7·4	·1351	·1350	·1348	·1346	·1344	·1342	·1340	·1339	·1337	·1335	0	0	1	1	1	1	1	1	2
7·5	·1333	·1332	·1330	·1328	·1326	·1325	·1323	·1321	·1319	·1318	0	0	1	1	1	1	1	1	2
7·6	·1316	·1314	·1312	·1311	·1309	·1307	·1305	·1304	·1302	·1300	0	0	1	1	1	1	1	1	2
7·7	·1299	·1297	·1295	·1294	·1292	·1290	·1289	·1287	·1285	·1284	0	0	0	1	1	1	1	1	1
7·8	·1282	·1280	·1279	·1277	·1276	·1274	·1272	·1271	·1269	·1267	0	0	0	1	1	1	1	1	1
7·9	·1266	·1264	·1263	·1261	·1259	·1258	·1256	·1255	·1253	·1252	0	0	0	1	1	1	1	1	1
8·0	·1250	·1248	·1247	·1245	·1244	·1242	·1241	·1239	·1238	·1236	0	0	0	1	1	1	1	1	1
8·1	·1235	·1233	·1232	·1230	·1229	·1227	·1225	·1224	·1222	·1221	0	0	0	1	1	1	1	1	1
8·2	·1220	·1218	·1217	·1215	·1214	·1212	·1211	·1209	·1208	·1206	0	0	0	1	1	1	1	1	1
8·3	·1205	·1203	·1202	·1200	·1199	·1198	·1196	·1195	·1193	·1192	0	0	0	1	1	1	1	1	1
8·4	·1190	·1189	·1188	·1186	·1185	·1183	·1182	·1181	·1179	·1178	0	0	0	1	1	1	1	1	1
8·5	·1176	·1175	·1174	·1172	·1171	·1170	·1168	·1167	·1166	·1164	0	0	0	1	1	1	1	1	1
8·6	·1163	·1161	·1160	·1159	·1157	·1156	·1155	·1153	·1152	·1151	0	0	0	1	1	1	1	1	1
8·7	·1149	·1148	·1147	·1145	·1144	·1143	·1142	·1140	·1139	·1138	0	0	0	1	1	1	1	1	1
8·8	·1136	·1135	·1134	·1133	·1131	·1130	·1129	·1127	·1126	·1125	0	0	0	1	1	1	1	1	1
8·9	·1124	·1122	·1121	·1120	·1119	·1117	·1116	·1115	·1114	·1112	0	0	0	1	1	1	1	1	1
9·0	·1111	·1110	·1109	·1107	·1106	·1105	·1104	·1103	·1101	·1100	0	0	0	1	1	1	1	1	1
9·1	·1099	·1098	·1096	·1095	·1094	·1093	·1092	·1091	·1089	·1088	0	0	0	0	1	1	1	1	1
9·2	·1087	·1086	·1085	·1083	·1082	·1081	·1080	·1079	·1078	·1076	0	0	0	0	1	1	1	1	1
9·3	·1075	·1074	·1073	·1072	·1071	·1070	·1068	·1067	·1066	·1065	0	0	0	0	1	1	1	1	1
9·4	·1064	·1063	·1062	·1060	·1059	·1058	·1057	·1056	·1055	·1054	0	0	0	0	1	1	1	1	1
9·5	·1053	·1052	·1050	·1049	·1048	·1047	·1046	·1045	·1044	·1043	0	0	0	0	1	1	1	1	1
9·6	·1042	·1041	·1040	·1038	·1037	·1036	·1035	·1034	·1033	·1032	0	0	0	0	1	1	1	1	1
9·7	·1031	·1030	·1029	·1028	·1027	·1026	·1025	·1024	·1022	·1021	0	0	0	0	1	1	1	1	1
9·8	·1020	·1019	·1018	·1017	·1016	·1015	·1014	·1013	·1012	·1011	0	0	0	0	1	1	1	1	1
9·9	·1010	·1009	·1008	·1007	·1006	·1005	·1004	·1003	·1002	·1001	0	0	0	0	0	1	1	1	1
	0	1	2	3	4	5	6	7	8	9	1	2	3	4	5	6	7	8	9

SI UNITS AND ABBREVIATIONS

International agreement has been reached on the adoption of a common system of units called 'le Système International d'Unités' and known as SI. This book has been written in this system.

There are seven basic SI units and two supplementary units as follows:

Name of physical quantity	Name of unit	Symbol
length	metre	m
mass	kilogramme	kg
time	second	s
electric current	ampere	A
thermodynamic temperature	kelvin	K
luminous intensity	candela	cd
amount of substance	mole	mol
plane angle	radian	rad
solid angle	steradian	sr

Other units derived from the above basic units are:

area	square metre	m^2
volume	cubic metre	m^3
density	kilogramme per cubic metre	$kg\ m^{-3}$
speed, velocity	metre per second	$m\ s^{-1}$
angular velocity	radian per second	$rad\ s^{-1}$
acceleration	metre per second per second	$m\ s^{-2}$
force	newton	N
pressure	newton per square metre	$N\ m^{-2}$
work energy	joule	J
power	watt	W

This table is not exhaustive and it does not give all the inter-connections between the units such as the following relations:

$$1\ newton = 1\ kg\ m\ s^{-2}$$
$$1\ joule\ \ = 1\ newton\ metre \quad \text{i.e. } 1J = 1N\ m$$
$$1\ watt\ \ = 1\ joule\ per\ second \quad \text{i.e. } 1W = 1J\ s^{-1}$$

The units which are named after people—for example, the joule, the newton and the watt—are written without a capital letter *but* their abbreviations are the capital letters J, N and W respectively.

Note that we never add s to an abbreviation to make a plural. Thus 3 kilogrammes will be abbreviated to 3 kg and *not* 3 kgs. Similarly a length of 10 metres is abbreviated to 10 m.

Units of Length

In full		Abbreviated	
1000 micrometres	= 1 millimetre	1000 μm	= 1 mm
10 millimetres	= 1 centimetre	10 mm	= 1 cm
1000 millimetres	= 1 metre	1000 mm	= 1 m
1000 metres	= 1 kilometre	1000 m	= 1 km
1000 kilometres	= 1 megametre	1000 km	= 1 Mm

Units of Mass

1000 microgrammes	= 1 milligramme	1000 μg	= 1 mg
1000 milligrammes	= 1 gramme	1000 mg	= 1 g
1000 grammes	= 1 kilogramme	1000 g	= 1 kg
1000 kilogrammes	= 1 megagramme	1000 kg	= 1 Mg

All the above abbreviations are in roman type. Other abbreviations for physical quantities in the text of the book will be printed in italics, e.g. in Chapter 3 we have s for second and *s* for distance.

ANSWERS

Exercise 1

1. (*a*) arrow *OQ*, (*b*) arrow *PO*, (*c*) arrow *QP*, (*d*)
2. The *total* displacement is from *A* to *B*: a displacement of magnitude 10 m from the original point *A*
3.

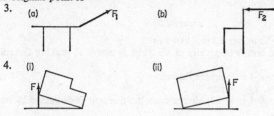

4.

Exercise 2

1. (*a*) 17 mm, (*b*) 23 mm
2. (*a*) 35 mm, (*b*) 41 pennies
3. (*a*) 48 pennies, (*b*) 12 pennies

Exercise 3

1. (i) *m*, (ii) *F*, (iii) *m*, (iv) *m*, (v) *F*, (vi) *F*
2. (i) 4 metres per second per second, (ii) 40 metres per second per second, (iii) 3 metres per second per second.
3. (i) 5 metres per second per second, (ii) 40 metres per second per second
4. 3·56 grammes
5. Copper, 8·475 grammes; nickel, 2·825 grammes

Exercise 4

1. If *L* extends 7 mm for every 6-penny load, then an extension of 21 mm arises from a load of 18 pennies; *R* is extended 11·25 mm
2. For *L* an extension of 4 mm arises from 3 N, therefore an extension of 20 mm arises from 15 N; *R* is extended 45 mm
3. 24 mm in *L* and 54 mm in *R*
4. The same force would be registered. If the point of contact had a greater force on one side than the other, it would move in the direction of the greater force.
5. 8 N in the upper spring, 6 N in the lower spring
6. 8 N in the upper spring, 6 N in the lower spring

Exercise 5

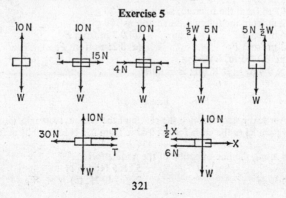

Exercise 6

1. The machine will register the same weight as if he had kept his coat on
2. The donkey is pulling the same load whether the farmer holds the buckets or puts them on the floor
3. Find the mass of the case when empty and the mass when full; subtract the two results

Exercise 7

1. (a) 0·24 m, (b) 0·12 m, (c) 0·2 m, (d) 0·1 m
2. The midpoint or centre of gravity of the ruler is above F, making the ruler 'top heavy'
3. On the other arm 0·4 m from the fulcrum
4. $100 \times 60 = 120x$, $x = 50$ mm
5. $10 \times 0·48 = y \times 0·4$, $y = 12$; alternatively the moment is decreased by one-tenth on either side
6. 5 N m, 6 N m, 30 N m, 3 N m
7. (i) 0, (ii) 2·5 N m, (iii) 0, (iv) 2·5 N m
8. 2 N, 0·02 m, 4 N, 0·04 m
   3 N, 0·03m, 6 N, 0·06 m
   2·5 N, 0·025 m, 5 N, 0·05 m
   4 N, 0·04 m, 8 N, 0·08 m
9. The moment of W_2 about F is still $W_2 \sin 30°$, FS is now below the beam
10. $3 \text{ N} \times l = 2 \text{ N} \times 0·3 \sin 45°$; $l = 0·1414$ m or $0·1 \times \sqrt{2}$ m

Exercise 8

1. Remove the body from pan P and replace by known masses till the pointer returns to the chosen point: the sum of the masses is equal to the unknown mass
2. Interchange the body and the 140 g; if they still balance, then the apparatus is accurate
3. (a) 8 g, 4 g, 2 g, 1 g; (b) Add 16 g to (a); (c) Add 32 g to (b)
4. $9 g + 1 g = 7 g + 3 g$
5. $1 g, 3 g - 1 g, 3 g, 3 g + 1 g, 9 g - 3 g - 1 g, 9 g - 3 g, 9 g + 1 g - 3 g,$
   $9 g - 1 g, 9 g, 9 g + 1 g, 9 g + 3 g - 1 g, 9 g + 3 g, 9 g + 3 g + 1 g$

Exercise 9

1. (a) 0·18 m from the fulcrum, on the other arm
   (b) As in (a)
   (c) 0·02 m from the fulcrum, same arm as 20 g
   (d) 0·2 m from the fulcrum, same arm as 100 g
2. (a) 0·4 m from F (b) 0·3 m from F
   (c) 0·3 m from F (d) 0·25 m from F
3. (a) 3 m, (b) 0·5 m, (c) 0·4 m
4. $10W \times 0·4 \text{ m} = B \times 0·1$; $B = 40W$

Exercise 10

1. Each vertical passes through the centre of the square, rectangle, circle
2. 300 N parallel to the given forces; 0·12 m from 200 N force, 0·24 m from 100 N force
3. 300 N along the line of action of the middle force
4. 40 N   5. 165 N; 415 N
6. 24 N; 16 N   7. (a) 40 N; (b) $W + W_1$

Exercise 11

1. At the intersection of the lines joining the midpoints of opposite sides
2. 0·4 m
3. C.G. of triangle *ABC* lies on *BD*, 20 mm from *B*; C.G. of triangle *ADC* lies on *DB*, 20 mm from *D*
4. The C.G. of any strip parallel to *AB* lies on the line joining the midpoints of *AB* and *DC*
5. See page 23
6. At the centre of the circles. A shape in the form of the letter L, C, etc.

Exercise 12

1. $NG = 53\frac{1}{3}$ mm   2. $OG = 0\cdot25$ m
3. The C.G. lies 50 mm from *E* on the line *EF*, where *F* is the midpoint of *AB* (consider $2W$ at the centre of the square and W at the C.G. of triangle *ABE*)
4. If $OX = \frac{1}{3} OD$ and $OY = \frac{1}{3} OB$, where *O* is the centre of the rectangle, then the C.G. lies at the midpoint of *XY*
5. 1·4 m from *A*
6. If *E* and *F* are midpoints of *DA* and *CB*, respectively, then $EG = \frac{2}{3}$ m
7. At the C.G. of the triangle, i.e. the intersection of the medians

Exercise 13

1. 300 N   2. 140 N
3. W; the points of attachment of the legs to the rim form an equilateral triangle whose centroid is at the centre of the circle
4. A three-legged stool does not rock on uneven ground
5. Splaying the legs makes the stool less likely to tip over

Exercise 14

1. (a) 18 km h^{-1}, (b) 72 km h^{-1}, (c) 9 km h^{-1}, (d) 90 km h^{-1}, (e) 54 km h^{-1}, (f) 3·6 km h^{-1}, (g) $3\cdot6x$ km h^{-1}
2. (a) 20 m s^{-1}, (b) 5 m s^{-1}, (c) $3\frac{1}{3}$ m s^{-1}, (d) $8\frac{1}{3}$ m s^{-1}, (e) 25 m s^{-1}, (f) $\frac{10}{36}$ m s^{-1}, (g) $\frac{10x}{36}$ m s^{-1}
3. $\frac{1}{360}$ m s^{-1}
4. (a) 36 km h^{-1}, (b) 24 km h^{-1}, (c) 30 km h^{-1} 5. 70 km h^{-1}
6. 6 km h^{-1}; walking 1 km uphill will take 15 min, and walking back to the starting point will take 5 min, therefore the time to travel 2 km is 20 min
7. 1660 m
8. The minute hand has an angular velocity of 1 revolution per hour or $\frac{\pi}{1800}$ rad s^{-1}, clockwise; the hour hand has an angular velocity of 1 revolution per 12 h or $\frac{\pi}{21\,600}$ rad s^{-1}, clockwise

Exercise 15

1.

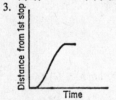

Time (h)

4.

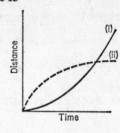

2. The uniform speed is represented by the dotted line in the answer to Question 1

3. Fig. 38. Starting from base at $t = 0$, we travel a distance of 20 km in 1 h with uniform speed. In the next hour we travel 10 km with uniform speed back towards base. We then rest for 1 h before finally travelling back to base in the next hour with the same uniform return speed of 10 kmh^{-1}.

Fig. 39. Starting (P) 30 m away from the base point, we travel with uniform speed and reach base in 2 seconds (Q). We continue on with the same speed for a further 2 seconds until (R) we are 30 m on the other side of the base point.

5. *Arthur.* Walks 5 km (OP) with uniform speed in 1 h; stops (PQ) for $\frac{1}{2}$ h; walks 5 km (QR) with the same uniform speed in 1 h; stops (RS) for $\frac{1}{2}$ h; walks 5 km (ST) with uniform speed in 1 h.

Bill. Rides 5 km (HJ) with uniform speed in $\frac{1}{2}$ h; stops (JK) for $\frac{1}{2}$ h; rides 5 km (KL) with uniform speed in 1 h; stops (LM) for $\frac{1}{2}$ h rides 5 km (MN) with uniform speed in $\frac{1}{2}$ h.

The distance between the towns is 15 km. Arthur takes 4 h and Bill 3 h to complete the journey. They meet midway between the towns, 2 h after Arthur started.

Exercise 16

1. (*a*) 11 m s^{-1}, (*b*) 28 m s^{-1}
2. (*a*) (i) 12 m s^{-1}, (*a*) (ii) 24 m s^{-1} (note: $s = 3t^2$); (*b*) 12 m s^{-1}
3.

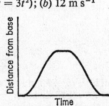

4.

Exercise 17

1. (*a*) (a) 40 m, (b) 4 km, (c) 20 m
   (*b*) (a) 90 m, (b) 12 km, (c) 87·5 m followed by 3·5 m on return journey.
2. Distance travelled $= \frac{1}{2}(520 + 120) \times 15 = 4800$ m

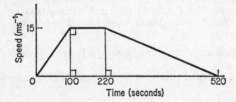

3.

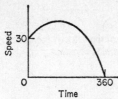

(b) $1 \text{ km h}^{-1} \times 3 \text{ seconds} \times 100 = 1 \times \dfrac{3}{60 \times 60} \times 100 = \frac{1}{12} \text{ km}$

(c) Approximately 47 complete (10 mm $\times$ 10 mm) squares, giving a distance of $\frac{1}{12} \times 47 = 3\frac{11}{12} \text{ km}$

4. (a) 60 s, (b) 12·5 m s^{-1} (c) 4 metres per second every 5 s, (d) 1 metre per second every second

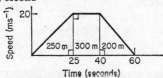

Exercise 18

1. (a) 0·2 m s^{-2}, (b) $100 + 600 + 1000 = 1700$ m

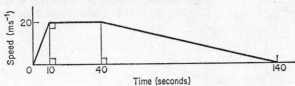

2.

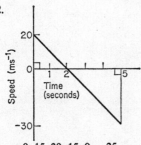

s: 0, 15, 20, 15, 0, −25,
t: 0, 1, 2, 3, 4, 5

3. When $t = 7$ seconds the body is $-75 + 12 = -63$ m from the point of first observation. The body is at rest when $t = 5$ seconds; after this time it begins to return to the point of first observation

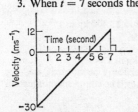

4. 10 seconds

Exercise 19

1. In any two equal intervals of time the magnitude of the velocity increases by the same amount and the direction of the acceleration remains constant
2.   (i) $s = 25$ m, $t = 5$ s
      (ii) $s = 125$ m, $t = 5$ s
      (iii) $a = -1.25$ m s^{-2}, $t = 40$ s
      (iv) $v = 100$ m s^{-1}, $a = 0$
      (v) $v = 32$ m s^{-1}, $s = 256$ m
      (vi) $u = 0$, $s = 225$ m
      (vii) $u = 100$ m s^{-1}, $a = -10$ m s^{-2}
3. 0.375 m s^{-2}; 300 m   4. 1.5 s; $8\frac{8}{9}$ m s^{-2}
5. 11 540 m; 13 510 m; 15 615 m

Exercise 20

1. (a) 2 s   (b) 19.6 m
  (c) $s = -24.5$ m; the ball is 24.5 m below the point of projection
  (d) 1 and 3 seconds
2. (a) At the point of projection
  (b) 4.9 m; this is the distance fallen in 1 s from the greatest height
  (c) 1 and 19 s; ($s = ut + \frac{1}{2}gt^2$ yields $93.1 = 98t - \frac{1}{2} \times 9.8t^2$ which reduces to $19 = 20t - t^2$ on division by 4.9)
3. Point B on the graph corresponds to the greatest height. Point C corresponds to the moment of return to the point of projection. Time of flight is 15 s, greatest height 276 m

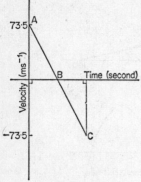

Exercise 21

1. 44.1 m; 29.4 m s^{-1}
2. 193.6 m; 63.4 m s^{-1}
3. $\sqrt{2} + 1 + 1 = 3.414$ s
4. 99.6 m; 59.6 m s^{-1}
5. 4.9 m; 14.7 m; 24.5 m

Exercise 22

1. 50 kg m s^{-1} 2. $620 \times 20 = 12\,400$ kg m s^{-1}
3. 4.5 m s^{-1} 4. 200 N
5. 12.5 m s^{-2}
6. Velocity of impact is 9.8 m s^{-1}; momentum 39.2 kg m s^{-1}
7. The momentum before impact is 20 kg m s^{-1} vertically downwards, and the momentum after impact is 10 kg m s^{-1} vertically upwards. The change in momentum is 30 kg m s^{-1}
8. 30 N
9. 40 kg m s^{-1} per second (the time of 5 s is irrelevant)   10. 48 kg

Exercise 23

1. 36 kg m s^{-1}; 18 m s^{-1} 2. 35 kg m s^{-1} 3. 1·2 kg m s^{-1}
4. Acceleration is 9·7 m s^{-2}; magnitude of force is 2500 × 9·7 kN
5. 2·825 kg m s^{-1} 6. (a) 400 m s^{-1}, (b) 240 m s^{-1}

Exercise 24

1. (a) 500 J, (b) 2000 J, (c) 1125 J
2. (a) 300 kg m s^{-1}, (b) 75 kg, (c) 3000 J, (d) 30 m
3. (a) $\frac{1}{5500}$ s, (b) 13·5 kg m s^{-1}, (c) 74·25 kN, (d) 3712·5 J

Exercise 25

1. (a) 6·4 m s^{-1}, (b) 1·6 m s^{-1}, (c) −4·4 m s^{-1}, (d) −7·6 m s^{-1}
2. 8 m s^{-1}; 16 J
3. The sum of the momenta before impact is zero, therefore the momentum of each mass after impact must be equal and opposite. Since the masses are equal, the velocities must be equal and opposite.
4. Yes
5. The truck will move back as a result of the impulse necessary to give the bag the velocity v. The momentum before and after the throw is zero. Therefore, if u is the velocity of the man and truck after the throw, we have $(m + M)u = m_1v$

Exercise 26

1. The rate of change of momentum of the hammer is 2 × 20 × 200 kg m s^{-1} in 1 min; the average force on the ground is therefore 133⅓ N
2. The rate of change of momentum of the drill is 1 × 50 × 500 kg m s^{-1} in 1 min; the average force on the wall is 416⅔ N.
3. The rate of change of momentum of the jet is 30 × 20 kg m s^{-1} in 1 s; force on the truck is 600 N; acceleration is 1·2 m s^{-2}
4. Length of jet in 1 s is 0·03 ÷ π 0·01^2 = 95·55 m; rate of change of momentum at the wall face is 30 × 95·55 kg m s^{-1} per second; force on the wall is 2866·5 N

Exercise 27

1. (i) 0·8, (ii) 0·6, (iii) 1⅓   2. 0·9397   3. 0·3420
4. 0·5000   5. 0·8660   6. 0·7071, 0·7071, 1
7. 0·5774   8. 0·4000
9. $\angle BCA = 53°$, $\angle CAB = 37°$ 10. Use Pythagoras' theorem

Exercise 28

1. (i) 0·5, (ii) $\frac{1}{2}\sqrt{3}$, (iii) $\frac{1}{3}\sqrt{3}$
2. $AC = \sqrt{2}$ m (i) $\frac{1}{2}\sqrt{2}$, (ii) $\frac{1}{2}\sqrt{2}$, (iii) 1
3. 300 tan 75° = 1120 m (nearest metre)
4. tan (angle of elevation) = 23·315 ÷ 100, angle TOH = 13° 08′; tan $\angle HOW$ = 8·815 ÷ 100, angle HOW = 5° 02′
5. 25 tan 85° = 285·8 m, 25 tan 80° = 141·8 m; average speed is 2·4 m s^{-1}

Exercise 29

1. sin 120° = sin 60° = 0·866, cos 120° = −cos 60° = −0·5,
tan 120° = −tan 60° = −1·7321
2. sin 170° = sin 10° = 0·1736, cos 170° = −cos 10° = 0·9848,
tan 170° = −tan 10° = 0·1763
3. sin 140° = sin 40° = 0·6428, cos 140° = −cos 40° = 0·7660, tan 140° = −tan 40° = −0·8391
4. sin 30° = 0·5 = sin 150°, x = 150°

5. $-\cos 30° = -0.3660 = \cos 150°, x = 150°$
6. $\tan 45° = 1, \tan 135° = -1; 2x = 135°, x = 67.5°$
7. $0.809 = \sin 54° = \sin 126°; 2x = 54°$ or $2x = 126°; x = 27°$ or $63°$
8. $0.5 = \cos 60°, -0.5 = -\cos 60° = \cos 120°; x = 120°$
9. $0.3640 = \tan 20°; -0.3640 = \tan 160°; x = 160°$
10. Dividing by $\cos x$ gives $\tan x = 1$, hence $x = 45°$

Exercise 30

1. $b = \dfrac{100 \sin 60°}{\sin 45°} = 122.45$ m; $c = \dfrac{100 \sin 75°}{\sin 45°} = 136.6$ m, $C = 75°$

2. $B = 120°; a = \dfrac{1000 \sin 40°}{\sin 120°} = 742.3$ m; $c = \dfrac{1000 \sin 20°}{\sin 120°} = 395$ m

3. $\sin B = \dfrac{300 \sin 20°}{200} = 0.513, B = 30° 52'; C = 129° 08'; c = 453.6$ m

4. 728 N, 1630·6 N

5. $b = \dfrac{1000 \sin 40°}{\sin 130°} = 1000 \tan 40° = 839.1$ N; $c = \dfrac{1000 \sin 10°}{\sin 130°} = 226.7$ N

Exercise 31

1. 40 N, 100 N, 119·8 N, 94·15 N, respectively
2. $\cos X = 0.35, X = 69° 31'$; therefore $\cos A = -0.35, A = 110° 29', \cos B = 0.95, B = 18° 12'$
3. 138·5 mm, 277 N
4. $AC = 130$ mm; $\angle ACB = 67° 23' = 67.4°, \angle CAB = 22° 37' = 22.6°$
$b^2 = c^2 + a^2 - 2ca \cos B; c^2 = a^2 + b^2 - 2ab \cos C$

Exercise 32

1. $\overrightarrow{AC}$, magnitude 176·75 N at 45° to AB
2. $\overrightarrow{BD}$, magnitude 960 N direction 51° 20′ to BA
3. The resultant passes through 0, of magnitude 500 N at angle θ to the 400 N force, where $\tan \theta = 0.75$
4. $\dfrac{x}{50} = \dfrac{2}{3}, x = 33\frac{1}{3}$ m

5. Resultant velocity $\sqrt{29}$ m s^{-1} at 68·2° to the bank; $\dfrac{x}{100} = \dfrac{2}{5}, x = 40$ m
6. 74·31 mm at N 61° 35′ E; 7·431 km

Exercise 33

1. 14 N at 21° 47′ to the force of 10 N
2. 14 m s^{-1} at 21° 47′ to the velocity of 10 m s^{-1}
3. 14 m s^{-1} (nearest metre) at 30° 21′ to the impulse of 10 kg m s^{-1}
4. 10 kN through A, bisecting the angle between the given forces
5. $3\sqrt{3}$ kN or 5·1963 kN, bisecting the angle between the given forces

Exercise 34

1. By Lami's Theorem $\sin A = \sin B = \sin C$, and since $A + B + C = 360°$ it follows that $A = B = C = 120°$
2. $T = \sqrt{3}$ kN
3. 150° between the horizontal string and the string with a tension of 200 N. Remaining angles 120° and 90°. Tension in horizontal string 100 $\sqrt{3}$ N
4. Inclined at 30° to the vertical and 90° to the first string. Tension 150 $\sqrt{3}$ N
5. $T = 2\sqrt{3}$ kN
6. Tension in the string is 115·5N. Horizontal force is 57·75 N
7. Tension in AB and CD is 115·5 N. Tension in BC is half the tension in the other two sections

Exercise 35

1. (i) $\overrightarrow{AC}$, magnitude $|a|\sqrt{2}$ or $|b|\sqrt{2}$, direction 45° to AB, above AB;

   (ii) $\overrightarrow{DB}$, magnitude $|a|\sqrt{2}$ or $|b|\sqrt{2}$, direction 45° to AB, below AB

2. (i) $\overrightarrow{AC}$, magnitude 5, direction 36° 50′ to AB, above AB;

   (ii) $\overrightarrow{DB}$, magnitude 5, direction 36° 50′ to AB, below AB

3. The surface of a sphere centre O and radius 3

4. $x = 0$ 5. $x = 0$

6. (i) A vector in the same direction as a but having twice the magnitude of a;
   (ii) a vector in the same direction as b but having treble the magnitude of b;
   (iii) the vector sum of (i) and (ii)

Exercise 36

1. (i) $b-2a$, (ii) $-2b$, (iii) $\overrightarrow{FD} = \overrightarrow{AC} = a + b$, (iv) $\overrightarrow{DC} = \overrightarrow{OB} = \overrightarrow{OC} + \overrightarrow{CB} = a-b$, (v) $\overrightarrow{BC} + \overrightarrow{CD} = b-a+b = 2b - a$

2. (i) $\overrightarrow{DQ} + \overrightarrow{QO} = -b + a$, (ii) $-2b + 2a$ or $2(a - b)$, (iii) $2b + a$, (iv) $2b + a$, (v) $\overrightarrow{DM} = \overrightarrow{PB} = a - 2b$

3. (i) $c + d$, (ii) $b + c + d$, (iii) $-c -b$, (iv) $-d -c -b$, (v) $-d -c -b -a$, (vi) 0

Exercise 37

1. (i) $2a + b$, (ii) $a + 2b$

2. (i) $\frac{1}{2}(b - a)$, (ii) $\frac{1}{3}(b - a)$, (iii) $\overrightarrow{BM} - \overrightarrow{BN} = \frac{1}{6}(b - a)$, (iv) $\overrightarrow{AM} = \frac{1}{2}(a + b)$, (v) $\overrightarrow{AN} = \frac{1}{3}(2a + b)$

3. (i) $\overrightarrow{AP} = \overrightarrow{AB} + \overrightarrow{BP} = b + \frac{1}{3}\overrightarrow{BC} = b + \frac{1}{3}(c - b) = \frac{1}{3}(c + 4b)$, (ii) $\frac{1}{3}(b + 4c)$

Exercise 38

1. $50\sqrt{2}$ N $= 70.7$ N in direction North-west

2. 359.8 N at 0° 34′ to direction AO

3. 206.5 kg m s^{-1} at 44° 16′ to direction AO

Exercise 39

1. $447 \times 2 = 894$ N in magnitude but having the same line of action.

2. $90 \times 3 = 270$ N in magnitude but having the same line of action where $AB = BD = 30$ mm

3. The resultant is $40\sqrt{2} = 56.56$ N at 45° to AB and passing through the mid-point of AB and BC

4. Magnitude of the resultant is $50\sqrt{2} = 70.7$ N; direction of the resultant is 45° to AB

5. The resultant is 145.3 N at 26° 34′ to AB; line of action is AC

6. The force polygon is closed indicating a zero resultant but the system is not in equilibrium

Exercise 40

1. (i) 22.4 N in same direction, (ii) 5.6 N in same direction

2. 5 at 53° 08′ to the positive x axis

3. The resultant has a magnitude of 26 along AC.

4. $\sqrt{61} = 7.81$, 50.2° to OA and on the same side as D

5. (i) 400 i + 300 j, (ii) 300 i + 400 j

6. 700 i + 700 j. Magnitude of resultant is 700 $\sqrt{2}$ N acting in the direction which bisects angle AOB

Exercise 41

1. In each case the magnitude of the resultant is doubled but the direction remains unchanged

2. In each case the magnitude of the resultant is unchanged but its direction is reversed

3. $X = \frac{37}{13}$, $Y = \frac{114}{13}$, R = 9·23 kN at 72° 01′ to AB

4. $X = 1$, $Y = 6\sqrt{3}$, R = 10·44 kN at 84·5° to AB

5. $X = 20$, $Y = 20$, R = 20 $\sqrt{2}$ kN at 45° to AB

6. $X = 20$, $Y = 20\sqrt{3}$, R = 40 kN at 60° to AB

7. Each pair has a resultant of 20 kN acting through the centre of the hexagon

8. 20 $\sqrt{2}$ kN in a direction NE.

9. $R = 2 + 8\sqrt{2}$ = 13·31 kN, in a direction parallel to AC

10. $R = 2 + 6\sqrt{2}$ = 10·48 kN, in a direction parallel to AC

Exercise 42

1. No change

2. $X = 60$, $Y = 100$, $\theta = 59°\ 02′$, $x = -0·4$; resultant is directed into the first quadrant

3. $X = 200$, $Y = 69·28$, $\theta = 19·1$; resultant will still pass through B no matter what the forces are in AB or BC

4. B

5. $X = 10$, $Y = 34·14$, R = 35·58 N, $\theta = 73°\ 40′$; line of action passes through P where $BP = PA$

6. The sum of the moments about the centre of the square is zero

7. $X = 0$, $Y = 273·2$, $R = 273·2$, $\theta = 90°$; line of action passes through P between A and B with $BP = 0·0683$ m

8. No change

Exercise 43

1. 53 km h^{-1} East both before and after

2. (i) A relative to C, 75 km h^{-1} East; B relative to C, 70 km h^{-1} East; (ii) B relative to A, 5 km h^{-1} West; C relative to A, 75 km h^{-1} West

3. (i) Against the current, 200 s; (ii) with the current, 66$\frac{2}{3}$ s. Velocity relative to the current is 1 m s^{-1} in either direction

4. 450 km h^{-1} North; 13$\frac{1}{3}$ min   5. 550 km h^{-1} South; 10$\frac{10}{11}$ min

6. (i) 2$\frac{1}{2}$ m s^{-1}; (ii) $-7\frac{1}{2}$ m s^{-1}, the negative sign indicating the opposite direction to that suggested in Fig. 119

7. (i) $\frac{2}{3}$ m s^{-1}; (ii) $(20 \pm \frac{2}{3})$ m s^{-1}

Exercise 44

1. New York: 5.05 p.m. Tuesday
   Tokyo: 7.20 a.m. Wednesday
   Sydney: 8.05 a.m. Wednesday
   Cape Town: 11.15 p.m. Tuesday
   Rio de Janeiro: 7.05 p.m. Tuesday

2. 20° W

3. Find the latitude by Pole Star at night; find longitude by G.M.T. for local noon

4. 105° 58′ E   5. When shadows point directly to the North

Exercise 45

1. (i) 177 km h^{-1}, N 7° W; (ii) 186 km h^{-1}, N 4° W
2. Course W 10° 11′ N; resultant velocity 232·1 km h^{-1} due West; time 25·8 min
3. Resultant velocity $\sqrt{5}$ m s^{-1} due West; course W 41° 48′ S; time 40 $\sqrt{5}$ = 89·44 s
4. Course is directly across at right angles to the current; resultant speed $\sqrt{2}$ m s^{-1}; time of crossing 150 s
5. B to C, N 10° W; C to D, W 10° S; D to A, S 10° W
6. Speed of train is 10 m s^{-1}. Speed of marble is 1 m s^{-1} relative to the train. Velocity of marble relative to track is $\sqrt{101}$ m s^{-1} at angle θ to direction of motion, where tan θ = 0·1

Exercise 46

1. 70 m s^{-1} East; 5 s   2. 2 km h^{-1} due West
3. Rate of change of $\overrightarrow{WM}$ is the rate of change of ($\overrightarrow{WA} + \overrightarrow{AM}$), which is 5 km h^{-1} at an angle θ to BA, where tan θ = $\frac{4}{3}$
4. 0·5016 m s^{-1} at an angle θ to direction carried, where tan θ = 0·08; 0·54 m s^{-1} or 0·46 m s^{-1} in direction being carried

Exercise 47

1. Velocity 10$\sqrt{41}$ km h^{-1} along BA; cars will collide at O
2. Velocity 100 $\sqrt{89}$ = 943·4 km h^{-1} in direction E 77° N; they do not collide
3. At 17° 21′ to the vertical, towards him
4. At 73° 54′ to the vertical
5. Velocity of B is 5 $\sqrt{13}$ = 18·03 m s^{-1} in direction N 56° 19′ E (See figure)

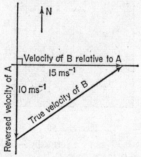

6. N 56° 19′ W at 18·03 km h^{-1}

Exercise 48

1. 100 m; A is 80 m past O and B is 60 m from O
2. 32 km h^{-1}, W 51° 20′ S; A is 5 km past O, B is 4 km from O
3. 86·02 km h^{-1}, N 54° 28′ E; A is 1·1 km from O, B is 1·5 km past O
4. E 42° 18′ N
5. tan α = $\frac{1}{60}$, α = 0° 57′ forward of the target
6. Result unchanged
7. 70 km h^{-1}, E 52° S; A is 1·7 km from O, B is 0·4 km past O (Graphical solution)

Exercise 49

1. The woman is midway between W_1 and W_2; the man is midway between M_1 and M_2
2. Given by R_4, M_4, and W_4
3. The woman is 36 m past O; the man is 12 m from O

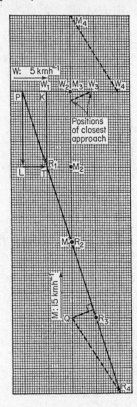

Exercise 50

1.   (i) Horizontally $-5 \cos 30°$ m s^{-1}; vertically $-5 \sin 30°$ m s^{-1}
    (ii) Horizontally -5 m s^{-1}; vertically 0
   (iii) Horizontally $-5 \sin 39°$ m s^{-1}; vertically $5 \cos 39°$ m s^{-1}
2.   (i) Horizontally $-10 \cos 70°$ m s^{-1}; vertically $10 \sin 70°$ m s^{-1}
    (ii) Horizontally -10 m s^{-1}; vertically 0
   (iii) Horizontally $-10 \cos 40°$ m s^{-1}; vertically $-10 \sin 40°$ m s^{-1}
3. 13 m s^{-1} at angle θ to the horizontal, where $\tan \theta = 2{\cdot}4$
4. Horizontally 5 m s^{-1}; vertically $5 \sqrt{3}$ m s^{-1}
5. At 40° to the horizontal, i.e. the direction of the tangent
6. The tension in the string and the weight of the body

Exercise 51

1. (i) 127·5 m, (ii) 10·19 s, (iii) 882·5 m, (iv) $x = 866$ m, $y = 9·5$ m; horizontal component 86·6 m s^{-1} (constant), vertical component $-38·1$ m s^{-1}, (v) 100 m s^{-1} in

direction 30° below the horizontal. (Since the curve is symmetrical about the greatest height the tangent at A is obtained from the tangent at 0 by symmetry.)

   2. Greatest height 255·1 m; range 1020 m; horizontal component of velocity is 50 $\sqrt{2}$ m s^{-1}, vertical component is (50 $\sqrt{2}$ − 9·8) m s^{-1}, hence velocity at $t = 1$ is 93·31 m s^{-1} at 40° 44′ to the horizontal

   3. 5 cos 30° m s^{-1}

   4. $x = 952·6$ m, $y = −43·5$ m (i.e. *below* the point of projection)

   5. 4·518 s

Exercise 52

   1. 319·7 m; tan θ = −0·3132, direction is 17° 24′ below the horizontal

   2. 9180 m

   3. Maximum range is 16 320 m; area covered is π (16 320)^2 m^2

   4. 329 m (to nearest metre)   5. 40·8 m

   6. sin α = 0·49, α = 29° 20′   7. $v^2 = u^2 − 2gy$

   8. $\sqrt{(u^2 −2gy)}$

   9. The magnitude of the velocity of the projectile at a point is the same as the magnitude of the velocity it would acquire at the same point after being released from rest at a height of $\dfrac{u^2}{2g}$

   10. 23·77 m s^{-1} 11. 2$\frac{4}{7}$ m; 1$\frac{3}{7}$ s, 0·22 s (nearest 0·01 s)

   13. sin 2α = 0·6125, 2α = 37° 46′

Exercise 53

   1. (i) g sin 60°, (ii) g sin 45°, (iii) g sin 30°

   2. 1·428 s; 7 m s^{-1} 3. 10·2 m; 10 m s^{-1} down the plane

Exercise 54

   1. $\sqrt{\dfrac{2d}{g}}$

   2. (i) On the vertical line through A. (ii) The circle will have A as its highest point and it will also touch the slope PQ; $OA = 10$ sin 15° = 2·588 m. (iii) $t^2 = \dfrac{2OA}{g} = 0·5280$; $t = 0·7266$ s

   3. Total time is $\dfrac{3\sqrt{2}}{2}$ s. (Note that the motion from B to C is not a chord of quickest descent related to the circumscribed circle of the hexagon.)

Exercise 55

   1. (i) $F = 0·8379$ N, $N = 2·3022$ N; (ii) $F = 1·225$ N, $N = 2·122$ N; (iii) $F = N = 1·732$ N

   2. Each answer would be doubled

   3. $F = 4·9$ N, $N = 8·4868$ N; 9·8 N vertically upwards, i.e. equal and opposite to the weight of the body

Exercise 56

   1. All results are 0·4; (i) 3·36 N, 3·2 N; (ii) 5·04 N, 4·8 N

   2. (i) 14 N, (ii) 16·8 N   3. (i) 10·08 N, (ii) 4·32 N   4. 480 N

Exercise 57

   1. (i) 45°, (ii) 20°, (iii) 58°

   2. (i) 0·5774, (ii) 1·7321, (iii) 0·404

   3. Normal reaction mg cos 30°; frictional force mg sin 30°

   4. 45°; mg sin 45°   5. 669·6 N   6. 369·6 N

   7. We know that the body will slide because the angle of slope is greater than 45°; velocity 7·18 m s^{-1}

   8. Result unchanged

Exercise 58

1. $KL = OK \sin 38°$; $\angle LOK = 38°$; $T = 603\cdot4$ N
2. $\dfrac{100g}{\sin \theta} = \dfrac{T}{\sin (180° - 38°)}$: force is least when $\theta = 90°$, i.e. $T = 100g \sin 38°$
3. Double the answer   4. $T = 100g \sqrt{2}$ 5. 783 N

Exercise 59

1. Slides first when angle of slope is given by $\tan \theta = 0\cdot9$; inclination would have to reach 45° before toppling takes place
2. (i) $\tan 40° = 0\cdot8391$: block topples when angle of slope is given by $\tan \theta = 0\cdot25$ with the 100-mm edge lowermost, or by $\tan \theta = 0\cdot5$ when the 50-mm edge is lowermost; hence block topples before sliding; (ii) Slides before toppling
3. 60°

Exercise 60

1. Stable   2. Neutral   3. Stable
4. Neutral   5. Stable   6. Unstable

Exercise 61

1. $\mu = 0\cdot2$; $0\cdot1414g = 1\cdot386$ N
2. $3\cdot27$ m s^{-2}; thrust on pulley is $T \sqrt{2} = 4\cdot624$ N at 45° below the horizontal, $T = 3\cdot27$
3. $0\cdot7833$ s,   4. $0\cdot0335g = 0\cdot3283$ m s^{-2} 5. $0\cdot1418g = 1\cdot39$ m s^{-2}
6. Acceleration $\frac{1}{3}g = 3\cdot27$ m s^{-2}; tension $1\frac{1}{3}g = 13\cdot08$ N
7. Acceleration $4\cdot9$ m s^{-2}; velocity of larger mass is $4\cdot9$ m s^{-1} vertically down, velocity of smaller mass is $4\cdot9$ m s^{-1} vertically up
8. $0\cdot5$ s; $1\cdot225$ m   9. 495 N   10. 485 N

Exercise 62

1. $577\cdot3$ N
2. For concurrency the lines of action meet on the perpendicular bisector of the bar and form an isosceles triangle.
3. 260 N, $1\cdot4$ m from B
4. $\triangle OLB$; $\overrightarrow{LB}$ for R, $\overrightarrow{BO}$ for S, $\overrightarrow{OL}$ for the weight 60 N
5. (i) $S \cos \angle OBL = R$; (ii) $S \sin \angle OBL = 60$ N
6. The diagram is that of Fig. 179, with $\angle ABK = 45°$, $LB = \frac{1}{2}OL$; reaction at the wall is 50 N, direction of force at B is 63° 26' to the horizontal
7. (i) $R \cos \angle KAO = 150$ N, (ii) $T = R \sin \angle KAO$, (iii) $150 \times KO = T \times AK$
8. 75 N ($2KO = AK$ and $\overrightarrow{AK}$ represents the weight in the triangle of forces AKO)
9. Tension in the cable at B is $50 \sqrt{2} = 70\cdot7$ N; reaction at A on the beam is $70\cdot7$ N at 45° to the horizontal (point of concurrency of the three forces is the midpoint of CB)

Exercise 63

1. (i) $R \cos \angle OAC = N \cos 70°$; (ii) $R \sin \angle OAC + N \sin 70° = 20$
2. $N = \dfrac{40 \cos 15°}{\tan 75°} = 40 \sin 15° = 10\cdot352$
3. $12\cdot5$ N at right-angles to the rod. (Take moments about the lower end of the rod)
4. (i) $R \cos \angle OBK = N \cos \angle OAL$; (ii) $R \sin \angle OBK + N \sin \angle OAL = 30$
5. The rod rests in a horizontal position, therefore reaction at each end of the rod passes through the centre of the bowl; resolving vertically, $2R \sin 60° = 10$, $R = 5\cdot773$ N

2. $R \times 0.05 = \dfrac{200}{11}$, $R = 363\frac{7}{11}$ N

3. (i) 0·1 m; (ii) 0·09 m

4. Common velocity of hammer + nail is 4·995 m s^{-1}. $R = 1249$ N

5. Common velocity of hammer + post is $\sqrt{4\cdot9}$ m s^{-1}. Work done by external forces $(12g - R)\, 0\cdot01$ J. Change in kinetic energy $-3g$. $R = 312g$ N

Exercise 73

1. 411·6 J s^{-1}

2. Gain in potential energy is $\dfrac{100 \times 16 \times 9\cdot8 \times 1}{60} = 261\frac{1}{3}$ J s^{-1}; gain in kinetic energy is $\frac{1}{2} \times 100 \times 16 \times \frac{1}{16} = 50$ J s^{-1}; power used is $311\frac{1}{3}$ J s$^{-1} = 311\frac{1}{3}$ W

3. 400×18 J s$^{-1} = 7200$ W   4. $R \times 0\cdot5 = 750$, $R = 1500$ N

5. Gain in potential energy is $\dfrac{600 \times 9\cdot8 \times 10}{60} = 980$ J s^{-1}; gain in kinetic energy is $\frac{1}{2} \times \dfrac{600 \times 10 \times 10}{60} = 500$ J s^{-1}; total work done on the water is 1480 J s^{-1}, represents 75 per cent of engine's output; power of engine is $1973\frac{1}{3}$ W

6. $300 \times 10\,\pi$ J per min $= 50\,\pi$ W

7. Tractive force is given by $T \times 15 = 1500$, whence $T = 100$ N; equating resultant force to mass × acceleration down the slope we have $T + 1000g \sin 30^\circ - R = 0$, whence $R = 5$ kN

8. At constant speed the resultant force on the car is zero; tractive force is $T = 1000 + 1000g \sin 30^\circ = 5900$ N; maximum speed is $\dfrac{10\,000}{5900} = 1\cdot695$ m s^{-1}

9. 30 km h^{-1} is $8\frac{1}{3}$ m s^{-1}; $R = 120$ kN

10. Work done is measured by the increase in the potential energy of the hammer, $7 \times 9\cdot8 \times 3 = 205\cdot8$ J for each lift; rate of working is 27·44 W

Exercise 74

1. (a) third type (*FEL*), (b) second type (*FLE*), (c) first type (*LFE*), (d) third type (*FEL*), (e) second type (*FLE*), (f) second type (*FLE*), (g) first type (*LFE*), (h) first type (*LFE*), (i) second type (*FLE*), (j) first type (*LFE*)

2. (a) 0·25, (b) 6, (c) 4, (d) 2/3, (e) 4, (f) 2·4, (g) 3·75, (h) 2·5, (i) 6, (j) 10

3.

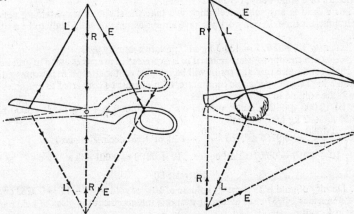

4. No, at the beginning of the lift it is of the second type (*FLE*)

5. 3200 N   6. $(1\cdot2 \times 500) \div 14 = 42\cdot86$ N

Applied Mathematics Made Simple

Exercise 75

1. 30 rev/min; V.R. $= 3$ 2. V.R. $= 3 \div 14$
3. V.R. $= \frac{1}{3}$, one revolution of the pedal wheel gives three revolutions of the back wheel; circumference of back wheel is 1·98 m, speed of bicycle is 5·94 m s^{-1}
4. M.A. $= 2\frac{1}{3}$, V.R. $= 2\frac{1}{2}$, efficiency $= 14 \div 15 = 93\frac{1}{3}$ per cent
5. (i) 62·5 N, (ii) 78·1 N   6. 1908 N   7. M.A. $=$ V.R. $= 5$

Exercise 76

1. M.A. $= 1·75$; V.R. $= 2$ 2. M.A. $= 0·45 \times 2 = 0·9$; $E = 333\frac{1}{3}$ N
3. Pitch 12·5 mm; $AC = 86·6$ mm $= 4 \times \pi d$, $d = 6·9$ mm
4. 80π 5. 26·5 N   6. 60π; $E = 125 \div \pi$

Exercise 77

1. (i) M.A. $= 1·9$; V.R. $= 3$; efficiency 0·63. (ii) M.A. $= 3$; V.R. $= 4$; efficiency 0·75. (iii) M.A. $= 3\frac{1}{3}$; V.R. $= 4$; efficiency 0·83
2. M.A. $= 1·75$; V.R. $= 8$, efficiency $\frac{7}{32} < 0·5$ therefore no fall-back of load
3. $\frac{1}{2}W$

Exercise 78

1. M.A. $= 150 \div 48 = 3·125$; efficiency $= 0·625$
2. $E = 0·5L + 10$; efficiency results 0, 0·61, 0·63, 0·65, 0·65; greatest efficiency $\frac{2}{3}$
3. $E = 0·2L + 20$; yes, since the maximum efficiency is 0·5
4. (i) $E = 0·3L + 20$; (ii) 0·15; (iii) $\frac{1}{16}$

Exercise 79

1. (i) 1·024, (ii) 8·93, (iii) 0·87, (iv) 13·6
2. (i) 15·72 kg, (ii) 22·74 kg, (iii) 21 kg, (iv) 38·64 kg
3. (i) 1000 kg m^{-3}, (ii) 1024 kg m^{-3}, (iii) 10 000 kg m^{-3}
4. Tin, iron, copper, silver, lead, gold   5. 2710 kg m^{-3}
6. (i) When set in the frame of a window putty cannot be called fluid because it will not yield to the smallest possible force; (ii) Fluid; (iii) Jelly when set is not a fluid since it will rest in any shape: it is solid like set putty in the sense that its various parts cannot move freely within, but it is not rigid; (iv) Grains of sand have the properties of a solid body; (v) Fluid
Thus a fluid is any substance which will take the shape of a containing vessel, given sufficient time. In our work we have always assumed a solid body to be a rigid one.
7. Density $10 \div 0·075 = 133·3$ kg m^{-3}; relative density 0·13
8. $m \div \rho$ in the appropriate units: if m is expressed in grammes and ρ in grammes per cubic centimetre then the result will be in cubic centimetres; if m is expressed in kg and ρ in grammes per cubic centimetre, then m must be converted to grammes to obtain the volume in cubic centimetres
9. (i) 10 000 g $\div$ 2 g cm^{-3} = 5000 cm^3, (ii) 10 000 g $\div$ 2 g m^{-3} = 5000 m^3, (iii) 10 kg $\div$ 2 kg m^{-3} = 5 m^3
10. (i) $\frac{1000}{1024} = 0·9766$ m^3, (ii) $\frac{1000}{830} = 1·21$ m^3 (two decimal places)
11. $10 \div 8930 = 0·001\ 12$ m^3 copper; $10 \div 7100 = 0·001\ 408$ m^3 zinc

Exercise 80

1. Density of gold is 19 g cm^{-3}; volume of 10 g of gold is $10 \div 19 = 0·5263$ cm^3; the difference is 1·0737 cm^3; assuming accurate measurement the piece of gold must contain a cavity
2. Volume of the mixture is $10 \div 0·5 = 20$ cm^3. Let mass of liquid A be m g, so the volume of A is $m \div 0·4 = 2·5\ m$; mass of liquid B is $(10-m)$ g and the volume of B is $(10-m) \div 0·8 = 12·5 - 1·25\ m$. Hence $20 = 2·5\ m + 12·5 - 1·25\ m$, $m = 6$ g. Mass of A is 6g, mass of B is 4g

3. $\dfrac{100}{8} = \dfrac{m}{6} + \dfrac{(100-m)}{10}$

$\therefore$ $m = 37.5$ kg of A; 62.5 kg of B

4. 8.9 kg of copper, 7.3 kg of tin

5. In A: $d = \dfrac{554 + 9 \times 746}{10\,000} = 0.7268$

in B: $d = 0.5732$

finally $\frac{1}{2}(0.8 + 0.5) = 0.65$

6. Total mass is unchanged at $5 + 16 \times 1.83 = 34.28$ g; final volume is 20 cm³, loss is 1 cm³

7. $19.3\,G + 10.5\,S = (G + S)12.7$, hence $S/G = 3$ or silver:gold $= 3:1$

Exercise 81

1. The bottle holds 25 cm³ of *any* liquid: mass of liquid is 20 g

2. Mass of turpentine is 21.75 g; R.D. $= 21.75 \div 25 = 0.87$; density is 870 kg m⁻³

3. Total mass + powder outside is $(33 + 1.6)$ g; total mass + powder inside is 34.1 g; mass of water displaced is 0.5 g; R.D. is $1.6 \div 0.5 = 3.2$

4. Mass of water displaced is $33 + 3 - 35.6 = 0.4$ g; R.D. $= 7.5$; density $= 7500$ kg m⁻³

5. (i) $25 \times 0.82 = 20.5$: bottle holds 20.5 g methylated spirit; (ii) Mass of bottle + methylated spirit is 28.5 g; mass of spirit displaced is $29.5 - 27.6 = 1.9$ g, which has a volume of $1.9 \div 0.82$; R.D. of the wax is $\dfrac{0.82}{1.9} = 0.432$

Exercise 82

1. (i) R.D. of body is $\dfrac{10}{10 - 8} = 5$

   (ii) R.D. of liquid is $\dfrac{10 - 7.5}{10 - 8} = 1.25$

2. $\dfrac{\text{Weight of liquid}}{2\,\text{N}} = 0.9$; weight registered is 4.2 N

3. Weight of equal volume of water is $0.25 \div 0.2 = 1.25$ N; $T + 0.25 = 1.25$, $T = 1$ N

4. Mass of water displaced is 134 g $= 0.134$ kg, therefore upthrust on the completely immersed ball is $0.134g$ N; weight of the ball is $0.056g$ N; $F + 0.056g = 0.134g$, $F = 0.078g$ N

5. $d = \frac{2}{3}$; force 2 N

Exercise 83

1. If the volume of the body is V cm³ then the mass of an equal volume of water is $V \times 1$ g. Since the body floats, the mass of the body = mass of water displaced = $0.75\,V \times 1$ g; therefore

$$\therefore\ d = \dfrac{0.75\,V \times 1}{V \times 1} = 0.75$$

2. $\frac{2}{3}$

3. Mass of liquid displaced is $\frac{2}{3}\,V \times 0.8$ g = mass of the body; therefore

$$d = \dfrac{\frac{2}{3}\,V \times 0.8}{V \times 1} = \dfrac{8}{15}\,(\text{or } 0.53)$$

4. 20 cm³ immersed in water, 24 cm³ immersed in least-dense liquid; $d = \frac{20}{24} = 0.83$, hence range is from 0.83 to 1.00

5. Volume immersed in water is 20 cm³, volume immersed in liquid $d = 1.2$ is $20 \div 1.2 = 16\frac{2}{3}$ cm³; volume immersed halfway up the stem is $18\frac{1}{3}$ cm³ and is the mark for a liquid with $d = \dfrac{20}{18\frac{1}{3}} = \dfrac{12}{11}$

6. 4 cm

7. The extra volume displaced is 3 cm³, so an extra mass of 3 g of lead shot would be needed in the tube; the force which this represents is $0 \cdot 003g$ N
8. 5 cm
9. The egg is suspended at the depth where fresh water and salt water meet. R.D. of salt water is greater than that of fresh water: R.D. of the egg is intermediate between the two. The egg therefore floats on the salt-water layer.

Exercise 84

1. The density of fresh water being 1000 kg m⁻³, it follows that the volume to be displaced is 100 m³. Imagine a barge in the form of a box which is 10 m × 4 m on the water-line with $2\frac{1}{2}$ m below the water-line. If the length is increased, less of the boat is below the water (e.g. 100 m³ = 25 m × 4 m × 1 m), so that the boat will be able to take a greater load. But to make the boat longer means spreading the same mass of iron more thinly with consequent risks.
2. Depth immersed = 3 cm; extra displacement required for sinking is 50 × 4·5 cm³ = 225 cm³, hence 225 g of lead shot required
3. Displacement required is 40 × 7 = 280 cm³; since the tin already has a mass of 100 g, extra water required is 180 g
4. The maximum displacement is 320 cm³ so the maximum mass of water which the tin can carry without sinking is 220 g; the tin will therefore sink and leave the hydrometer floating behind
5. The smaller cube has a volume of 8 = 2 × 2 × 2 cm³ and will float in water with 0·75 of its volume immersed. Here the minimum depth of water required is 1·5 cm. Minimum volume of water required is (2·5 × 2·5 − 4) 1·5 = 3·375 cm³. (Notice that this is less than the volume immersed.)
6. 100 t = 100 × 1000 kg displaces a volume of 100 m³ and since this lowers the boat through 0·1 m the area of cross section of the boat is 1000 m². To float a mass of 1000 kg we require a displacement of 1 m³ in fresh water and $\dfrac{1}{1 \cdot 024}$ m³ in salt water.

To float 1100 t we require $1100 \left(1 - \dfrac{1}{1 \cdot 024}\right)$ m³ less displacement in salt water than in fresh water. Since the area on the water-line is 1000 m² the ship will rise 1·1 (1 − 0·9766) = 0·025 74 m. (This result is given to four figures simply because four-figure tables were used for calculations.)

Exercise 85

1. 235·5 kN 　　　　　 2. 450 N ÷ 0·0004 m² = 1 125 000 N m⁻²
3. Weight of the body is $0 \cdot 04g$ N; increase in pressure is $8g$ N m⁻²
4. Jack supports a load of 500 kg on 0·0008 m²; pressure is $625 \, 000g$ N m⁻²

Exercise 86

1. (i) 36 N to B, 24 N to C, 12 N to D; (ii) 12 N to A, 6 N to C, 3 N to D; (iii) 120 N to A, 90 N to B, 30 N to D; (iv) 4 N to A, 3 N to B, 2 N to C
2. The pressure supplied by the cork is $\dfrac{100}{\pi \times 0 \cdot 01 \times 0 \cdot 01 \times 1000}$ kN m⁻² = 318·5 kN m⁻²; thrust on base is 25 × 100 N, i.e. the ratio of the areas is 25:1
3. $4a$ 　　　　 4. 1024 kg m⁻³ × g m s⁻² × 500 m = $512g$ kN m⁻²
5. Thrust of the base on the water is given by $F - mg = ma$; $F = 1 \cdot 5g + 1 \cdot 5 \times 2$ = 17·7 N; pressure on base 177 kN m⁻²
6. $p = 830$ kg m⁻³ × g m s⁻² × 0·06 m = $49 \cdot 8g$ N m⁻²
7. $16g$ N m⁻²
8. Pressure at a depth of 10 m is 1024 kg m⁻³ × g m s⁻² × 10 m = $1024g$ N m⁻²; Thrust on the cork is $4 \cdot 096g$ N

Exercise 87

1. (i) M.A. of complete machine is 120; load is 12 000 N
   (ii) 10 N
   (iii) $\dfrac{10 \times 0.1}{30} = \dfrac{1}{30}$ m
2. Thrust required is 48 N; pressure is 24 N mm^{-2} = 24 000 kN m^{-2}
3. 1000 N
4. Force on piston is 200 kN m^{-2} $\times$ 0.03 m^2 = 6 kN = 6000 N; work done is 6000 $\times$ 0.1 = 600 J
5. 0.002 $\times$ R = 600, R = 300 000 N   6. Between A and P

Exercise 88

1. Thrust is 1 m $\times$ 1.5 m $\times$ 1000 kg m^{-3} $\times$ g m s^{-2} $\times$ 10 m = 1 470 00 N = 147 kN
2. Thrust is 10 m $\times$ 4 m $\times$ 1000 kg m^{-3} $\times$ 9.8 m s^{-2} $\times$ 5 m = 1.96 MN
3. Thrust on other side is 3 $\times$ 4 $\times$ 1000 $\times$ 9.8 $\times$ 1.5 = 0.1764 MN; resultant thrust is 1.7836 MN
4. Thrust on the top face is 625g N; thrust on the side face is 632.8g N
5. (i) 36 000g N, (ii) 12 000g N, (iii) 24 000g N

Exercise 89

1. No vacuum would have been formed and the tube would have remained full of mercury after being inverted
2. By tilting the tube into an inclined position until the mercury fills the tube: if air has entered the space, the bubble will always become obvious as the inclination of the tube is increased.
3. $h = \dfrac{13.6 \times 0.75}{1.25} = 8.16$ m; advantage is greater sensitivity, e.g. 1 mm change in the mercury barometer will appear as a change of approximately 11 mm in the glycerine barometer
4. The pressure is equivalent to 20 mm of mercury.
$$p = 13\,600g \times 0.02 = 2665.6 \text{ N m}^{-2}$$
5. 1000 $\times$ g $\times$ 20 + 13 600 $\times$ g $\times$ 0.75 = 30 200g N m^{-2}
6. Atmospheric pressure holds the card in place

Exercise 90

1. No, because there will be an upthrust on the inflated balloon equal in magnitude to the weight of air displaced
2. Upthrust is 650g N; weight of gas in the balloon is 50g N, hence total weight which could be lifted is less than 600g N. (The average person has a weight of 500–600 N.)
3. The balloon stops rising when the upthrust is 400g + 50g = 450g N; density of the air in the layer is 450 $\div$ 500 = 0.9 kg m^{-3}
4. (i) 1.4, (ii) $\frac{1}{14}$ m
5. 0.15 m, diameter of tube makes no difference
6. Relative density $= \frac{4}{3}$; density = 1333.3 kg m^{-3}

Exercise 91

1. 100 $\times$ 0.01 = 150 v, $v = \frac{1}{150}$ m^3
2. 300 kN m^{-2} $\times$ 4 m^3 = 1200 kN m
3. 300 $\times$ 4 = p $\times$ 1.5, p = 800 kN m^{-2}; $\dfrac{p_1}{p_2} = \dfrac{\rho_1}{\rho_2} = \dfrac{3}{8}$
4. 10.2

5. $p_1 = 13\,600g \times 0.76, v_1 = v$
 $p_2 = (13\,600g \times 0.76 + 1000g \times 30), v_2 = ?$
 $\therefore 13\,600g \times 0.76 = (13\,600g \times 0.76 + 1000g \times 30)v_2$
 $\therefore 136 \times 0.76 = (136 \times 0.76 + 300)v_2$, hence $v_2 = 0.25$ m^3

6. The pressure due to 0·02 m of mercury is $13\,600g \times 0.02$ N m^{-2} = 2665·6 N m^{-2}

7. 0·76 m

8. 0·76 m (given by the intersection of the graph with the pressure axis)

Exercise 92

1. (i) The siphon will continue to work for as long as the hole is below the level of the water in the beaker (ii) The siphon will no longer work

2. The bell is lifted by pulling the chain and then allowed to fall under its own weight. The water then floods over into the opening H and, as soon as the water has passed K, the siphon is complete

3. $p_1 = 10, v_1 = V; p_2 = 30, v_2 = ?$; hence $v_2 = \frac{1}{3}V$

4. We find the volume of air at atmospheric pressure which will compress to the volume of the bell at a pressure equivalent to the weight of a 20-m column of water ($p_1 = 10, v_1 = ?; p_2 = 20, v_2 = 30$; hence $v_1 = 60$ m^3). An extra 30 m^3 is required.

5. Atmospheric pressure of 10·2 m on the water barometer. (i) $p_1 = 10.2, v_1 = 5a$; $p_2 = ?, v_2 = 4a$; hence $p_2 = 12.75$; the pressure is therefore 1000 kg m^{-3} × g m s^{-2} × 12·75 m = 12 750g N m^{-2} (ii) The depth of the water-line inside the bell is $12.75 - 10.2 = 2.55$ m below the free surface; bottom of the bell is 3·55 m below the surface (iii) $p_1 = 10.2, v_1 = ?; p_2 = 13.75, v_2 = 5a$; hence $v_1 = 6.74a$; an extra 0·348 of the volume of the bell is required

INDEX

Absolute unit of force, 8, 69
Acceleration, 8, 46, 50, 165
   due to gravity, 59
Aneroid barometer, 289
Angle, of depression, 88
   of elevation, 88
   of friction, 187
Angular velocity, 37
Archimedes' principle, 267
Area of triangle, 92
Area, under v/t graph, 48
   positive and negative, 52
Atmospheric pressure, 287
Average speed, 36

Balance,18
Barometer, Aneroid, 289
   Fortin, 289
   Torricelli, 288
Base point, 42
Belt drive, 244
Body, rigid, 1, 201
Boyle's Law, 294
Brakes, 182
Bramah press, 282
Buoyancy, 257, 273

Calculation of moments, 20
Capstan, 246
Cassiopeia, 142
Centre of buoyancy, 274
Centre of gravity, 13, 21, 23, 28
   of composite figures, 31
   of uniform lamina, 23, 29, 30
   of uniform thin rod, 28
Centre of mass, 29
Centroid, 29
Characteristic of a force, 2
Closest approach, positions of, 156
Coefficient of friction, 185
Components, 121, 163
Concurrent forces, 201
Connected particles, 197
Conservation of energy, 227
Conservation of linear momentum, 79
Coplanar forces, 20, 107
Cosine rule, 91, 94
Course, 148

Dead reckoning, 141
Density, 257, 258
   of mixtures, 262
   relative, 257, 260
Directed line segment, 96
Disc brakes, 182
Displacement, 2, 36
   rate of change of, 36
Displacement–time graph, 53
Distance–time graph, 39
Distortion, 1, 78
Diving bell, 299
Dynamics, 1
Dynamometer, 9

Efficiency, 247
Elastic bodies, impact of, 78
Elastic limit, 5
Elasticity of a body, 78
Energy, 217
   conservation of, 227, 228
   kinetic, 75, 218
   potential, 225
Equations of motion, 55, 166, 169
Equilibrant, 98
Equilibrium, 11, 212, 273
   limiting, 180
   neutral, 195
   stable, 195
   unstable, 195
Extension of a spring, 4

Falling bodies, 62
Fluid, 258
Force, 2
   characteristics of, 2
   line of action of, 3
   turning effect of, 13
   unit of, 8, 69
Force diagram, 118
Forces, action and reaction, 9
   coplanar, 20
   parallel, 22, 25
   simple diagrams thereof, 10
Fortin barometer, 289
Friction, 179
   angle of, 185
Fulcrum, 14, 20, 21

Gear wheel, 245
Gradient, 39
Gravitational unit of force, 69
Gravity, 59

Great Bear, 142
Greatest height, 168
Greenwich Mean Time, 146
Greenwich, Meridian of, 142

Hare's apparatus, 292
Hooke's Law, for a spring, 5
Hydraulic press, 282
Hydrometer, 271
Hydrostatics, 1, 257

Impact, 78
Impulse, 72
Inclined plane, 175, 249
International Date Line, 147
International nautical mile, 137

Jack, 250
Joule, 75

Kilogramme, 7
Kinetic energy, 75, 218
   change in, 218
Knot, 137
Kinematics, 35

Lamina, 23, 29, 30
Lami's Theorem, 105
Latitude, 141
Law of the machine, 254
Lever, 237
   double, 239
   first type, 237
   second type, 239
   third type, 240
Light string, 10
Like parallel forces, 25
Limiting equilibrium, 180, 187
Line of action, of a force, 3
   of a resultant, 131
   of a weight, 15
Line of quickest descent, 178
Liquid, pressure, 276
   thrust, 284
Longitude, 141

Machine, law of, 254
Mass, of a body, 6
Mechanical advantage, 236
Mechanics, 1
Medians of a triangle, 30

Meridian, 141
Metacentre, 274
Moment of force, 13
  in three dimensions, 33
Moment of a force, about
    a fulcrum, 15
  a point, 20
  clockwise, anticlock-
    wise, 16, 25
Momentum, linear, 66,
    67, 232
  conservation of, 79
Motion, 1
  along chord of circle,
    177
  equations of, 55
  in vertical circle, 177
  laws of, 66
  of projectile, 163
  on inclined plane, 175
Moving reference point,
    153

Newton's, First Law, 67
  Laws of Motion, 66
  Second Law, 68
  Third Law, 67
Newton, the absolute unit
    of force, 8
Non-concurrent vectors,
    118
Normal, 179, 206
Normal reaction, 180

Parabola, 167
Parallel forces, 22, 25
Parallelogram law, 99
Particle, 13
Pascal's Law, 278
Pendulum, period of, 60
Pitch, 250
Pivot, 14, 20
Plimsoll mark, 275
Plough, the, 142
Plumb-line, 60
Pole Star, 142
Polygon of forces, 115
Point of concurrency,
    201
Positions of closest
    approach, 156
Potential energy, 225
Power, 217, 233
Pressure, 276
  at a point, 279
  average, 277
  of a gas, 292
  uniform, 277

Principle of conservation
    of energy, 228
Principle of work, 239
Projectile, 162
  general equations, 169
  greatest height, 168, 171
  horizontal range, 168,
    170
  maximum horizontal
    range, 173
  time of flight, 170
Pulley system, 252
Pumps, 297

Radians, 37
  per second, 37
Reaction of a fulcrum, 22
Relative density, 265
Relative density bottle,
    265
Relative motion, 135
Relative velocity, 139
Resolution, 120, 128
Rest, 1
Restitution, 78
Resultant, of parallel
    forces, 22, 25
  of three vectors, 104
Retardation, 50
Rigid body, 1
Rough surface, 179

Sand-glass, 136
Scalars, 101, 224
Screw, 249
Screw press, 250
Simple force diagrams, 10
Sine rule, 91
Siphon, 298
Smooth, 179
Space diagram, 118
Specific gravity, 260
Speed, 36
  average, 36
  non-uniform, 44
Speed-time graph, 46
Spring, 4
  elastic limit, 5
Spring balance, 8, 9
Static friction, 183, 185
Submarine, 274
Sundial, 145

Taking moments, 16
Tangent and normal to a
    curve, 45
Tension, 10
The newton, 69

Thrust on a plane area,
    284
Toppling or sliding, 193
Torricelli's experiment,
    287
Track, 148
Trajectory, 167
Triangle of forces, 104
  law, for vectors, 98
Trigonometric functions,
    84
  acute angle, 84
  obtuse angle, 90
Turning effect of a force,
    13, 15

Uniform body, 28
Uniform pressure, 277
Uniform speed, 36
Uniform velocity, 36
Upthrust, 268

Vacuum, motion in, 59
Vector algebra, 107
Vector components, 163
Vector quantities, 96, 97
Vectors, addition, 101,
    108
  associative law for
    addition, 109
  distributive law, 109
  product with scalar, 109
  subtraction, 108
  sum, 97
  zero, 108
Velocity, 36
  angular, 37
  formulae, 38
  positive, negative, 50
  ratio, 244
Velocity-time graph, 51
Vertical, 10
Vertically downwards
    direction, 10
Vertically upwards direc-
    tion, 10
Viscous fluid, 258

Watt, 233
Weight, of a body, 6
  of a mass, 15
Weston differential pul-
    ley, 252
Wheel and axle, 243
Windlass, 246
Work, 217

Zero potential energy, 226